FLORA OF TROPICAL EAST AFRICA

SCROPHULARIACEAE

S.A. GHAZANFAR[1], F.N. HEPPER[2] & D. PHILCOX[3]

Herbs, undershrubs, sometimes climbers or shrubs, rarely trees; many partial root parasites, a few wholly parasitic. Leaves exstipulate, simple, alternate or opposite (especially lower ones), rarely all basal or the upper-most whorled, often variously lobed or dissected, toothed or entire. Flowers hermaphrodite, zygomorphic (sometimes very slightly) usually in bracteate spikes or racemes, less often solitary in leaf-axils, or in cymes. Calyx persistent; calyx-tube campanulate, tubular or almost none; calyx-teeth usually 4 or 5, rarely 3, sometimes calyx 2-lipped, valvate, imbricate or open in bud. Corolla united; tube campanulate, cylindric or ventricose or enlarged above, straight or curved or bent, sometimes very short, sometimes with 1 or 2 spurs or sacs at the base, usually 4- or 5-lobed with the lobes more or less equal and all spreading, or distinctly 2-lipped, with the upper lip entire or 2-lobed, lower lip usually 3-lobed. Stamens 4, didynamous or 2, rarely 3, 5 or 6–8; staminodes sometimes present. Ovary superior, 2-locular, ovules numerous. Fruit usually capsular, septicidal or loculicidal, rarely a berry and indehiscent, or fruit separating into 2 cocci.

About 220 genera with ± 3000 species worldwide; mostly herbaceous, especially in temperate regions.

Recent molecular study of a limited number of species of Scrophulariaceae sensu lato by Olmstead *et al.* (Amer. J. Bot. 88(2): 348–361 (2001)and Oxelman *et al.* in Taxon 54(2): 411–425 (2005)) have shown that the family consists of 5 monophyletic groups with *Mimulus* not belonging to any of the groups. The family-level classification accepted by them include Scrophulariaceae sensu stricto, Orobanchaceae (including parasitic and semiparasitic genera), Veronicaceae, Stilbaceae and Calceolariaceae. Beardsley & Olmstead (Amer. J. Bot. 89: 1093–1102 (2002)), later segregated Phyrmaceae to include *Mimulus* and other genera derived from within *Mimulus*, and more recently two further families, Linderniaceae and Gratiolaceae have been segregated. Fischer in his treatment of the Scrophulariaceae in Kubitzki (ed. J.W. Kadereit), Families and Genera of Vascular Plants vol. 7 (2004), does not formally segregate any families from the traditional Scrophulariaceae sensu lato, but lists the families to be recognised and their genera arranged according to the available morphological and molecular data. In our treatment for FTEA, we have included all genera except *Orobanche* and *Cistanche*; these have been published seperately as Orobanchaceae sens. str. (3 carpels and parietal placentation) (1957): both genera are however keyed out in the generic key.

[1] All genera have been revised and updated by S.A. Ghazanfar. S.A.G. thanks Bernard Verdcourt for going over the final draft.

Alectra, Bartsia, Cycniopsis, Diclis (in part), *Dopatrium, Ghikaea, Hebenstretia, Jamesbrittenia, Lindenbergia, Mecardonia, Melasma, Micrargeria, Nanorrhinum, Parastriga, Pseudosopubia, Selago, Torenia* (in part), *Verbascum* (in part), and cultivated species by S.A. Ghazanfar

[2] *Anticharis, Aptosimum, Artanema, Bacopa, Buttonia, Craterostigma, Cycnium, Diclis, Freylinia, Glossostigma, Halleria, Harveya, Kickxia, Limosella, Mimulus, Misopates, Rhamphicarpa, Scoparia, Sibthorpia, Stemodia, Stemodiopsis, Striga, Thunbergianthus, Torenia, Verbascum, Zaluzianskya* and *Sopubia* species key by F.N. Hepper; *Veronica* revised by D. Albach, Johannes Gutenberg-Universität Mainz

[3] *Buchnera, Gerardiina, Graderia, Hedbergia, Limnophila, Lindernia* and *Sopubia* by (late) D. Philcox

Conservation assessments of species by S.A. Ghazanfar. Assessments follow Categories and Criteria as listed for the IUCN Red List Version 3.1; all assessments are preliminary and are made from herbarium material.
Vernacular names and uses given where available on herbarium sheets.

Cultivated species

Several species of Scrophulariaceae are, or have been in cultivation in the Flora area.

Alonsoa acutifolia Ruiz. & Pav.

Straggling erect herb up to 50 cm tall. Leaves ovate to ovate-lanceolate, 20–30 mm long, margins dentate. Flowers in terminal racemes. Corolla orange to vermillion with a green throat; anthers yellow. Capsule ovoid.

KENYA. Naivasha District: South Kinangop, 22 Nov. 1953, *Verdcourt* 1040!

A. linearis Ruiz. & Pav.

Erect herb up to 50 cm. Leaves appearing vertillate, linear, 10–15 mm long. Flowers in terminal racemes. Corolla tubular, yellow/orange; anthers yellow, large, coherent. Capsule ovate, beaked.

KENYA. Nairobi District: Nairobi, 28 Aug. 1953, *R.H.D. Wood* in EA 10329!

Antirrhinum majus L. "Snapdragon".

Erect herb up to 50 cm. Leaves ovate to lanceolate, up to 40 mm long. Flowers red to yellow, tubular, strongly 2-lipped. Capsule ovoid, velvety green.

KENYA. Nairobi District: Chiromo Estate, *Mathenge* 730!

TANZANIA. Lushoto District: Hotel, *Shunda* 7!; also recorded as an escape in Kenya, Takaungu, near Gedi ruins, 1945, *Bally* 5200!

Calceolaria mexicana Benth.

Annual herb, up to 30 cm. Leaves pinnatisect, up to 6 cm long, lobes ± dentate. Flowers in cymose clusters, pale yellow. Stamens reddish at the base. Capsule ovoid.

KENYA. Nakuru District: E of Hyrax Hill at Historical House, 9 Aug. 1967, *Mwangangi* 173!

Digitalis purpurea L. "Foxglove".

Biennial herb. Leaves of first year rosette, oblong-lanceolate, about 20 cm long and 10 cm wide, acute, long-cuneate and with petiole winged, margins crenate-dentate, softly grey pubescent. Inflorescence 0.5–1.8 m tall, in one-sided raceme by twisting of pedicels. Calyx lobes ovate, acute. Corolla pink (or sometimes white in gardens) with darker spots inside among the long hairs, tube ± 45 mm long; lobes 5, short, rounded, spreading. Capsule ± 8 mm long, acute, longer than the calyx.

KENYA. Naivasha District: Kinangop, Dec. 1948, *Bally* 17895!; Nairobi, Oct. 1942, *Jex-Blake* 11633!

A common western European plant of open woodland and hedges on acid soil, also cultivated in gardens. Although grown occasionally as an ornamental plant in gardens in upland East Africa, it seems to have become naturalised in a few places.

Maurandya purpusii Brandegee

Climbing herb with tough wiry stems, glabrous. Leaves triangular-hastate, 3–4 cm long, up to 4 cm wide at base, coarsely and sharply toothed; petioles often curled and assisting in climb. Flowers solitary, with long pedicels subtended by leafy bracts. Calyx broadly 5-lobed. Corolla purple to carmine, 4–5 cm long, funnel-shaped, lobes shallow, with golden hairs inside and near the base of stamens and in 2 rows down the lower lip. Capsule rotund.

KENYA. Nairobi City Park, 12 Aug. 1970, *Mwangi* in EA 14444

M. barclaina Lindley

Woody or herbaceous climber or creeper, glabrous. Leaves triangular-hastate, up to 35 mm long, and up to 25 mm wide at base, coarsely toothed; petioles 2–3 cm long. Flowers solitary, with long pedicels up to 5 cm subtended by leafy bracts. Calyx with lanceolate lobes, villous. Corolla deep rich blue, 3 cm long, funnel shaped, lobes shallow. Capsule rotund, ± included in the calyx.

KENYA. Nairobi City Park, 20 Aug. 1977, *Gillett* 21526! & 24 Sept. 1951, *Sangai* 142!

TANZANIA. Arusha District: Arusha Hotel, 18 May 1933, *Ritchie* H29/33!

Paulownia tomentosa Steud. "Princess tree".

Small tree or shrub up to 8 m tall with a rough grey-brown bark. Leaves cordate, 12–30 cm long, velvety green above, brownish beneath. Flowers in conical inflorescences, up to 50 cm long. Calyx to 1.5 cm, tomentose. Corolla blue, crinkly ventrally, glandular outside. Fruits 3–4 cm in diameter, yellowish green, viscid-glandular.

TANZANIA. Lushoto District, in private garden, May 1959, *Mgaza* s.n.! & Feb. 1960, *Gardner* 449/60!

Russelia equisetiformis Schltdl. & Cham. (Syn.: *R. juncea* Zucc.). "Coral plant".

Shrub up to 1.5 m high with numerous lax whorled branches sometimes arching and rooting, ribbed, and *Equisetum*-like. Leaves usually totally absent, if present up to 7 in a whorl, obovate, 1–2 cm long, ± 1 cm broad, or very small. Flowers in 1–2(–3)-flowered fascicles and slender peduncles. Corolla scarlet, tubular, slightly saccate, up to 2.5 cm long. Capsule broadly ovoid, beaked by persistent style.

KENYA. Nairobi City Park, *Sangai* 414!

TANZANIA. Uzaramo District: Dar es Salaam, *Ruffo* 1062!; Lushoto District: Amani, 12 Aug. 1932, *Geilinger* s.n.!

Frequently cultivated as a garden ornamental; occasionally escaping and naturalizing by roadsides.

R. sarmentosa Jacq. "Red Rocket".

Shrub with spreading branches, up to 1.5 m tall. Leaves ovate, up to 7 cm long and 4 cm wide, margins dentate. Flowers in axillary cymose clusters, somewhat nodding. Corolla scarlet, tubular, up to 3 cm long.

KENYA. Nairobi City Park, *Mwangi* in EA14447!

TANZANIA. Uzaramo District: Dar es Salaam, State House grounds, *Ruffo* 518!; Dar University Campus, *Wingfield* 2274!

In addition, the ornamentals *Freylinia tropica, Hebe* spp., *Penstemon* × *gloxinioides* and *Phygelius aequalis* are also reported to be cultivated in Kenya (fide Ann Robertson, pers. comm., 2005).

KEY TO THE GENERA

Most semiparastic genera turn black on drying due to the presence of orobanchin and iridoid compounds in their leaves and stems. This character (blackening on drying) is helpful in identification of herbarium specimens, and includes most species of *Alectra, Buchnera, Cycnium, Ramphicarpa* and *Striga.* The number of stamens, spurred, bent or entire filaments and the presence of mono- or bithecal anthers are important in the identification of genera, as is the lobation of the calyx.

1.	Plants holoparasitic, lacking chlorophyll, never green	2
	Plants autotrophic or semiparasitic, with chlorophyll	5
2.	Corolla tube bent at right angles below throat	34. **Striga** (p. 138)
	Corolla tube straight or curved, but not with a right angle bend	3
3.	Corolla distinctly bilabiate; limb with upper lip 2-lobed and lower lip 3-lobed	**Orobanche**
	Corolla with limb ± equally lobed	4
4.	Stem thick and fleshy, often swollen towards base; anthers with both thecae fertile	**Cistanche**
	Stem often not thick and fleshy or swollen towards base; anthers with one fertile theca, the other sterile, longer and subulate-acuminate	32. **Harveya** (p. 117)

5. Plants climbing 6
   Plants not climbing 7
6. Leaves toothed or 3-lobed; calyx inflated in fruit; flowers carmine, rose-lilac to violet (in Flora area) .... 39. **Buttonia** (p. 170)
   Leaves with margins coarsely toothed near base or entire; calyx not inflated in fruit; flowers pink to mauve .... 29. **Thunbergianthus** (p. 106)
7. Leaves in a basal rosette, cauline leaves present or absent 8
   Leaves not in basal rosette 9
8. Stems very reduced; leaves entire; filaments sharply angled about the middle .... 21. **Craterostigma** (p. 56)
   Stems not reduced; leaves toothed or pinnatifid; filaments not angled .... 3. **Verbascum** (p. 12)
9. Aquatic herbs or of marshy habitats; leaves submerged and/or aerial, when submerged then different from aerial 10
   Terrestrial herbs and shrubs or small trees; leaves only aerial 13
10. Leaves of two types; submerged leaves finely divided, aerial leaves undivided, deeply lobed or variously divided .... 16. **Limnophila** (p. 41)
    Leaves of one type 11
11. Stems ± fleshy; corolla bilabiate; calyx deeply 5-fid; fertile stamens 2, anterior stamens reduced to staminodes .... 19. **Dopatrium** (p. 50)
    Stems not fleshy; corolla regular; calyx 4–5-divided; fertile stamens 4 or 2, staminodes absent 12
12. Calyx (4–)5-toothed; corolla 5(rarely 4)-lobed; stamens 4 .... 25. **Limosella** (p. 93)
    Calyx deeply 4(–5)-lobed, the fifth when present usually smaller; corolla with 4 (rarely 5) lobes; stamens 2 .... 28. **Veronica** (p. 100)
13. Corolla saccate at base, lower lip 3-lobed, produced into a large palate closing the hairy throat .... 7. **Misopates** (p. 22)
    Corolla not saccate at base 14
14. Corolla tube extending to form a distinct spur 15
    Corolla not spurred or spur obscure (1–2 mm long) 16
15. Filaments of anterior pair of stamens bent at base; anthers monothecal by confluence, cohering in pairs .... 5. **Diclis** (p. 18)
    Filaments not bent at base; anthers bithecal, marginally coherent forming a ring-like structure .... 6. **Nanorrhinum** (p. 22)
16. Lower lip of corolla with a cavity .... 4. **Angelonia** (p. 18)
    Lower lip of corolla without a cavity 17
17. Fertile stamens 2, with or without 2 staminodes 18
    Fertile stamens 4 or 5 22
18. Small creeping moss-like herbs; calyx 3–4-toothed; anther thecae divergent at base .... 24. **Glossostigma** (p. 91)
    Erect or prostrate herbs; calyx 4–5-lobed or -toothed 19

19. Anterior filaments each with a distinct spur arising at or near base .... 23. **Lindernia** (p. 67)
    Anterior filaments without spur 20
20. Stems often fleshy; posterior 2 stamens fertile, anterior 2 reduced to minute staminodes .. 19. **Dopatrium** (p. 50)
    Stems not fleshy; staminodes absent 21
21. Leaves opposite; calyx divided into 4(–5) lobes, the fifth when present usually smaller; corolla with 4 (rarely 5) lobes; capsule more or less compressed at right angles to the septum .... 28. **Veronica** (p. 100)
    Leaves alternate; calyx equally 5-lobed; corolla 5-lobed; capsule not compressed .... 2. **Anticharis** (p. 10)
22. Anthers hispid or ciliate; capsule compressed at the top at right angles to the septum, emarginate at apex .... 1. **Aptosimum** (p. 8)
    Anthers glabrous; capsule not compressed at top 23
23. Fruit separating into 2 cocci 24
    Fruit not separating into 2 cocci 25
24. Calyx with 1 lobe, spathaceous; corolla tubular, divided almost to the base forming a large, flattened or concave 4(–5)-lobed limb .... 48. **Hebenstretia** (p. 197)
    Calyx 3–5-lobed, lobes equal or unequal with dorsal lobes small and anterior lobes deeply cleft, calyx thus appearing 2- or 3-lobed; corolla tube not divided to base, limb 2-lipped, 5-lobed .... 49. **Selago** (p. 198)
25. Calyx unequally 4-lobed, median clefts usually deeper than lateral ones; corolla limb separated into galea (hood) and lip ...... 46. **Bartsia** (p. 191)
    Calyx 2–5-lobed; corolla limb not separated into galea and lip 26
26. Stamens 4 with 1 staminode; filaments of the longer pair twisted; fruit reflexed when mature .... 15. **Stemodiopsis** (p. 38)
    Stamens 4 or 5, staminode usually absent; filaments not twisted; fruit not reflexed when mature 27
27. Flowers placed above the axils of leaves, with 2 small glands between the axil and pedicel ... 42. **Ghikaea** (p. 183)
    Flowers placed in the axil of leaves or terminal 28
28. Shrubs or small trees; flowers red, corolla tube inflated; fruit baccate, indehiscent .... 8. **Halleria** (p. 24)
    Herbs or shrubs; corolla tube usually not inflated; fruit capsular, dehiscent, rarely baccate (baccate in *Cycnium adonense* only) 29
29. Calyx 2–3-lobed (in Flora area); corolla tube exceeding the calyx (≥ 2× calyx) .... 10. **Zaluzianskya** (p. 30)
    Calyx (3–)4–5-lobed; corolla tube equalling or exceeding the calyx, but usually ≤ 2× calyx 30
30. Corolla tube bent sharply or curved above the middle or gibbous near the apex 31
    Corolla tube straight or sometimes curved 33

31. Mat-forming prostrate herbs; flowers solitary; corolla tube curved and gibbous near throat; limb 4-lobed .... 36. **Cycniopsis** (p. 152)
Erect herbs; flowers in spikes; corolla tube bent at right angles; limb 5-lobed 32
32. Plants hirsute or scabrous; corolla strongly 2-lipped, tube not widening below throat ... 34. **Striga** (p. 138)
Plants glandular; corolla not strongly 2-lipped, tube widening below throat .... 35. **Parastriga** (p. 150)
33. Anther thecae with distinct connective (stipitate) 34
Anther thecae not stipitate 36
34. Upper pair of anthers without appendage, lower pair with filiform connective, the upper branch of which bearing one anther, the lower curving upwards and ending in a rounded appendage representing the sterile theca .... 40. **Pseudosopubia** (p. 172)
Connective not prolonged 35
35. Leaves opposite or whorled; calyx lobes deeply divided; corolla white or blue ...... 14. **Stemodia** (p. 36)
Leaves opposite or the upper alternate; calyx lobes divided to middle; corolla yellow to brownish yellow .... 13. **Lindenbergia** (p. 34)
36. Calyx deeply 4(–5)-lobed; corolla 4-lobed .... 27. **Scoparia** (p. 98)
Calyx 3–5-toothed or -lobed; corolla 5-lobed 37
37. Capsule strongly compressed longitudinally, appearing unilaterally winged .... 43. **Graderia** (p. 184)
Capsule globose to ovoid, not compressed 38
38. Calyx inflated or inflated in fruit .... 30. **Melasma** (p. 106)
Calyx not inflated 39
39. Anterior pair of filaments or all with a distinct spur or appendage at or near the base or sharply bent at base 40
Filaments without spur or appendages 43
40. Filaments of anterior pair sharply bent at the base; anthers monothecal by confluence, cohering in pairs .... 5. **Diclis** (p. 18)
Filaments of anterior pair not sharply bent at the base; anthers bithecal 41
41. Calyx 3–5-winged or prominently 5-ridged or angled .... 22. **Torenia** (p. 62)
Calyx not winged, ridged or angled 42
42. Flowers $>$ 2 cm long; anterior filaments filiform and arched with an appendage at the base of each; anthers connivent at the apex .... 20. **Artanema** (p. 54)
Flowers usually $\leq$ 2 cm long, if $>$ 2 cm long then filaments without spur or appendages; anterior filaments each with distinct spur arising at or near base; anthers free or contiguous .... 23. **Lindernia** (p. 67)
43. One cell of each anther perfect, the other empty or nearly so 44
Both anther cells perfect or anthers monothecal 45

44\. Corolla > 2 cm long, tube much enlarged above, lobes with entire or undulate-crenate margins ..... 32. **Harveya** (p. 117)
Corolla < 2 cm long, lobes with entire margins 41. **Sopubia** (p. 175)

45\. Calyx lobed to the base, lobes ± imbricate ... 11. **Jamesbrittenia** (p. 30)
Calyx not lobed to the base, lobes not imbricate 46

46\. Anthers monothecal 47
Anthers bithecal 49

47\. Corolla salver-shaped; corolla tube with throat and inside of tube bearded; limb with upper lobes united, emarginate, lower lip 3-lobed .. 38. **Cycnium** (p. 156)
Corolla tube and throat not bearded; limb with lobes equal or nearly so 48

48\. Leaves pinnatisect; filaments bearded at base or above middle; capsule obliquely beaked .. 37. **Rhamphicarpa** (p. 154)
Leaves not pinnatisect; filaments glabrous; capsule ± beaked but not obliquely so ..... 33. **Buchnera** (p. 123)

49\. Stamens 5, filaments with long pilose hairs above middle ..... 45. **Gerardiina** (p. 189)
Stamens 4, filaments not pilose 50

50\. Calyx lobes markedly unequal, the outer broadest, laterals narrow 51
Calyx lobes ± equal 52

51\. Corolla yellow, 2-lipped, upper lobe obovate, emarginate; lower lobe shallowly 3-lobed with rounded margins ..... 17. **Mecardonia** (p. 46)
Corolla white or bluish, almost regular, with all lobes equal or the upper 2 connate .... 18. **Bacopa** (p. 48)

52\. Corolla markedly bilabiate; tube cylindric ... 12. **Mimulus** (p. 32)
Corolla not markedly bilabiate; tube campanulate or broadly tubular 53

53\. Prostrate creeping herbs; leaves reniform .... 26. **Sibthorpia** (p. 96)
Erect, tufted or woody herbs or undershrubs; leaves various but not reniform 54

54\. Calyx 4-lobed; corolla subrotate; seeds ovoid, longitudinally winged ..... 47. **Hedbergia** (p. 195)
Calyx 5-lobed; corolla regular; seeds linear, clavate or obovoid or discoid with a marginal wing 55

55\. Shrubs or small trees; leaves often verticillate; seeds with a membranous testa and winged margin ..... 9. **Freylinia** (p. 27)
Annual or perennial herbs; leaves alternate or opposite, lower leaves often reduced to scales 56

56\. Leaves ovate to lanceolate or reduced to scales; corolla persistent in fruit; anthers coherent or connivent in pairs; plants blackening on drying ..... 31. **Alectra** (p. 108)
Leaves linear to filiform; corolla deciduous; anthers free; plants remaining green on drying ..... 44. **Micrargeria** (p. 187)

1. APTOSIMUM

Burch., Travels S. Africa 1: 217 (1822); Skan in F.T.A. 4(2): 267 (1906); Hepper in F.W.T.A. ed. 2, 2: 354 (1963); *nom. conserv.*

Chilostigma Hochst. in Flora 1: 372 (1841)

Undershrubs or woody herbs, densely branched. Leaves alternate, congested, narrow, often spinescent, 1-veined. Inflorescence axillary, 1-several-flowered; flowers ± sessile, 2-bracteolate. Calyx usually deeply 5-segmented, lobes subulate. Corolla tube veined, long, lower portion narrow, expanded above into a tubular throat; corolla-lobes 5, nearly equal, rounded, the two upper outside in aestivation. Stamens 4, didynamous, attached near the middle or base of the corolla, included; anthers transverse, hispid or ciliate, unilocular by confluence, those of the two lower smaller, posticous ones often empty; staminodes absent; style filiform; stigma large or small, emarginate. Capsule globose, apex emarginate or rounded, compressed at the top at right angles to the septum, septicidally 2-locular. Seeds numerous. Testa reticulate.

About 40 species in tropical Africa, mainly southern Africa.

Aptosimum pumilum (*Hochst.*) *Benth.* in DC., Prodr. 10: 345 (1846); Skan in F.T.A. 4(2): 268 (1906); F.P.S. 3: 132 (1956); Hepper in F.W.T.A. ed. 2, 2: 355 (1963); U.K.W.F.: 254 (1994); Wood, Handb. Fl. Yemen: 259 (1997); Collenette, Wildflow. Saudi Arabia: 673 (1999); P. Cuccuini & C. Nepi in Thulin (ed.), Fl. Somal.: 3: 269 (2006); Fischer in Fl. Ethiop. & Eritr. 5: 280 (2006). Type: Sudan, Kordofan, *Kotschy* 227 (B†, holo.; K!, P!, iso.)

Tough annual or biennial herb up to 15 cm high, with a single taproot, densely branched, forming a tuft 6–20 cm in diameter. Leaves held erect, lamina breaking off leaving the long petiole as a weak spine; lamina linear to linear-oblanceolate, (1.5–)2.5–7 cm long, 2–4 mm wide, apex acute, scabrid on the margins and midrib beneath, more or less finely pubescent, ciliate along narrowly winged petiole. Flowers sessile, congested in the leaf axils. Bracts linear, long-ciliate. Calyx 5-segmented, segments subulate, 10–12 mm long, strongly 3-veined, long-ciliate, calyx-tube very short. Corolla tube white with pink spots, 15 mm long, pubescent; corolla-lobes 5, almost equal, pale lilac, pink or blue, 2 mm long. Stamens 2 + 2, pairs very unequal, lower pair minute, sterile; filaments inserted about the middle of the corolla tube. Capsule round, both compartments inflated, apex emarginate and slightly winged, venose. Seeds black, 1 mm long, subquadrangular, minutely papillose. Fig. 1, p. 9.

KENYA. Northern Frontier District: Mt Kulal, July 1958, *Verdcourt* 2224! & Laisamis, Nov. 1977, *Carter & Stannard* 732!; Masai District: km 24 from Magadi to Nairobi, 14 Aug. 1961, *Greenway & Polhill* EA 12409!

TANZANIA. Arusha District: Arusha, June 1928, *Haarer* 1426!; Masai District: Olbalbal, foot of escarpment, 22 July 1966, *Greenway & Kanuri* EA 12605!

DISTR. **K** 1, 3, 6; **T** 2; Mauritania, Niger, Sudan, Ethiopia, Somalia; SW Arabia

HAB. Bare sandy and rocky places; 500–1200 m

USES. None recorded on specimens from our area

CONSERVATION NOTES. Least Concern (LC); common, but not widespread in the Flora area

SYN. *Chilostigma pumilum* Hochst. in Flora 24, 1: 372 (1841)

Fig. 1. *APTOSIMUM PUMILUM* — **1**, habit × $^{2}/_{3}$; **2**, leaf underside × 1; **3**, flower × 4; **4**, flower dissected × 4; **5**, capsules × $1^{1}/_{2}$; **6**, capsule × 6; **7**, seed × 20. 1, 2 from *Carter & Stannard* 732; 3, 4 from *Greenway & Kanuri* 12605; 5–7 from *Gilbert et al.* 5650. Drawn by Juliet Williamson.

2. ANTICHARIS

Endl., Nov. Stirp. Dec.: 22 (1839) & Gen. Pl.: 682 (1839) & Iconogr.: t. 93 (1839)

Distemon Aschers. in Monatsberg., Akad. Wiss. Berlin 1866: 880 (1867)
Gerardiopsis Engl., P.O.A. C: 359 (1895) & in E.J. 23: 507 (1897)

Small herbs, erect, with glandular pubescence. Leaves alternate, entire. Flowers axillary, solitary, pedicellate, often 2-bracteolate. Calyx 5-lobed, lobes lanceolate. Corolla 5, lobes sub-equal. Stamens 2, anterior, included; filaments inserted above the base of the corolla tube; anthers at first crescent-shaped, after dehiscence straight, free or cohering; staminodes absent; style filiform; stigma slightly bifid. Capsule 2-locular, loculicidal and septicidal. Seeds numerous.

10 species in dry zones, from SW Africa to Arabia and India.

Bracteoles inserted well above the middle of the pedicel .... 1. *A. senegalensis*
Bracteoles inserted at or below the middle of the pedicel ... 2. *A. arabica*

1. **Anticharis senegalensis** (*Walp.*) *Bhandari* in Bull. Bot. Survey India 6: 327 (1964) & in Fl. Indian Desert: 279 (1978); P. Cuccuini & C. Nepi in Thulin (ed.), Fl. Somal.: 3: 267 (2006); Fischer in Fl. Ethiop. & Eritr. 5: 278 (2006). Type: Senegal, *Leprieur* s.n. (K!, iso.)

Erect annual, 7–33 cm high, drying blue-black, almost simple or more usually profusely branched, glandular-pubescent throughout; root simple. Leaves linear, 1–3.5 cm long, ± 2 mm wide, sessile, base attenuate, apex acute, margins entire. Flowers blue with white throat, solitary; pedicels 8–12 mm long, with a pair of subulate bracteoles 1–2 mm long, inserted well above the middle. Calyx deeply divided, lobes ± 3 mm long, joined only at the base, linear. Corolla tube 8–10 mm long; lobes blue with darker veins, ± 2 mm long, rounded at the apex; anthers free or ± cohering, ciliate; style as long as the corolla tube; stigma slightly bifid. Capsule ovoid, 7–8 mm long, acutely beaked, dehiscing into 4 segments, minutely pubescent. Seeds oblong, 1 mm long, longitudinally ridged, minutely striated. Fig. 2, p. 11.

KENYA. Turkana District: Lokori–Sigor, 19 May 1970, *Mathew* 6292!; Masai District: near Ol Lorgosailic [Olorgesailie], 14 Aug. 1961, *Polhill & Greenway* 448!; Kilifi District: Malindi–Lali Hills, 15 Nov. 1967, *D. Wood* 1350!
TANZANIA. Masai District: between Ketete R. and Mto wa Mbu, 12 Mar. 1964, *Greenway & Hunter* 11359! & road to Engaruka, 21 Feb. 1970, *Richards* 25483!; Arusha District: Lower Nduruma R., June 1928, *Haarer* 1434!
DISTR. **K** 1–3, 6, 7; **T** 2; extending across the southern Sahara to furthest West Africa, and in the drier parts of South West Africa; Arabia and India
HAB. In semi-desert, an ephemeral; 200–1200 m
USES. None recorded on specimens from our area
CONSERVATION NOTES. Least Concern (LC)

SYN. *Doratanthera senegalensis* Walp., Repert Bot. 3: 305 (1844)
D. linearis Benth. in DC., Prodr. 10: 347 (1846). Type as for *A. senegalensis*
Anticharis linearis (Benth.) Aschers. in Monatsber. Akad. Wiss. Berlin 1866: 882 (1867); Hiern, Cat. Afr. Pl. Welw. 1: 756 (1898); Hemsley & Skan in F.T.A. 4(2): 276 (1906); F.P.S. 3: 34 (1956); Hepper in F.W.T.A. ed. 2, 2: 354 (1963); U.K.W.F.: 254 (1994); Philcox in F.Z.: 8(2): 9 (1990); Wood, Handb. Fl. Yemen: 260 (1997)
Gerardiopsis fischeri Engl., P.O.A. C: 359 (1895) & in E.J. 23: 507 (1897). Type: E Africa, locality not known, *Fischer* 119 (HBG, holo.)

2. **Anticharis arabica** *Endl.*, Nov. Stirp. Dec.: 23 (1839); Wood, Handb. Fl. Yemen: 260 (1997); Miller & Morris, Ethnofl. Soqotra Archipel.: 679 (2004); P. Cuccuini & C. Nepi in Thulin (ed.), Fl. Somal.: 3: 268 (2006). Syntypes: Yemen, Mts Sidr & Kasr, Geddae, *S. Fischer* 17 & near Ferihe, *Schimper* 748 (M; syn.)

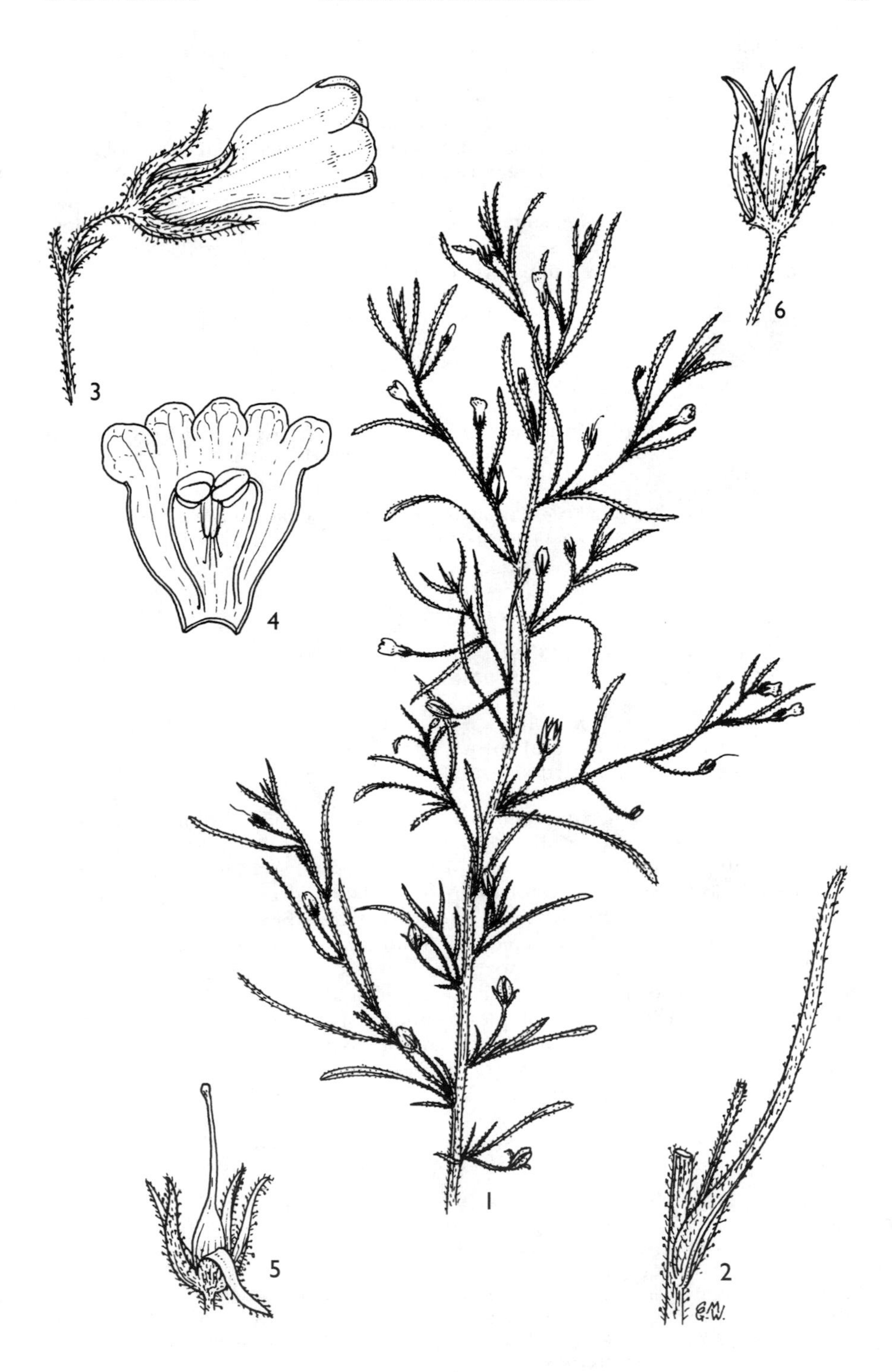

FIG. 2. *ANTICHARIS SENEGALENSIS* — **1**, habit × $^1/_3$; **2**, leaf × 4; **3**, flower × 4; **4**, corolla opened to show stamens × 4; **5**, carpel and calyx × 4; **6**, dehisced capsule × 4. 1, 2 from *Bally* B5217; 3–6 from *Glover & Samuel* 2984. Drawn by Christine Grey-Wilson. Reproduced with permission from F.Z.

Erect annual, up to 20 cm high, simple or with a few branches, glandular-pubescent throughout. Leaves lanceolate to linear-lanceolate, 1–3.5 cm long, ± 2 mm wide, sessile, base attenuate, apex acute, margins entire. Flowers blue with white throat, solitary; pedicels 8–12 mm long, with a pair of subulate bracteoles 1–2 mm long, inserted at or below the middle of the pedicel. Calyx deeply divided, lobes ± 3 mm long, joined only at the base, linear. Corolla tube 8–10 mm long; lobes blue with darker veins, ± 2 mm long, rounded at the apex; anthers free or ± cohering, ciliate; style as long as the corolla tube; stigma slightly bifid. Capsule ovoid, 7–8 mm long, acutely beaked, dehiscing into 4 segments, minutely pubescent. Seeds oblong, 1 mm long, longitudinally ridged, minutely striated.

KENYA. Northern Frontier District: Derrati Wells, 30 km E of Lake Turkana, 11 Aug. 1974, *G. Jackson* 43 (photo!)
DISTR. **K** 1; Sudan, Eritrea, Ethiopia, Somalia; Arabia, Socotra
HAB. In deserts; ± 400 m
USES. None recorded on photo
CONSERVATION NOTES. Not much material from Kenya; possibly at threat from over-grazing and habitat degradation in the Flora area, but not uncommon elsewhere; assessed as of Least Concern (LC)

NOTE. A short-lived annual in dry desertic areas, here at its southernmost limit in northern Kenya; apparently not common but perhaps overlooked.

3. **VERBASCUM**

L., Sp. Pl.: 177 (1753) & Gen. Pl. ed. 5: 83 (1754); Murb. in Lunds Univ. Arsskr. 29(2): 1–630 (1933); Huber-Morath in Bauhinea 5: 7 (1973)

Celsia L., Sp. Pl.: 621 (1753) & Gen. Pl. ed. 5: 272 (1754); Skan in F.T.A. 4(2): 280 (1906); Murb. in Lunds Univ. Arsskr. 29(2): 1–237 (1925)
Rhabdotosperma Hartl in Beitr. Bio. Pflanze. 53: 57 (1977)

Biennial or perennial herbs, sometimes undershrubs, often woolly-pubescent; stems erect. Leaves cauline and alternate and in a basal rosette, toothed or sometimes pinnatifid, leaves of basal rosettes usually larger than cauline. Inflorescence racemose or spicate or often paniculate and tall. Flowers solitary or fasciculate (accessory flowers of Hartl), subtended by a bract, pedicellate. Calyx deeply 5-lobed or 5-partite, lobes ± equal, imbricate. Corolla subrotate, tube very short, 5-lobed, lobes broad, slightly unequal, the posterior outside. Stamens 4 or 5, inserted at the base of the corolla; filaments of anterior 2 or 3 bearded or all bearded; anthers transverse or oblique, monothecal by confluence; style entire, laterally compressed, dilated at the apex. Capsule globose or ovoid, septicidally 2-locular. Seeds numerous, minute, longitudinally 6–9-grooved and ridged, sometimes (not in Flora area) with a cellular network between the ridges.

A large genus of 300–400 species, mainly in the temperate regions of the Northern Hemisphere, from Eurasia to Indo-China, Cape Verde, N Africa to Congo-Kinshasa and S Tanzania.

Verbascum, with 5 stamens, and *Celsia* with 4 stamens have been previously recognised as distinct genera, but it is now generally agreed that this status is untenable (see Ferguson in B.J.L.S. 64: 229 (1971) and Huber-Morath in Bauhinia 5: 7 (1973)). Fischer in Kubitzki (ed. J.W. Kadereit), Fam. & Gen. of Vasc. Pl.: vol. 7, p. 361 (2004) has followed Hartl and separated *Rhabdotosperma* Hartl from *Verbascum* on the basis of its (*Rhabdotosperma*) seed endosperm morphology and accessory flowers. For the purpose of this Flora I have retained all species in the genus *Verbascum*, but species 3 & 4 can be referred to *Rhabdotosperma*. A taxonomically difficult genus, showing vast variations in some species.

Verbascum longifolium Ten., a native of Italy and Greece has been introduced at Muguga, Kenya (**K** 4) where it is recorded to grow well. It has grey tomentose leaves about 60 cm long and a large panicle 2 m high with yellow flowers.

1. Whole plant densely stellate-pubescent; flowers in fascicles on the rhachis; stamens 5 .... 1. *V. sinaiticum*
   Plants almost glabrous to floccose-pubescent; flowers solitary in the axils of each bract along the rhachis; stamens 4 or rarely 5 (in *V. scrophularifolium*) 2
2. Leaves lyrato-pinnate with numerous lobes .... 2. *V. interruptum*
   Leaves simple or with a few pinnae along the petiole 3
3. Stem often leafy; bracts about as long as the fruiting pedicel .... 3. *V. scrophulariifolium*
   Stem not leafy; bracts much shorter than fruiting pedicel .... 4. *V. brevipedicellatum*

1. **Verbascum sinaiticum** *Benth.* in DC., Prodr. 10: 236 (1846); Murb. in Lunds Univ. Arssk. n.f. 29(2): 234 (1933); Wickens, Fl. Jebel Marra: 147, Map 150 (1976); Blundell, Wild Fl. East Afr.: 379, pl. 394 (1987); U.K.W.F.: 254 (1994); Wood, Handb. Fl. Yemen: 260 (1997); Fischer in Thulin (ed.), Fl. Somal.: 3: 264 (2006) & in Fl. Ethiop. & Eritr. 5: 246 (2006). Types: Egypt, Sinai, *Schimper* 357 (K!, syn.) & between Suez and Gaza, *Bové* 77 (P, syn.)

Biennial herb, 1–2 m or more high; stems simple, densely stellate-pubescent, with a whitish or pale rusty colour. Juvenile plants with a rosette of leaves oblanceolate to oblong, up to 30 cm long, 7–11 cm wide, broadly petiolate, coarsely serrate, acute to acuminate; lower stem leaves similar, smaller, upper ones becoming ovate, ± sessile; all densely stellate-pubescent on both surfaces. Inflorescence well-branched, rarely simple; lower bracts ovate, acuminate, ± 2 cm long, upper bracts inconspicuous. Flowers 2–7 in rather distant groups; pedicels 3–5 mm long. Calyx deeply lobed, lobes lanceolate, 4–9 mm long, densely stellate-pubescent. Corolla yellow, 20–28 mm in diameter. Filaments and anthers orange, filaments bearded with white and purple hairs. Capsule ovoid, 6–7 mm long and 4–5 mm broad, pubescent.

KENYA. Northern Frontier District: Huri Hills, July 1957, *Joy Adamson* 623!; Nakuru District: Njoro, 17 Oct. 1953, *Drummond & Hemsley* 4811!; North Nyeri District: Nanyuki, 31 July 1954, *Verdcourt* 1137!

DISTR. **K** 1, 3, 4; Sudan, Ethiopia, Somalia, Sinai mountains, eastwards through Iran to Afghanistan

HAB. Dry grassy places, roadsides; 1500–2400 m

USES. None recorded on specimens from our area

CONSERVATION NOTES. Least Concern (LC)

SYN. *V. ternacha* A.Rich., Tent. Fl. Abyss. 2: 108 (1851); Engl., Hochsgebirgsfl. Trop. Afr.: 375 (1896); Skan in F.T.A. 4(2): 279 (1906). Type: Ethiopia, Tchélikoté, *Petit* s.n. (P!, syn.) & near Tchelatchekanné, *Schimper* 621 (BM!, K!, syn.)

V. somaliense Baker in K.B. 1895: 222 (1895) Type: Somalia, Golis-range, *E. Cole* (K!, syn.) & *Lort Phillips* (K!, syn.)

2. **Verbascum interruptum** (*Fresen.*) *Kuntze*, Rev. Gen.: 469 (1891); Fischer in Fl. Ethiop. & Eritr. 5: 251 (2006). Type: Ethiopia, Simien, *Rüppell* s.n. (Location unknown; not at FR). Neotype: Ethiopia, Schire Dschogati, *Schimper* 514 (K!, neo., chosen by Fischer (2006))

Biennial, 40–100 cm high; stems slender, simple or sparsely branched, villous to glabrous towards the base, ± glandular-hairy above. Lower leaves 4–15(–21) cm long and 1–4 cm wide, petiolate, lyrato-pinnate, with several to many small oblong or tooth-like segments on each side, terminal lobe much larger broadly ovate or ovate-elliptic, coarsely toothed, pubescent to glabrous; upper stem leaves few, decreasing markedly in size, triangular-ovate, acute at apex, semi-amplexicaul, coarsely toothed; petiole 6–13 cm. Bracts 2 to 4 times shorter than the fruiting pedicels, glandular.

Racemes lax, many-flowered; pedicels in fruit 10–22 mm long. Calyx deeply divided, lobes lanceolate, 2–3 mm long, glandular. Corolla yellow with purple centre, 12–20 mm in diameter. Stamens purple, all bearded with white hairs; style 8–12 mm long, glabrous. Capsule subglobose, 5–6 mm in diameter, glabrous. Seeds numerous, ± 0.5 mm long.

KENYA. Northern Frontier District: Marsabit, 16 Feb. 1953, *Gillett* 15121! & 14 Feb. 1972, *Bally & Smith* 14796! & Marsabit forest edge, 6 Aug. 1968, *Faden* 68/403!
DISTR. **K** 1; Ethiopia, Eritrea
HAB. Montane mist forest and grassland with *Croton* and *Olea*; 1450–2000 m
USES. None recorded on specimens from our area
CONSERVATION NOTES. Not much material from Kenya; possibly at threat from habitat degradation in the Flora area, but not uncommon elsewhere. Assesed here as of Least Concern (LC)

SYN. *Celsia interrupta* Fresen. in Flora 21: 605 (1838); Skan in F.T.A. 4(2): 281 (1906); Murb. in Lunds Univ. Arsskr. N.F. Avd. 2, 22(1): 158 (1925)
C. valerianifolia A.Rich., Tent. Fl. Abyss. 2: 112 (1851); Skan in F.T.A. 4(2): 282 (1906). Type: Ethiopia, Tigre near Memsah, 1839, *Quarton Dillon* & *Petit* s.n. (P!, holo.; K! photo)

NOTE. The type of *Verbascum interruptum* is not traceable at the Fresenius herbarium at FR.

3. **Verbascum scrophulariifolium** (*A.Rich.*) *Huber-Morath* in Bauhinia 5: 15 (1973); U.K.W.F.: 254 (1994). Type: Ethiopia, Semien, Mt Taber, *Schimper* 1417 (B†, holo.; BM!, iso.)

Biennial herb 0.8–2 m tall, ± hairy, usually glandular in the upper parts. Leaves lanceolate, at base up to 30 cm long and 6.5 cm wide, coarsely doubly toothed; stem leaves shortly petiolate, sometimes with very small pinnae or sessile and cordate, decreasing in size upwards. Inflorescence elongated, up to 50 cm, ± simple; bracts triangular, acute, coarsely toothed, almost as long as the pedicel or leafy and longer than pedicel; pedicels in fruit remaining short or sometimes elongating, 5–20 mm long. Calyx deeply 5-divided; lobes lanceolate oblong, 4–8 mm long, toothed or entire. Corolla yellow with purple centre, 14–28 mm in diameter. Filaments all bearded. Capsule globose to ovoid, 8 mm long and wide. Seeds striate.

UGANDA. Ruwenzori, May, *Scott Elliot* 7723!; Mt Elgon, Bulambuli, 4 Sept. 1932, *A.S. Thomas* 525! & 11 Nov. 1933, *Tothill* 2298!
KENYA. West Suk District: Cherangani Hills, Kaibwibich, Aug. 1968, *Thulin & Tidigs* 44!; Trans-Nzoia District: Mt Elgon, Jan. 1931, *Lugard* 491!; Kiambu District: Limuru, Sigoni, *P.L.D. Wood* s.n.!
TANZANIA. Masai District: Ngorogoro Crater, 14 July 1963, *Verdcourt* 3685A!
DISTR. **U** 2, 3; **K** 2–4/6; **T** 2; Cameroon, Burundi, Sudan, Ethiopia
HAB. Montane grassland; 1800–3600 m
USES. None recorded on specimens from our area.
CONSERVATION NOTES. Least Concern (LC); widespread

SYN. *Celsia scrophulariifolia* A.Rich., Tent. Fl. Abyss. 2: 112 (1851); Engl. in Abhandl. Akad. Wissensch. Berlin 1891(2): 376 (1892); Skan in F.T.A. 4(2): 284 (1906); Murb. in Lunds Univ. Arsskr. 22: 72 (1925); A.V.P.: 166 (1957)
Verbascum schimperi Skan in F.T.A. 4(2): 280 (1906); Murb. in Lunds Univ. Arssk. 29: 293 (1933). Type: Ethiopia, Begemeder, *Schimper* 1398 (K!, holo.; BM!, iso.)
Celsia foliosa Chiov. in Ann. Bot. Roma 9: 84 (1911). Type: Ethiopia, Semien, *Chiovenda* 931 (FT, holo.)
C. scrophulariifolia A.Rich. subsp. *foliosa* (Chiov.) Murb. in Lunds Univ. Arsskr. 22(1): 74 (1925)
Rhabdotosperma scrophulariifolia (A.Rich.) D.Hartl in Beitr. Biol. Pfl., 53 (1): 58 (1977); Fischer in Fl. Ethiop. & Eritr. 5: 252 (2006)
R. scrophulariifolia (A.Rich.) D.Hartl subsp. *foliosa* (Chiov.) D.Hartl in Beitr. Biol. Pfl., 53 (1): 58 (1977)

FIG. 3. *VERBASCUM BREVIPEDICELLATUM* — **1**, habit × $^1/_3$; **2**, flower × $^1/_2$; 3; flower detail × 4; **4**, capsule × 4; **5**, seed × 8. 1, 2, 3 from *Hepper & Jaeger* 7195; 4, 5 from *Greenway & Kanuri* 14572. Drawn by Juliet Williamson.

NOTE. A polymorphic species not clearly defined from *V. brevipedicellatum* (see note under that species), but worth recognising on account of the larger bracts. Typically in Ethiopia it has toothed calyx lobes as long as the capsule, a character which is seldom linked with the longer bracts elsewhere. The length of the fruiting pedicel is sometimes no longer than the capsules, and it would be interesting to study the correlation of length of the fruiting pedicel with other characters.

A collection from Kenya, Chyulu Hills, *Luke* 1104, has a laxer inflorescence and smaller flowers, but falls within the range of this taxon.

4. **Verbascum brevipedicellatum** (*Engl.*) *Huber-Morath* in Bauhinia 5: 11 (1973); Blundell, Wild Fl. East Afr.: 378, pl. 393 (1987); U.K.W.F.: 254 (1994). Type: Tanzania, Kilimanjaro [Kilimandscharo], *Meyer* 286 (B†, holo.). Neotype: Tanzania, Kilimanjaro, SE of Bismark hut, *Bigger* 2012 (K!, chosen here)

Biennial herb, somewhat woody at the base, 0.6–2.5 m high; stems almost glabrous to densely floccose, often tinged deep red. Lower leaves petiolate, petiole 1–2.5 cm long; blade simple or occasionally with small pinnate lobes on the petiole, oblong-lanceolate, 3–12 cm long, 1.5–4 cm wide, obtuse to acute, truncate at the base, ± glabrous or pubescent above, pubescent or floccose beneath, margins biserrate; upper leaves ± sessile, sometimes with base clasping. Inflorescence simple or sparingly branched; bracts lanceolate, 3–5(–12) mm long, sessile, serrate; pedicels in fruit 1.3–2.5 cm long, ± glandular-pubescent. Calyx deeply divided, 5–6 mm long in fruit, $^{2}/_{3}$ as long as the mature capsule. Corolla yellow with purple streaks at centre, about 2 cm diameter. Stamens reddish-purple bearded; anthers orange; style purplish. Capsule ovoid, 6–8 mm, glabrous. Seeds striate. Fig. 3, p. 15.

UGANDA. Karamoja District: Moroto Mt, June 1963, *J. Wilson* 1487!; Kigezi District: Kigezi, April 1949, *Purseglove* 2721!; Mt Elgon, April 1930, *Liebenberg* 1668!

KENYA. Mt Kenya, April 1975, *Hepper, Field & Townsend* 4879!; Kericho District: Sotik, 15 June 1953, *Verdcourt* 974!; Masai District: Enesambulai valley, 12 Sept. 1970, *Greenway & Kanuri* 14572!

TANZANIA. Mt Kilimanjaro, 21 July 1968, *Bigger* 2012!; Masai District: Ngorogoro Crater, 5 Sept. 1964, *Richards* 19130!; Lushoto District: Mkuzi, 1 Sept. 1971, *Greenway* 8697!

DISTR. **U** 1–3; **K** 1–6; **T** 2, 3, 6; Congo-Kinshasa, Sudan, Ethiopia, Somalia

HAB. Margins of upland forest and rocky montane grassland; 1200–3600(–4100) m

USES. An important medicinal herb whose dried leaves mixed with water are used to prevent (cure) kwashiorkor and fresh leaves used to help with afterbirth.

CONSERVATION NOTES. Least Concern (LC); common and widespread

SYN. *Celsia brevipedicellata* Engl. in Abhandl. Akad. Wissensch. Berlin 1891(2): 376 (1892); Skan in F.T.A.: 4(2): 285 (1906); Murb. in Lunds Univ. Arsskr. 22, 1: 65 (1925); A.V.P.: 164 (1957); F.P.U. ed. 2: 133 (1971)

C. brevipedicellata Engl. var. *homostemon* Murb. in Lunds Univ. Arsskr. 22, 1: 67 (1925). Type: Kenya, Aberdare Mts, 13 March 1922, *R.E. & T.C.E. Fries* 2262 (UPS!, holo.)

C. brevipedicellata Engl. var. *heterostemon* Murb. in Lunds Univ. Arsskr. 22, 1: 68 (1925). Type: East Africa, unspecified

C. keniensis Murb. in Lunds Univ. Arsskr. 22, 1: 70, t. 2 (1925); Glover, Prov. Check-List Brit. & It. Somal.: 244 (1947). Type: Kenya, Mt Kenya, *R.E. & T.C.E. Fries* 458 (K!, UPS, syn.), 1869 (UPS!, syn.)

[*Celsia floccosa* sensu Agnew UKWF: 550 (1974), *non* Benth.]

Rhabdotosperma brevipedicellata (Engl.) D.Hartl in Beitr. Biol. Pfl. 53(1): 58 (1977); Fischer, F. A.C. Scrophulariaceae: 12, pl. 2 (1999) & in Fl. Ethiop. & Eritr. 5: 252 (2006)

R. keniensis (Murb.) D.Hartl in Beitr. Biol. Pfl., 53 (1): 58 (1977); Fischer in Thulin (ed.), Fl. Somal.: 3: 266 (2006)

NOTE. This is an extremely variable species; field studies would be necessary to study the morphological variations (if constant) amongst populations. With the material that I (S.A.G.) have studied, it is difficult to recognise the species (and even less the infraspecific ranks) that have been described around this species.

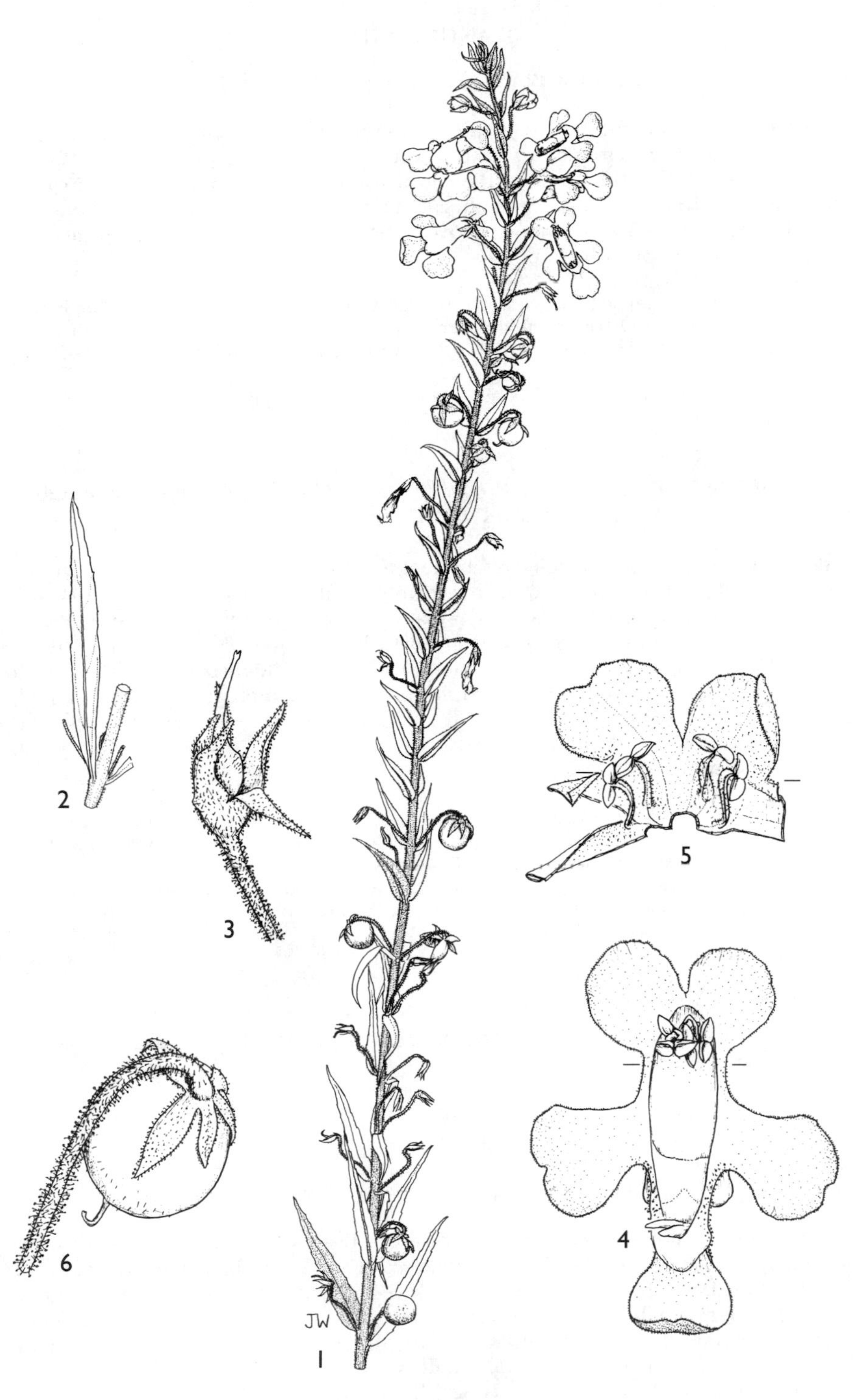

FIG. 4. *ANGELONIA BIFLORA* – **1**, habit × 2/3; **2**, leaf × 1/2; 3, calyx with young fruit × 4; **4**, corolla × 3; **5**, flower section with stamens × 3; **6**, capsule × 3. 1, 3–6 from *Ntemi, Sallu & Madege* 517; 2 from *Cadet* 5309. Drawn by Juliet Williamson.

4. ANGELONIA

Humb. & Bonpl., Pl. Aequim. 2: 92 (1812)

Herbs or subscandent shrubs. Leaves opposite or rarely subopposite, margins usually serrate. Flowers single or paired in the upper axils, often racemose. Calyx 5-lobed. Corolla 5-lobed, lobes spreading and rounded, the lower lip with a cavity. Stamens 4, didynamous; anthers bithecal. Ovary globose or ellipsoid, 2-locular; stigma acute. Fruit capsular, globose or ovoid, bivalvate, loculicidal. Seeds numerous, testa with a hyaline pitted membrane.

Some 30 species endemic to the New World, especially Brazil. Two species (*A. biflora* Benth. and *A. angustifolia* Benth.) have been introduced into cultivation in the Old World tropics for the value of the beautiful flowers, where *A. biflora* has become naturalized in some areas.

The very characteristic cavity or pocket in the lower lip of the corolla makes this an easily distinguished genus.

Angelonia biflora *Benth.* in DC., Prodr. 10: 254 (1846); Hepper in Fl. Masc. 129: 5 (2000). Type: Brazil, Ceara, *Gardner* 1795 (K!, holo.)

Perennial herb 40–60 cm high, viscid-pubescent throughout; stems several, simple, ± quadrangular. Leaves linear to narrowly linear-ovate, 3–9 cm long, 5–9(–14) mm wide, sessile, ± cuneate at base, very acute at apex, margins shallowly serrate. Bracts similar to leaves diminishing in size upwards, longer than the pedicels; pedicels 1–2 cm long, curved in fruit. Flowers blue-purple, with white throat, usually 2 in each upper axil. Calyx ± 4 mm long, 5-lobed, lobes ± equal, acuminate. Corolla 2-lipped, ± 1 cm long and 2 cm across, with small glands all over; lobes 5, ± equal; lower lip with a central cavity and thrust forward; style and stigma glabrous. Capsule rotund, ± 5 mm in diameter. Fig. 4, p. 17.

TANZANIA. Lushoto District: Amani Nature Reserve, road from Kwamkoro Forest Reserve, Ngua area, *Ntemi, Sallu & Athumani* 479! & Amani Nature Reserve, 10 Mar. 2000, *Ntemi Sallu & Madege* 517! & Amani Nursery, 22 Apr. 1930, *Greenway* 2226!

DISTR. **T** 3; South Africa, Madagascar; Reunion, Mauritius, Rodrigues. Apparently introduced recently, but now naturalised. Native to Brazil, now widely cultivated and often naturalized throughout tropical Asia and parts of tropical Africa

HAB. Waste and disturbed land, near streams; ± 950 m

USES. None recorded on specimens from our area

CONSERVATION NOTES. Least Concern (LC)

NOTE. This species has been misidentified as *A. salicariifolia* Humb. & Bonpl., *A. goyazensis* Benth. and *A. grandiflora* Morren.

5. DICLIS

Benth. in Hooker, Comp. Bot. Mag. 2: 23 (1836); Philcox in F.Z.: 8(2): 12 (1990)

Annual herbs, stems slender, prostrate or erect. Leaves opposite or the upper ones alternate, petiolate or sessile. Flowers axillary, solitary, pedicellate, ebracteate. Calyx deeply 5-lobed. Corolla 2-lipped, tube produced in front as a spur, sometimes spur obscure or very small; posterior lip 2-fid, anterior 3-fid. Stamens 4, didynamous; filaments of the anterior pair sharply bent at the base; anthers monothecal by confluence, cohering in pairs. Style small. Capsule subglobose, septicidal, valves reflexing, entire or partly bifid. Seeds ovoid, reticulate-fulveolate.

Nine species in tropical and South Africa, and Madagascar; usually in mountains.

1. Plant ± erect, sometimes decumbent; corolla spur 4–7 mm long; upland species .... 1. *D. tenella*
   Plant procumbent or erect; corolla spur 0–3 mm long 2
2. Stems rooting at the nodes; leaf margin with 4–8 coarsely broadly dentate-serrate apiculate teeth on each side; pedicels 3–4.5 cm long, exceeding the leaves; a high altitude bamboo forest species .... 2. *D. bambuseti*
   Stems not rooting at the nodes; leaf margin dentate or obscurely dentate; pedicels 1–2.5(–4) cm long, not exceeding the leaves; upland species .... 3. *D. ovata*

1. **Diclis tenella** *Hemsl.* in K.B. 1896: 163 (1896); Hemsley & Skan in F.T.A. 4(2): 287 (1906); Brenan in Mem. N.Y. Bot. Gard. 9(1): 10 (1954); Philcox in F.Z. 8(2): 12 (1990). Type: Malawi, Mt Chiradzulu, *Whyte* s.n. (K!, holo.)

Slender erect or weakly decumbent herb, 6–25 cm long, rooting at some nodes, usually rather densely hirsute with weak white hairs. Leaves opposite, petiolate; petiole 8–23 mm long; lamina rounded-ovate, 9–41 mm long, 6–38 mm wide, rounded to subcuneate at the base, acute at apex, margins coarsely serrate, slightly glutinous-hirsute. Flowers axillary, solitary; pedicels ± 2 cm long, slender to filiform, reflexed in fruit, shortly villous to ± glabrous. Calyx lobes 1–1.5 mm long, unequal, upper lobe much longer and ovate-oblong. Corolla white or pale mauve with darker spur, 7–12 mm long, lower lip 3-lobed, ± 3 mm long, 2 mm wide, emarginate, upper lip deeply bi-lobed; spur 4–7 mm long. Capsule 2.5–3 mm in diameter, laterally compressed, valves spreading, minutely glandular hairy. Fig. 5, p. 21.

TANZANIA. Songea District: Matengo Hills, Luwiri Kitesa, Mar. 1956, *Milne-Redhead & Taylor* 9033! & Mbinga, 4 Mar. 1987, *Congdon* 165!
DISTR. **T** 8; Malawi, Mozambique, Zimbabwe
HAB. Shady places under rock ledges; 1500–2100 m
USES. None recorded on specimens from our area
CONSERVATION NOTES. Least Concern (LC), but known only from a few specimens in the Flora area

2. **Diclis bambuseti** *R.E.Fries* in Act. Hort. Berg. 8: 47, fig. 1 (1925); Agnew, U.K.W.F.: 254, pl. 110 (1994); Fischer in Fl. Ethiop. & Eritr. 5: 256 (2006). Type: Kenya, Mt Kenya, *R.E. & T.C.E. Fries* 689 & 1231 (UPS, syn; K!, isosyn.), Mt Aberdare, *R.E. & T.C.E. Fries* 2323 (UPS, syn; K!, isosyn.). Lectotype: Kenya, Mt Kenya, *R.E. & T.C.E. Fries* 689 (K!, chosen here)

Diffusely branched decumbent herb 50 cm or more long, rooting at the nodes; internodes 3–5 cm long, hirsute with glandular hairs. Leaves opposite, petiolate; petiole 1.5–4 cm long; lamina broadly ovate, 1–2 cm long, 1.5–3.5 cm wide, truncate or sometimes subcordate at the base, acute at the apex, margins with 4–8 coarsely broadly dentate-serrate apiculate teeth on each side, laxly hirsute on both sides. Flowers axillary, solitary; pedicels slender, 3–4.5 cm long, glandular-hairy, exceeding the leaves. Calyx lobes linear-lanceolate or lanceolate, 1.5–2.5 mm long. Corolla white, pink towards the base, 11–12 mm long, covered with minute sessile glands, without a spur or with a small obtuse saccate spur 1–2 mm long, lobes of upper lip linear-oblong, 3 mm long, lower lobe longer and wider, subacute at apex. Capsule ± 5 mm in diameter, laterally compressed, glandular-hairy.

UGANDA. Mbale District: Benet Sabei, Dec. 1938, *A.S. Thomas* 2639!
KENYA. Trans-Nzoia District: Mt Elgon, May 1950, *Tweedie* 842!; Aberdare Mts, May 1922, *R.E. & Th.C.E. Fries* 689!; Mt Kenya, April 1975, *Hepper & Field* 4872!

TANZANIA. Morogoro District: Uluguru Mts, Hululu Falls, Mgeta R., Aug. 1951, *Greenway & Eggeling* 8668!
DISTR. **U** 3; **K** 3, 4; **T** 6; not known elsewhere
HAB. Montane moist forest in the bamboo (*Arundinaria alpina*) zone; 2000–3100 m
USES. None recorded on specimens from our area
CONSERVATION NOTES. Least Concern (LC)

NOTE. This species typically lacks a spur but some specimens have a small spur.

3. **Diclis ovata** *Benth.* in Hook. Comp. Bot. Mag. 2: 23 (1836); Engl. in P.O.A. C.: 356 (1895); Hemsley & Skan in F.T.A. 4(2): 287 (1906); Cufodontis in B.J.B.B. 33, Suppl.: 888 (1963); Hepper in F.W.T.A. ed. 2, 2: 354 (1963); Philcox in F.Z.: 12 (1990); U.K.W.F.: 254 (1994); Fischer, F.A.C. Scrophulariaceae: 24, pl. 6 (1999) & in Fl. Ethiop. & Eritr. 5: 256 (2006). Type: "Madagascar, Herb. Hooker" (K!, holo., see note)

Erect annual herb 3–17 cm high, ± branched, decumbent or ascending, rooting at nodes; stems subquadrangular, ± pubescent and glandular. Leaves mostly opposite below and alternate above, petiolate; petiole 5–19 mm long, glandular-hairy; lamina broadly ovate, 9–20(–41) mm long, 6–15(–26) mm wide, cuneate or truncate at the base, margins dentate sometimes obscurely so, sparsely pubescent especially on the veins. Flowers axillary, solitary; pedicels filiform, 1–2.5(–4) cm long, not exceeding the leaves, glandular. Calyx lobes unequal, 1.2–2 mm long, glandular-hairy. Corolla white with upper lip brownish with spur, 3.5 mm long, 1 mm wide, lower lip 3-lobed, upper lip emarginate. Capsule 4 mm in diameter, 4-valved, valves spreading at dehiscence. Fig. 5, p. 21.

UGANDA. Kigezi District: Kachwekano Farm, Mar. 1950, *Purseglove* 3342!; Mengo District: Entebbe, *E. Brown* 317!; Masaka District: Bugala I., Sese, Oct. 1958, *Symes* 498!
KENYA. Naivasha District: S Kinangop, June 1961, *Polhill* 430!; Kiambu District: Kabete, Aug. 1947, *Bogdan* 1065!; Masai District: km 48 on main road – Narok, 16 June 1956, *Verdcourt* 1502!
TANZANIA. Lushoto District: Mkuzi–Kwai road, May 1953, *Drummond & Hemsley* 2675!; Mbeya District: Mbeya, May 1975, *Hepper & Field* 5274! & Matagoro Hills, Feb. 1956, *Milne-Redhead & Taylor* 8859!
DISTR. **U** 2–4, 6; **K** 2–6; **T** 1–4, 6–8; Ethiopia, Angola, Zambia, Malawi, Mozambique, Zimbabwe; Madagascar, Mauritius
HAB. Usually in damp places in upland grassland, montane forest where it may be mat-forming, or in cultivated ground; 1300–2300 m
USES. None recorded on specimens from our area
CONSERVATION NOTES. Least Concern (LC); widespread

SYN. *Linaria veronicoides* A.Rich., Tent. Fl. Abyss. 2: 114 (1851). Type: Ethiopia, Adoua, September, *Quartin Dillon* s.n. (?FT)
Anarrhinum veronicoides (A.Rich.) O.Kuntze in Jahrb. Bot. Gart. Berlin 4: 269 (1886)
Simbuleta veronicoides (A.Rich.) O.Kuntze, Rev. Gen. Pl. 2: 465 (1891)

NOTE. There are 3 plants on a sheet at K with 'type sheet' written on it possibly in Bentham's handwriting. All are from Madagascar: the first collected by G.W. Parker in Aug. 1880, the middle one without a collector's name on a cut out label giving the name of the country only (Madagascar), and the third collected by R. Baron in July 1880. As the new species was published in 1836, neither of the two collections dated 1880 can be the holotype of *Diclis ovata*. In the protologue Bentham cites a specimen fom Madagascar which he has seen, but does not give the name of the collector or a collection number. It is quite probable that the middle specimen is one of Bojer's collections from Madagascar, of which a couple are at P, and I have seen them. The specimen at K is definitely the holotype, however it cannot be said if those of Bojer's at P are isotypes.

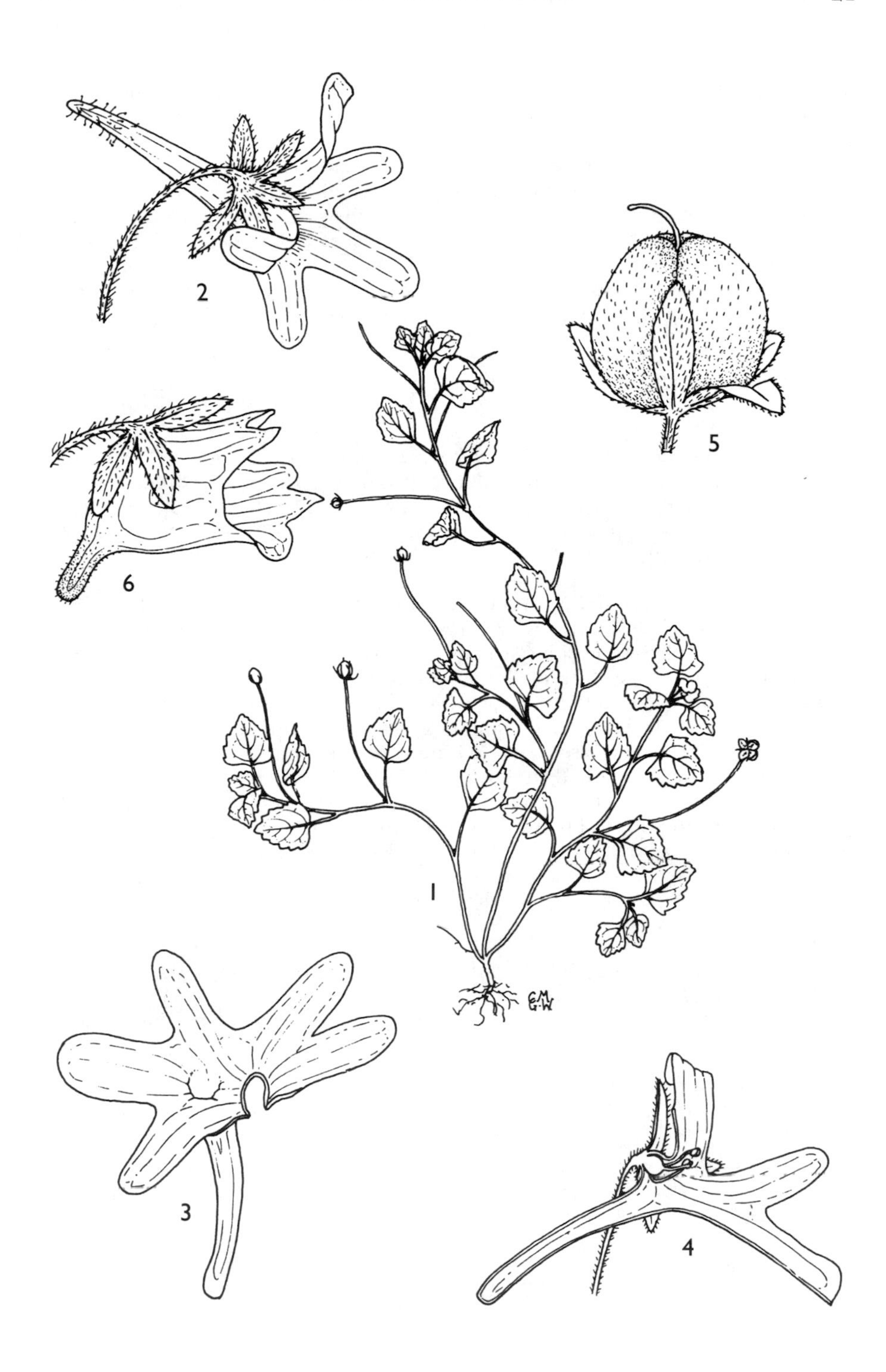

Fig. 5. *DICLIS TENELLA* — **1**, habit × $^2/_3$; **2**, flower × 4; **3**, corolla × 4; **4**, corolla dissected to show spur × 4; **5**, capsule × 8. *DICLIS OVATA* — **6**, flower × 8. 1–5 from *Brummit* 9673; 6 from *Fanshawe* 5773. Drawn by Christine Grey-Wilson. Reproduced with permission from F.Z.

6. NANORRHINUM

Betsche in Cour. Forsch. Inst. Senckenberg 71: 131 (1984); Ghebrehiwet in Nord. J. Bot. 20(6): 668 (2000)

Annual or perennial herbs; stems procumbent or climbing, glabrous or glandular or pubescent. Leaves, at least the upper ones, usually alternate. Flowers axillary, solitary, pedicellate. Calyx deeply 5-lobed, lobes subequal. Corolla bilabiate; upper lip 2-lobed, lower lip 3-lobed with a prominent hairy palate between closing the throat; corolla tube extended at the base into a long slender acute spur. Stamens 4, didynamous; anthers bithecal, glabrous, marginally coherent forming a ring-like structure; style simple; stigma capitate. Capsule subglobose, with 2 equal locules; dehiscence valvate. Seeds oblong-ovoid, tuberculate.

Ten species distributed in the tropical and subtropical regions of Africa and Asia.

Nanorrhinum ramosissimum (*Wall.*) *Betsche* in Cour. Forsch. Inst. Senckenberg 71: 132 (1984); Ghebrehiwet in Nord. J. Bot. 20(6): 669 (2000) & in Thulin (ed.), Fl. Somal.: 3: 275 (2006); Fischer in Fl. Ethiop. & Eritr. 5: 242 (2006). Lectotype: India, Uttar Pradesh, "near Mirzapur on Ganges R.", *Wallich* 3911c (K!; BM!, isolecto., selected by Sutton, 1988)

Much-branched perennial up to 60 cm long with wiry prostrate or decumbent branches, hanging or appressed against vertical cliffs, glabrous to sparsely pubescent. Leaves heteromorphic usually decreasing in size along the stems, well spaced, petiolate; petioles 1–6 mm long, patent. Basal leaves ovate to oblong-ovate to elliptic, 5–7 mm long, 2–4 mm wide, rounded to cuneate at base, obtuse, usually entire; upper leaves ovate, 4–10 mm long, 2–5 mm wide, usually hastate and truncate or cordate at the base, acute to obtuse, entire, pubescent. Flowers axillary; pedicels filiform, much longer than most of the leaves. Calyx lobes lanceolate, 3–5 mm long with narrow scarious margins, acuminate, glabrous. Corolla pale yellow, violet on the inside of the upper lip and near the base of the lower lip, ± 3.5 mm long including the spur, spur forming an obtuse angle with corolla tube, slender, longer than the corolla tube. Abaxial filaments pubescent. Capsule globose to subglobose, 2.5–5 mm, slightly shorter than the calyx. Seeds numerous, oblong-ovoid, tuberculate. Fig. 6, p. 23.

KENYA. Northern Frontier District: Choba Goff, Aug. 1959, *Archer* 348! & Mar. 1963, *Bally* 12569! & Mt Kulal, Nov. 1978, *Hepper & Jaeger* 6898! 6960!

DISTR. **K** 1; Sudan, Ethiopia, Somalia; widespread from Myanmar westwards through India, Pakistan, Iran and Arabia to NE Africa

HAB. Moist cliff crevices in dry country; 1000–1700 m

USES. None recorded on specimens from our area

CONSERVATION NOTES. Known only from a few collections in the Flora area, but widespread elsewhere. Here assessed as of Least Concern (LC)

SYN. *Linaria ramosissima* Wall., Pl. As. rar. 43, t. 153 (1831)
Kickxia ramosissima (Wall.) Janchen in Osterr. Bot. Zeitschr. 152 (1933); Sutton, Rev. Antirrhineae: 190, fig. 46 (1988)

7. MISOPATES

Rafin., Autikon Botanikon: 158 (1840)

Annual herbs. Leaves opposite but those in the upper part of the stem alternate, entire, minutely veined, shortly petiolate. Flowers axillary or in a leafy terminal raceme, sessile or shortly pedicellate. Calyx 5-lobed, with long narrow segments, at

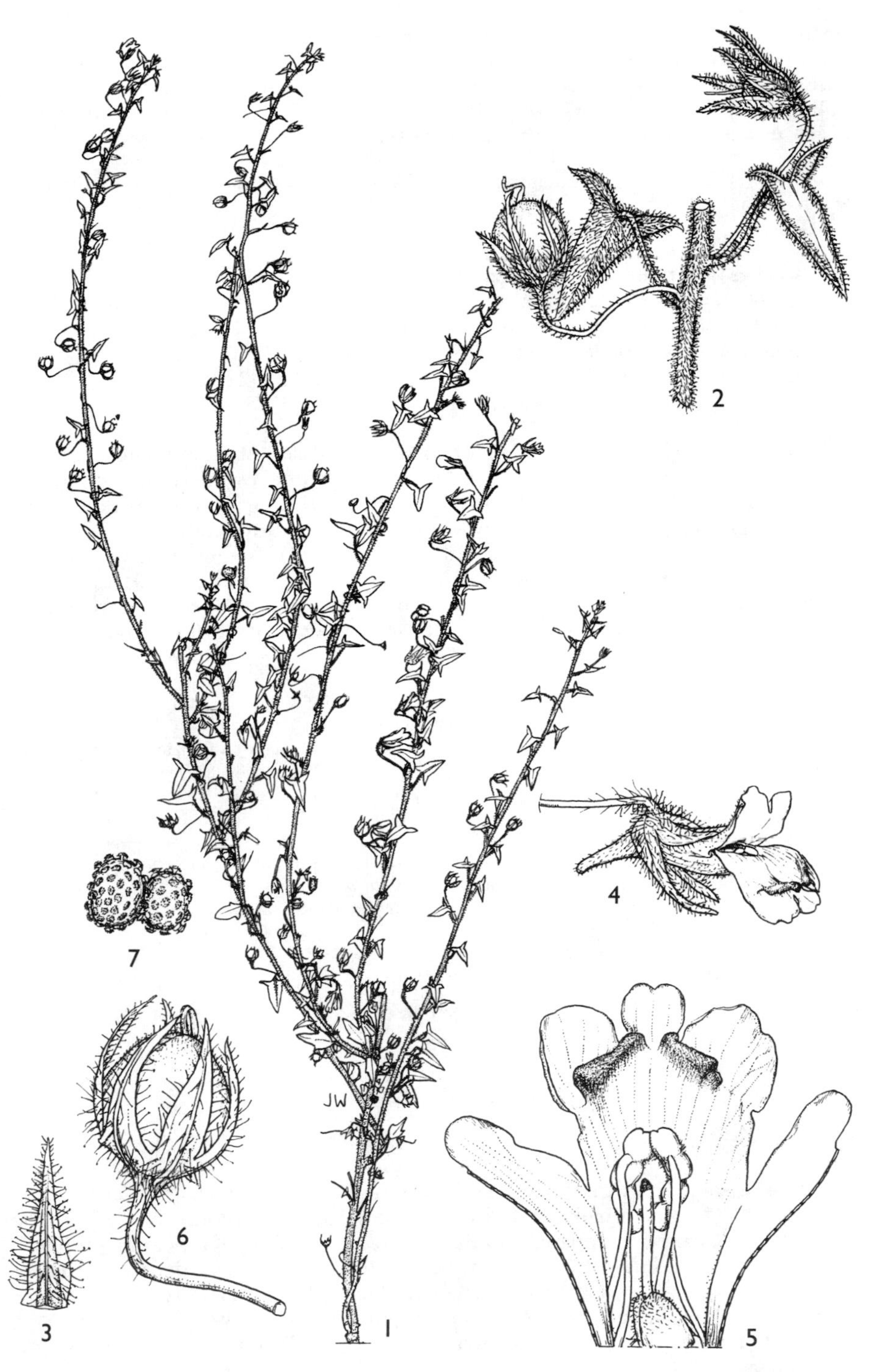

FIG. 6. *NANORRHINUM RAMOSISSIMUM* — **1**, habit × ⅔; **2**, leaf and capsule × 5; **3**, sepal × 8; **4**, flower × 6; **5**, flower opened × 10; **6**, capsule × 8; **7**, seeds × 20. All from *Hepper & Jaeger* 6960. Drawn by Juliet Williamson.

least as long as the corolla tube, the posterior exceeding the others. Corolla bilabiate, small or medium, saccate at the base, not spurred; upper lip 2-lobed; lower lip 3-lobed produced into a large palate closing the hairy throat; stamens 4, inserted at the base of the tube; staminode very small, scale-like. Capsule ovoid, oblique, thick walled, locules unequal, the anterior locule broad and short, dehiscing by two apical pores, posterior locule small with a single pore.

A small genus of 3 species extending from Cape Verde to NW India and from Europe to South Africa.

Misopates orontium (*L.*) *Rafin.*, Autikon Botanikon: 158 (1840); Wickens, Fl. Jebel Marra: 146, map 147 (1977); Philcox in F.Z. 8(2): 16, t. 5 (1990); U.K.W.F.: 254, pl. 110 (1994); Wood, Handb. Fl. Yemen: 262 (1997); Fischer, F.A.C. Scrophulariaceae: 11 (1999)& in Fl. Ethiop. & Eritr.: 5: 239 (2006); Thulin in Fl. Somal.: 3: 272 (2006). Lectotype: W Europe, (in fields), Herb. Clifford: 324. *Antirrhinum* 11 (BM), designated by Fischer in F.R. 108: 113 (1997)

Annual herb (10–)20–60 cm high, often simple or with erect branches, more or less pubescent with weak hairs or glabrescent. Leaves linear to narrowly lanceolate, 1–5 cm long, 2–3(–7) mm wide, base gradually narrowing into the petiole, apex acute to obtuse, entire, glabrous or slightly ciliate. Inflorescence few- to many-flowered raceme; flowers well separated; bracts leaf-like, longer than the flowers; pedicels 1–2 mm long, glandular-hairy. Calyx lobes linear, 8–10 mm long, lengthening in fruit to ± 14 mm, sparingly hispid and glandular-hairy. Corolla pale pink or white with purple veins, 8–21 mm long, hairs on the lip yellow. Capsule obliquely ovoid, 6–12 mm long, hispid. Seeds cylindrical, 1 mm long. Fig. 7, p. 25.

UGANDA. Karamoja District: Napak, June 1950, *Eggeling* 5886! & Mt Morongole, July 1965, *J. Wilson* 1953!; Kigezi District; Maziba, Dec. 1944, *Purseglove* 1616!

KENYA. Naivasha District: Longonot, July 1961, *Verdcourt* 3202!; Nakuru District: Elburgon, July 1951, *Bogdan* 3171!; Teita District: Taita Hills, Sept. 1953, *Drummond & Hemsley* 4296!

TANZANIA. Bukoba District: Kagera, Minziro Forest Reserve, 15 Aug. 2001, *Festo & Bayona* 1774!; Masai District: Ngorongoro Crater Wall, Feb. 1962, *Newbould* 5966!; Kondoa District: Kondoa, Jan. 1962, *Polhill & Paulo* 1206!

DISTR. **U** 1, 2; **K** 1, 3, 4, 6, 7; **T** 1, 2, 5; widespread in the mountains of eastern tropical Africa, North and South Africa, Europe, Arabia and parts of Asia

HAB. Weed of cereal crops and cultivation, generally on sandy soil and open grassy places, by roads; 1600–2850 m

USES. None recorded on specimens from our area

CONSERVATION NOTES. Least Concern (LC); widespread

SYN. *Antirrhinum orontium* L., Sp. Pl.: 617 (1753); Skan in F.T.A. 4(2): 294 (1906)

8. HALLERIA

L., Sp. Pl.: 625 (1753) & Gen. Pl. ed. 5: 274 (1754)

Shrubs or small trees, glabrous. Leaves opposite, petiolate, thick and glossy, entire or toothed. Flowers usually in fascicles, cauliflorous or among leaves, axillary, pedicellate. Calyx campanulate, with 3–5 shallow lobes, persistent. Corolla tubular, curved or straight, inflated in the upper part, obscurely 2-lipped, the upper shortly bilobed, the lower shortly 3-lobed. Stamens 4, didynamous; filaments inserted at the middle of the corolla tube; anther cells diverging, dehiscing by slit and ± confluent; staminode absent; style filiform; stigma minute. Ovules numerous in both locules. Fruit a berry.

A genus of 7 species distributed in eastern and southern Africa, and Madagascar.

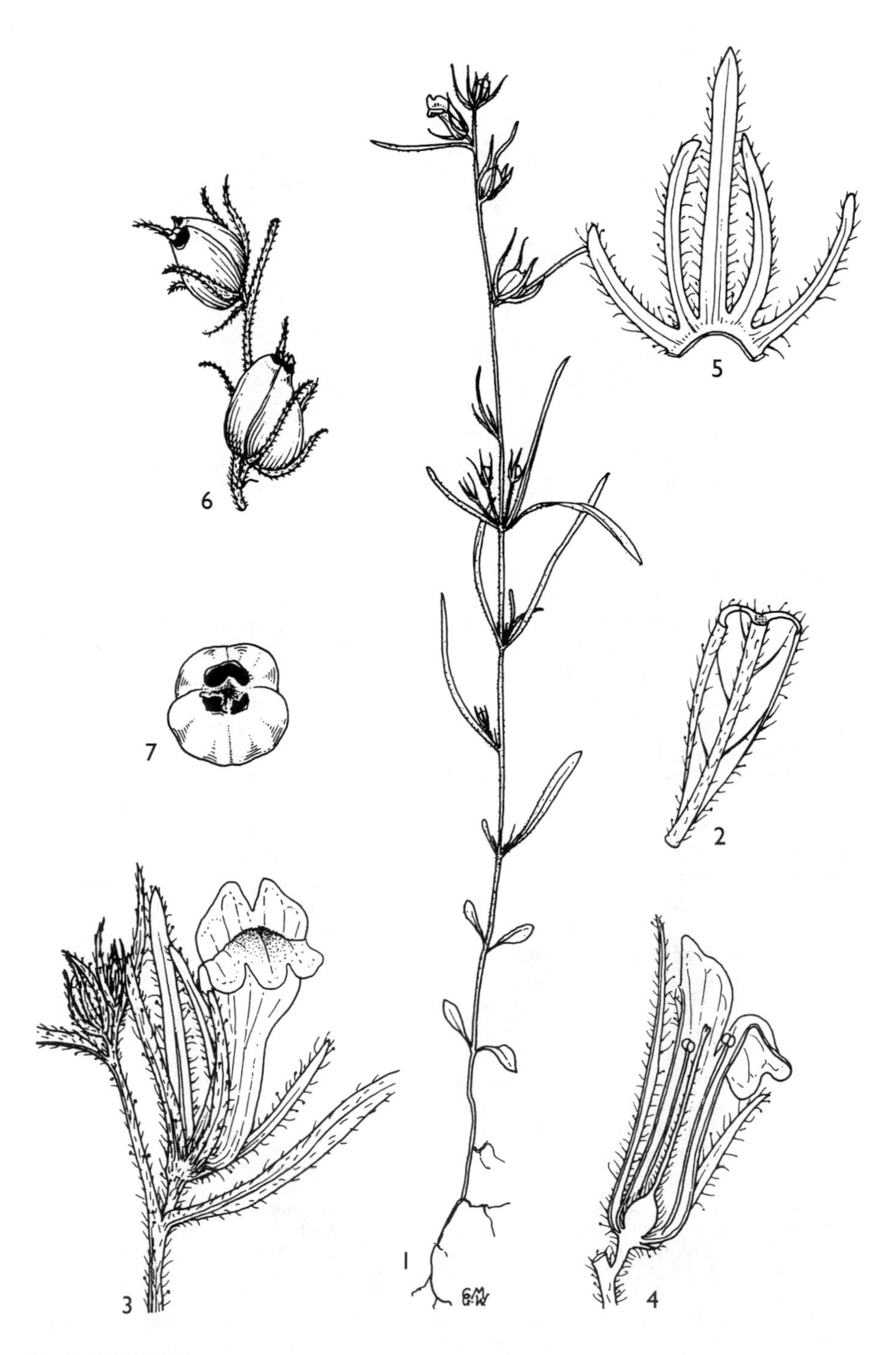

FIG. 7. *MISOPATES ORONTIUM* — **1**, habit × $^2/_3$; **2**, leaf base × 2; **3**, flowering branch × 4; **4**, flower dissected × 4; **5**, calyx opened × 4; **6**, fruiting branch × 2; **7**, capsule (apical view) × 4. 1–5 from *Polhill & Paulo* 1206; 6, 7 from *Jackson* 2074. Drawn by Christine Grey-Wilson. Reproduced with permission from F.Z.

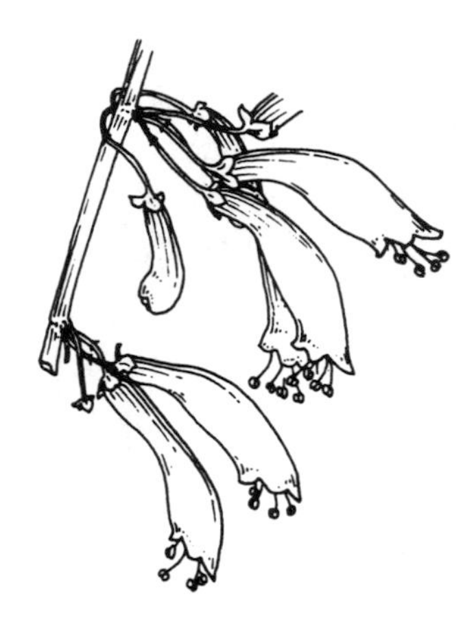

FIG. 8. *HALLERIA LUCIDA* — Flowering branch (part) × ²/₃, from Phillips 3987. Drawn by Christine Grey-Wilson. Reproduced with permission from F.Z.

Halleria lucida *L.*, Sp. Pl.: 625 (1753); A. Rich., Tent. Fl. Abyss. 2: 116 (1851); Sims in Bot. Mag.: t. 1744 (1815); Benth. in Hook., Comp. Bot. Mag. 2: 54 (1854) & in DC., Prodr. 10: 301 (1846); Engl., E.J. 30: 401 (1901); Hiern in Dyer, Fl. Cap. 4(2): 207 (1904); Skan in F.T.A. 4(2): 295 (1906); K.T.S.: 530 (1961); F.F.N.R.: 377 (1962); Cufodontis in B.J.B.B. 33: Suppl. 892 (1963); Philcox in F.Z. 8(2): 16 (1990); U.K.W.F.: 255 (1994); K.T.S.L.: 589, illustr.: (1994); Wood, Handb. Fl. Yemen: 262 (1997); Collenette, Wildflow. Saudi Arabia: 674 (1999); Fischer, F.A.C. Scrophulariaceae: 21, pl. 5 (1999) & in Fl. Ethiop. & Eritr. 5: 255 (2006). Type: "Aethiopia" (= South Africa), *Burmann*, Pl. Afr. Dec.: 244, t. 89, f. 2. (Herman, Cat. Pl. Afr.: 17)

Shrub or small tree, 2–5 m, rarely up to 10 m, well-branched with drooping habit, bark grey. Leaves ± coriaceous, ovate, 4–9 cm long, 2–5 cm wide, base rounded to shortly cuneate, apex acuminate, margins serrate to irregularly serrulate to almost entire, usually entire in the apical part, glabrous, minutely glandular-punctate beneath; petiole 4–10 mm long. Flowers pendulous, usually 2–4 in leaf axils, many more aggregated on leafless stems. Pedicel 10–11 mm; bracteoles lanceolate, ± 2 mm, inserted below the middle of the pedicel. Calyx cup-shaped, lobes 4–5 mm, rounded at apex, spreading. Corolla tube orange-red, curved, 2–2.5(–3.5) cm, lobes rounded, 2 mm, obliquely zygomorphic. Stamens exserted ± 8 mm beyond the corolla lobes; style exserted. Fruit a berry, purple-black when ripe, ovoid, 1.5–2 × 0.7–1.5 cm, cupped by the calyx and often topped by the style. Seeds ellipsoid, somewhat plano-convex; testa thin, punctate, with a narrow marginal wing. Fig. 8.

UGANDA. Acholi District: Agoro, Nov. 1954, *A.S. Thomas* 4376!; Karamoja District: Illipath, Kadam, Jan. 1957, *Philip* 814!; Mbale District: Sipi to Suam River, NW Elgon, Oct. 1939, *Dale* 56!

KENYA. Northern Frontier District: Mt Nyiru, July 1960, *Kerfoot* 1975!; Meru District: Nyambeni Hills, Oct. 1960, *Verdcourt & Polhill* 2968!; Narok District: Olokurto to Olenguerone, May 1961, *Glover, Gwynne & Samuel* 1139!

TANZANIA. Arusha District: Mt Meru, Dec. 1966, *Mrs Richards* 21713!; Lushoto District: Western Shagai forest, May 1953, *Drummond & Hemsley* 2566!; Handeni District: Maunga valley above Mzinga, Sept. 1970, *Harris, Pocs & Csontos* 5160!

DISTR. **U** 1, 3; **K** 1, 3–6; **T** 2–7; extending from Ethiopia to South Africa (Table Mountain) and to Angola; SW Arabia

HAB. Montane forest, especially common along forest margins where it may be dominant; 900–2700 m
USES. None recorded on specimens from our area
CONSERVATION NOTES. Least Concern (LC)

SYN. *H. lucida* L., Sp. Pl.: 625 (1753) var. a. Type: "Aethiopia", Burmann, Pl. Afr. Dec.: 243, t. 89, f. 1.
H. abyssinica Jaub. & Spach, Illustr. Pl. Orient. 5: 65, 66, t. 459 & 460 (1855); Engl., Hochgebirgsfl. Trop. Afr.: 377 (1892); P.O.A. C.: 356 (1895). Type: Ethiopia, *Schimper* 858 (K!, P!, syn.)
H. elliptica Thunb. in Nov. Act. Upsal. 6: 39 (1879); P.O.A. C.: 356 (1895); Skan in F.T.A. 4(2): 296 (1906). Type: as for *H. lucida* var. a

9. **FREYLINIA**

Colla, Hort. Rupul.: 56 (1824); Hiern in Thiselton-Dyer, Fl. Capensis 4(2): 214 (1904); Phillips, Gen. S. Afr. Fl. Pl. ed. 2: 667 (1951)

Shrubs. Leaves opposite, verticillate or alternate, ± sessile, margins entire or toothed in the upper half. Inflorescence cymose, terminal or axillary. Calyx 5-lobed, persistent. Corolla 5-lobed; tube funnel-shaped or subcylindric, straight; lobes much shorter than the tube. Stamens 4, didynamous, occasionally a rudimentary fifth stamen present, inserted about or above the middle of the corolla tube or deep in the throat, included or scarcely exserted; anthers bithecal. Ovary 2-locular; style thick, linear, scarcely exserted; stigma capitate or globose. Fruit an ovoid, obtuse capsule, septicidal. Seeds linear or quadrate in outline or ellipsoid with membranous testa and a marginal wing.

A small genus of 5 species in southern Africa with a single species reaching the tropics.

Freylinia tropica *S.Moore* in J.L.S. 40: 152 (1911); Dyer in Fl. Pl. Afr. 33: pl. 1320 (1959); Philcox in F.Z.: 8(2): 20 (1990). Type: Zimbabwe, Chimanimani [Melsetter], *Swynnerton* 608 (BM!, holo.; K!, iso.)

Well-branched shrub (0.6–)1.5–2.5(–4) m high; stems quadrangular, glabrous, young growth viscid. Leaves very shortly petiolate, lamina coriaceous, oblong-oblanceolate, 8–31 mm long, 4–13 mm wide, base cuneate, apex obtuse, margins with a few coarse teeth on each side in the upper half, midrib and marginal vein conspicuous beneath, glabrous, glandular-punctate. Inflorescences mostly axillary, 2–5-flowered, viscid, very shortly pedunculate. Calyx campanulate, 2–4 mm long; lobes 2–2.5 mm, oblong, with minute glands. Corolla bluish purple or rarely white; tube 10–11 mm long, 2–3.5 mm in diameter, minutely glandular-pubescent; lobes suborbicular, 5 mm wide. Stamens ± 2 mm long; anthers 1 mm long. Ovary oblong-ovoid, 2 mm long; style thick, glabrous. Capsule 4–5 × 3 mm, ellipsoid, bilocular. Fig. 9, p. 28.

TANZANIA. Usambara District: Magamba forest near Lushoto, Feb. 1920, *Swynnerton* 1382!
DISTR. **T** 3; Mozambique, Zimbabwe, South Africa
HAB. Forested valleys, stream beds and grassy hillsides; 1500–1900 m
USES. None recorded on specimens from our area
CONSERVATION NOTES. In Flora area known from a single collection (see note below), otherwise assessed as of Least Concern (LC)

NOTE. It is surprising that only a single specimen is present for the Flora area, collected by Swynnerton in 1920 who also collected the type specimen from Zimbabwe in 1909. The species has not been collected since. It is possible that the species is now extinct in the Flora area or has somehow been overlooked. Collectors should look out for this species.

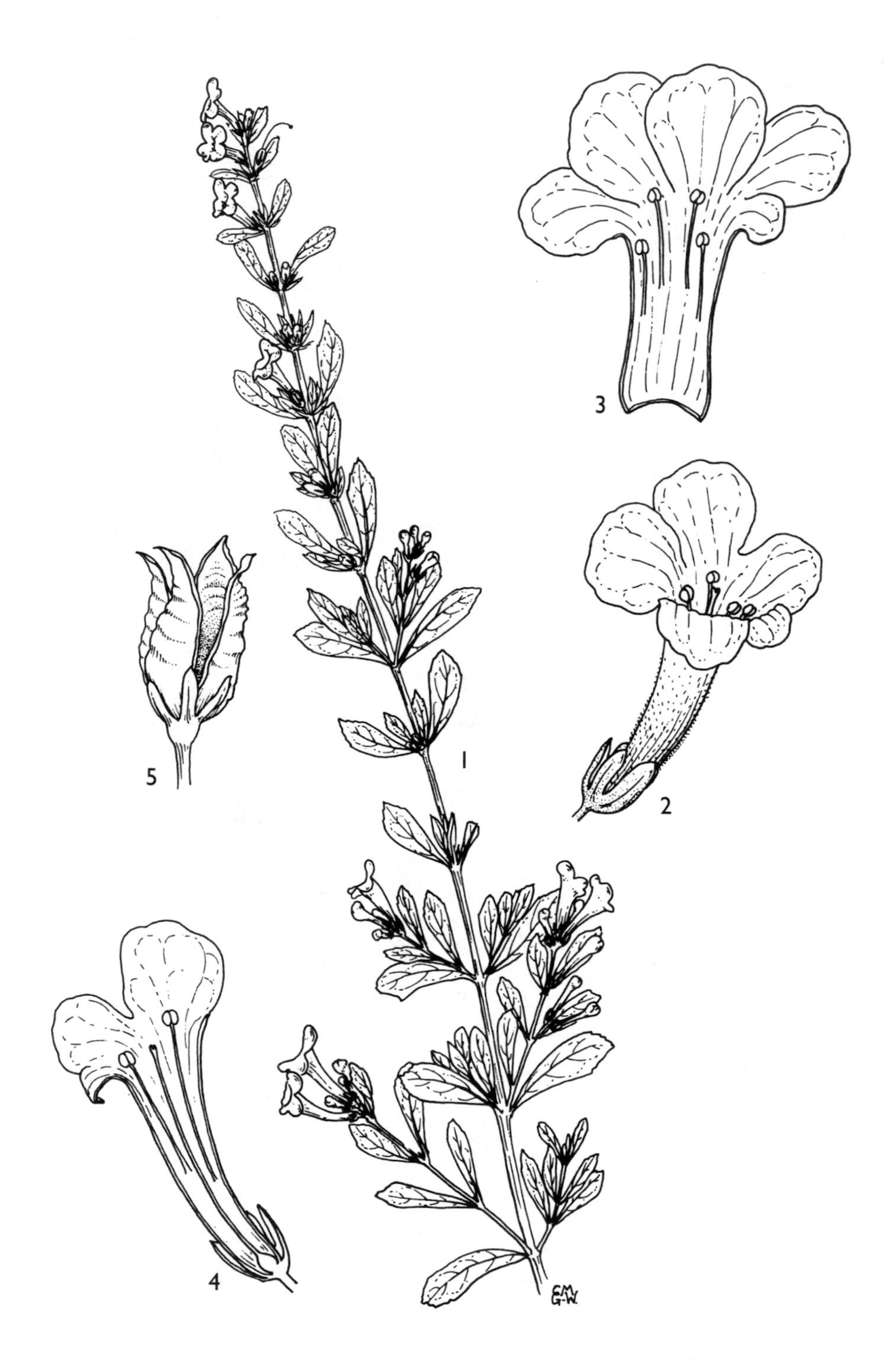

FIG. 9. *FREYLINIA TROPICA* — **1**, flowering branch × $^{2}/_{3}$; **2**, flower × 3; **3**, flower opened × 3; **4**, flower dissected × 3; **5**, capsule dehisced × 4. All from *Crook* M46. Drawn by Christine Grey-Wilson. Reproduced with permission from F.Z.

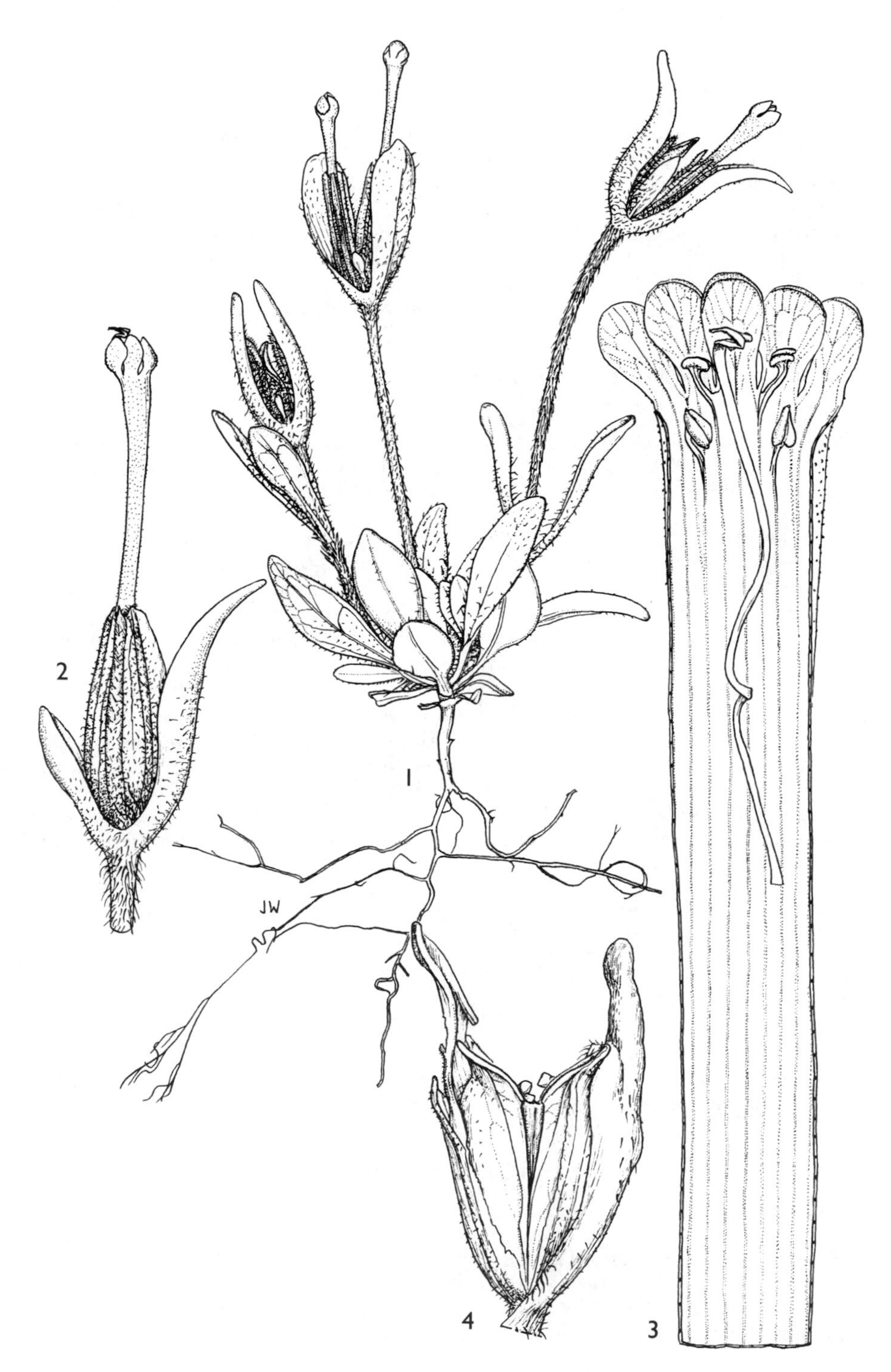

FIG. 10. *ZALUZIANSKYA ELGONENSIS* — **1**, habit × 3; **2**, flower with bract × 6; **3**, corolla tube opened × 12; **4**, capsule dehisced × 6. 1, 2, 4 from *Hedberg* 4478; 3 from *Friis* 241. Drawn by Juliet Williamson.

10. ZALUZIANSKYA

F.W.Schmidt, Neue U. Selt. Pfl.: 11 (1783); Walpers, Repert. Sp. Nov.: 3: 306 (1844); Hilliard, Rev. Manuleae: 460 (1994), *nom. cons., non Zaluzianskia* Necker

Annual or perennial herbs. Leaves simple, toothed or entire; lower leaves opposite; upper leaves usually alternate. Inflorescence a terminal spike or more rarely flowers axillary. Bract adnate to calyx. Calyx bilabiate or 5-lobed, 5-ribbed, sometimes 2- or 5-winged, persistent. Corolla bilabiate, 2–3-toothed or lobed or regularly 5-lobed. Stamens 4 and didynamous, or unusually 2 or 5, inserted near the apex of the corolla tube or mouth; filaments decurrent to base of tube forming a channel for the style; anthers monothecal. Ovary 2-locular with a conspicuous terete gland or a small globose swelling at the base; ovules many; stigma often exserted. Capsule septicidal. Seeds obscurely angled.

About 55 species, mainly confined to southern Africa.

Zaluzianskya elgonensis *Hedberg* in Bot. Not. 123: 512, fig. 2 (1970); Hilliard, Rev. Manuleae: 511 (1994). Type: Uganda, Bugishu District: Elgon, W slopes above Butadiri, *Hedberg* 4478 (UPS, holo.; K!, EA, MHU, PRE, iso.)

Small annual herb, 2–7 cm high (in cultivation up to 15 cm), turning black on drying; stem simple or branched from the base with 2–5 ascending or erect scape-like branches, with lax pubescence of patent to reflexed soft hairs. Leaves opposite, crowded towards the base and usually absent from the upper $^1/_2$–$^2/_3$ of the stem; lamina ovate to lanceolate, 3–15 mm long, 1.5–5 mm wide, entire, slightly pubescent especially along the margins, lower leaves ± petiolate, upper leaves sessile. Inflorescence terminal, capitate with (1–)2–4(–6) densely crowded flowers. Bracts sessile, erect, elliptic-lanceolate, 6–11 mm long, 1–2.5 mm wide. Calyx ± 5 mm long, hyaline with 5 purplish pubescent ribs, 2–3-toothed. Corolla purplish brown outside and cream inside with 5 orange spots round the throat; tube slender, 10–12 mm long, ± glandular outside, glabrous inside, greatly exceeding the calyx; limb cup-shaped and regular with 5 oblong lobes 1.5–2 mm long with entire, rounded apex. Stamens 4, didynamous. Capsule ± 6 mm long. Seeds irregularly prismatic, ± 0.4–0.5 mm in diameter, light brown. Fig. 10, p. 29.

UGANDA. Uganda, Bugishu District: Elgon, W slopes above Butadiri, Dec. 1967, *Hedberg* 4478!
TANZANIA. Kilimanjaro, Shira Plateau, July 1970, *Friis* 241!
DISTR. **U** 3; **T** 2; not found elsewhere
HAB. Afro-alpine zone, rocky ground with thin soil; ± 3800 m
USES. None recorded on specimens from our area
CONSERVATION NOTES. Vulnerable, VU D2 with less than 5 locations and the species has not been re-collected for almost 40 years

NOTE. Since both collections are from the same altitude and zone, the species is expected to occur in the same zone on Mt Kenya, but there are no collections from there so far.

11. JAMESBRITTENIA

(Roth.) Hilliard in Edinb. J. Bot. 49: 231 (1992) & Rev. Manuleae: 84 (1994)

Sutera Roth, Bot. Bemerk.: 172 (1807), *non* Roth, Nov. Pl. Sp.: 291 (1821); E.A. Bruce in K.B. 1940: 63 (1940)
Sutera Roth, Nov. Pl. Spec. 291 (1821), *non* Roth 1807
Lyperia Benth. in Hook., Comp. Bot. Mag. 1: 377 (1836) pro parte, excl. lectotype

Herbs or undershrubs, pubescent or ± viscid-pubescent or more rarely glabrous, becoming black on drying. Leaves opposite, at least the lower ones, often alternate above, petiolate or sessile, dentate, dissected or entire. Flowers solitary in axils of

FIG. 11. *JAMESBRITTENIA MICRANTHA* — **1**, habit × ²/₃; **2**, leaf × 2; **3**, flower side view × 6; **4**, flower opened × 6; **5**, capsule × 4; **6**, seed × 40. All from *S.O. Aleljung* 172. Drawn by Juliet Williamson.

leaves, in terminal racemes or cymes. Calyx 5-lobed almost to the base, lobes ± imbricate. Corolla tubular, expanding near the apex, 5-lobed or 2-lipped; lobes imbricate, spreading, entire or shortly 2-fid or emarginate at the apex. Stamens 4, didynamous, inserted in swollen part of corolla tube; anthers all perfect. Ovary 2-locular, ovules numerous; stigma obtuse or rarely 2-fid. Capsule septicidal. Seeds minute, numerous.

A mainly southern African genus of some 130 species; one species endemic to the Canary Islands and another extending to India.

Jamesbrittenia micrantha (*Klotzsch*) *Hilliard* in Edinb. J. Bot. 49: 231 (1992) & Rev. Manuleae: 208 (1994). Type: Mozambique, 'Rios de Sena', *Peters* s.n. (B†, holo.); near Lupata, Oct. 1858, *Kirk* s.n. (K!, neotype)

Perennial herb up to 60 cm high with a slender tap-root; stems several from base, erect, ascending or procumbent, wiry, subterete, minutely glandular-pubescent. Leaves ovate or ovate-oblong, 8–23 mm long, 4–12 mm wide, coarsely toothed to pinnatifid-lobed, the teeth or lobes entire or crenate-serrate; minutely pubescent and glistening glandular-punctate; petiole 2–5 mm long, usually narrowly winged on both sides. Flowers axillary in elongated racemes. Lower bracts similar to the leaves, upper ones decreasing in size and less lobed, all shorter than the pedicels. Pedicel 6–29 mm long (in fruit). Calyx-lobes oblong-spatulate or linear, ± 3 mm long, 1 mm broad, sometimes serrate towards the apex, subacute, minutely glandular-pubescent. Corolla yellow (or rarely white); tube 3–4 mm long, cylindric; upper lobes ± 1 mm long, lateral lobes ± 1.5 mm long, lower lobe longer, all lobes very obtuse to emarginate. Longer stamens almost exserted. Capsule ellipsoid-ovoid, ± 3 mm long. Seeds rugose. Fig. 11, p. 31.

TANZANIA. Mbeya District: Uyole, 11 km E of Mbeya, 25 Dec. 1974, *Aleljung* 172!
DISTR. **T** 7; Zambia, Malawi, Mozambique, Zimbabwe, Botswana, South Africa, Swaziland
HAB. Wet grassland and clayey or sandy places, near water; 800–1800 m
USES. None recorded on specimens from our area
CONSERVATION NOTES. Least Concern (LC) due to its wide geographical range of distribution; however there is only a single collection from Tanzania

SYN. *Lyperia micrantha* Klotzsch in Peters, Reise Mossamb. Bot.: 222 (1861)
Chaenostoma micranthum (Klotzsch) Engl., P.O.A. C: 356 (1895); Diels in E.J.: 23: 489 (1897)
Sutera fissifolia S.Moore in J.B. 38: 467 (1900). Type: Zimbabwe, Bulawayo, Jan. 1898, *Rand* 155 (BM!, holo.; BR, iso.)
S. micrantha (Klotzsch) Hiern in Dyer, Fl. Capensis 4(2): 263 (1904); Skan in F.T.A. 4(2): 303 (1906); Compton in J.S. Afr. Bot. Suppl. 2: 524 (1976); Philcox in F.Z.: 8(2): 29 (1990)
S. blantyrensis Skan in F.T.A. 4(2): 304 (1906). Syntypes: Malawi, Buchanan in Herb. *Wood* 6630 (K!), *Sharpe* 96 (K!)

12. MIMULUS

L., Sp. Pl.: 634 (1753) & Gen. Pl. ed. 5: 283 (1754)

Annual or perennial herbs, rarely undershrubs, glabrous, pilose or viscid. Leaves opposite, simple, entire or toothed. Flowers axillary, solitary or in terminal racemes. Calyx tubular, as long as the corolla tube, 5-lobed or 5-partite, 5-ribbed. Corolla bilabiate, tube cylindric or swollen, lobes normally shorter than the tube. Stamens 4, didynamous, inserted on the corolla tube at various heights, usually low down or near the base; filaments filiform, included; anthers bithecal; style filiform, terete; stigma flattened, shortly 2-lobed. Fruit a loculicidal capsule. Seeds minute, ellipsoid.

FIG. 12. *MIMULUS GRACILIS* — **1**, habit × $^{2}/_{3}$; **2**, flower × 4; **3**, flower dissected × 4; **4**, fruiting calyx × 4; **5**, capsule with calyx removed × 4; **6**, seeds × 44. All from *Robinson* 737. Drawn by Christine Grey-Wilson. Reproduced with permission from F.Z.

Over 100 species, mainly temperate America; a single species native to the mountains of tropical and southern Africa.

Mimulus gracilis *R.Br.*, Prodr. Fl. Nov. Holl.: 439 (1810); Hiern in Cat. Afr. Pl. Welw. 1: 758 (1898) & in Thiselton-Dyer, Fl. Cap.: 4(2): 354 (1904); Skan in F.T.A.: 4(2): 310 (1906); Wickens, Fl. Jebel Marra: 146 (1977); Philcox in F.Z.: 8(2): 36 (1990); U.K.W.F.: 255 (1994); Wood, Handb. Fl. Yemen: 262 (1997); Fischer, F.A.C. Scrophulariaceae: 74, pl. 28 (1999) & in Fl. Ethiop. & Eritr. 5: 267 (2006). Type: Australia, Hunters R., 1804, *R. Brown* s.n. (BM!, syn.)

Erect annual or perennial herb 20–50 cm, glabrous, branching from the base or simple; stems quadrangular; roots fleshy, white. Leaves opposite, sessile, ± amplexicaul, lanceolate to narrowly oblong, 2–5 cm long, 4–15 mm wide, decreasing in size upwards as bracts, apex obtuse to subacute, margins denticulate, midrib prominent, flanked by 2 lesser veins from the base. Flowers solitary, axillary, loosely racemose; pedicels 15–25 mm long, reaching up to 33 mm in the fruiting stage. Calyx 5–7 mm long, 5-angled, with 5 subequal short teeth 1–1.5 mm long, obscurely ciliate, persistent. Corolla white (with pale mauve tube) and yellow spots on lower lobe, unequally bilabiate; tube ± 8 mm long, upper lobe ± 2 mm, emarginate, lower lobe longer, broadly 3-lobed. Capsule ovoid, 5–6 mm long. Fig. 12, p. 33.

KENYA. Trans-Nzoia District: Kitale, McCoy's bridge, June 1969, *Tweedie* 3647!; Fort Hall District: Nairobi to Makuyu–Sagana, Mar. 1969, *Napper & Haines* 1979!; South Nyeri District: Kirinyaga, 27 Jan. 1972, *Mrs Robertson* 1743!

TANZANIA. Arusha District: Ndurumu, Jan. 1930, *Haarer* 2012! & Sakila, Mar. 1968, *Greenway & Kanuri* 13218!

DISTR. **K** 3, 4; **T** 2; Sudan, Ethiopia, Nigeria, Angola, Zambia, Mozambique, Zimbabwe, Botswana, South Africa; extending to Yemen, India, China and Australia

HAB. Wet pastures and drainage channels especially on black cotton soil; 1150–1800 m (–2650 m fide U.K.W.F.)

USES. None recorded on specimens from our area

CONSERVATION NOTES. Least Concern (LC)

NOTE. *M. puniceus* (Nutt.) Steud., a small shrub with red and orange flowers, American in origin, is recorded to be cultivated in Nairobi.

13. LINDENBERGIA

Lehm., Sem. Hort. Bot. Hamburg. 8 (1829); Hjertson in J.L.S. 119: 265 (1995)

Undershrubs or perennial herbs, branched, usually glandular-villous. Leaves opposite or the upper alternate, dentate. Flowers axillary, solitary or in terminal spikes or racemes, shortly pedicellate. Bracts similar to leaves but smaller. Calyx campanulate, 5-lobed, lobes ± equal, divided to middle. Corolla bilabiate, tube narrowly cylindric; upper lip emarginate or 2-lobed, lower lip longer, 3-lobed. Stamens 4, didynamous, included; filaments filiform, anther thecae stipitate, all fertile. Ovary glabrous (in our plants) or pilose; style filiform; stigma sub-clavate. Capsule ovoid, dehiscence loculicidal, valves entire. Seeds numerous, minute.

12 species distributed from NE Africa through Arabia and India to the Philippines; 4 species in Africa.

Lindenbergia indica (*L.*) *Vatke* in Osterr. Bot. Zeits. 25: 10 (1875); Hjertson in J.L.S. 119: 304 (1995); Wood, Handb. Fl. Yemen: 262 (1997); Hjertson in Thulin (ed.), Fl. Somal.: 3: 282 (2006); Fischer in Fl. Ethiop. & Eritr. 5: 263 (2006). Lectotype: Herb. Linn. No. 800.3 (LINN), designated by Prijanto in Reinwardtia 7: 550 (1969)

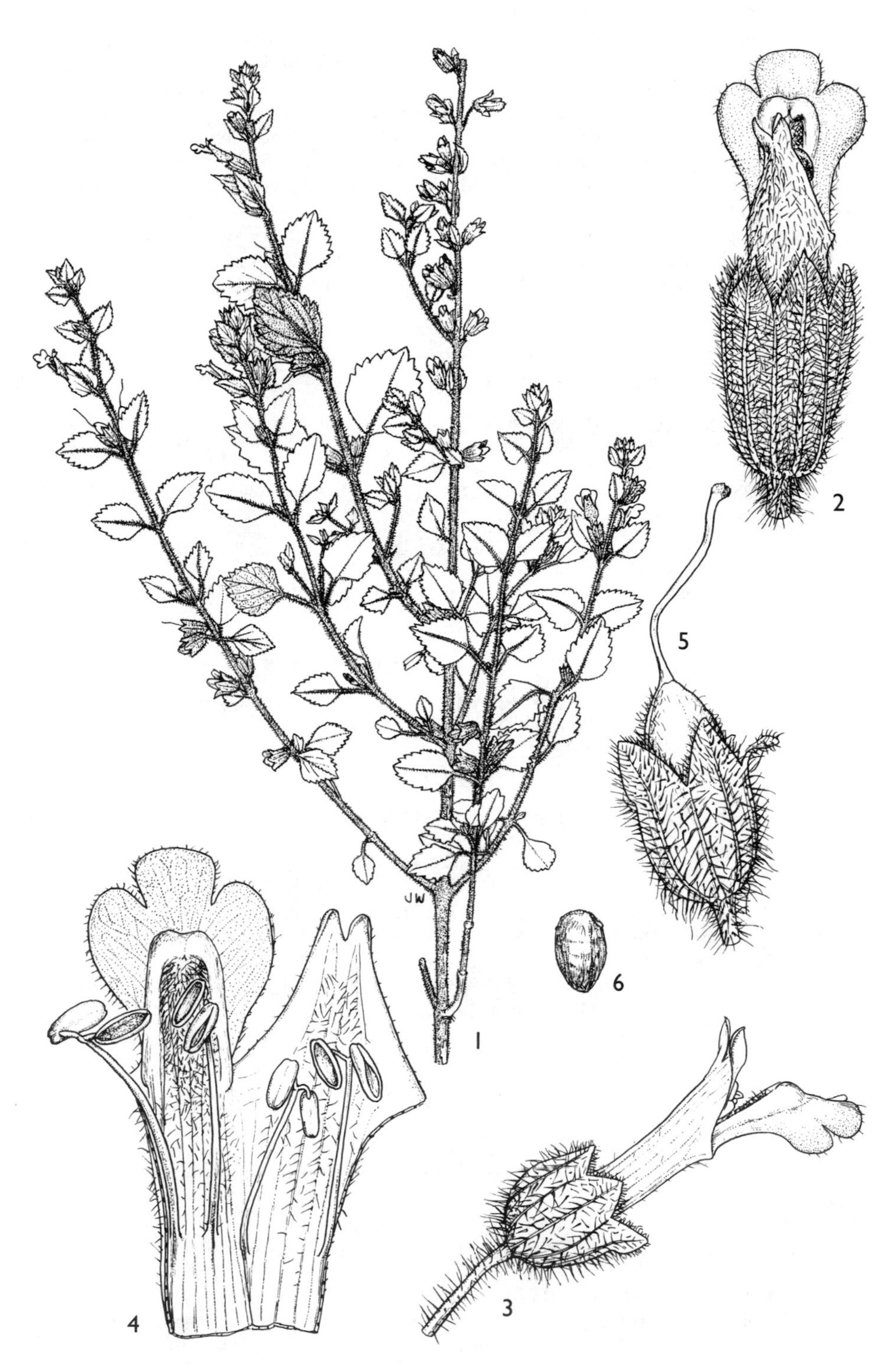

FIG. 13. *LINDENBERGIA INDICA* — **1**, habit × $^{2}/_{3}$; **2**, flower anterior view × 6; **3**, flower side view × 6; **4**, flower opened × 6; **5**, capsule and calyx × 6; **6**, seed × 40. 1, 2, 4 from *B. Mathew* 6671; 3 from *Gilbert & Thulin* 1541; 5, 6 from *Hepper & Jaeger* 6959. Drawn by Juliet Williamson.

Annual or perennial herb, 15–40 cm high, often profusely branched, densely or sparsely pilose with glandular and eglandular hairs (fresh material sticky to touch and fragrant). Leaves broadly ovate to ovate, 1–4 cm long, up to 2.5 cm wide, widely cuneate at the base, obtuse or acute at apex, margins coarsely crenate-dentate; lower leaves petiolate with petiole (5–)10–30 mm, upper leaves with shorter petioles or sessile. Flowers axillary; pedicel 1–5 mm. Bracts 3–4 mm. Calyx 5–6 mm long, lobes triangular, ± 1 mm, 10-ribbed, pubescent with glandular and eglandular hairs, pilose within. Corolla yellow to brownish yellow, often with brown spots, 12–19 mm long, tube narrow, upper lip villous inside, lower lip pubescent along median line. Stamens with the upper filaments inserted 2–5 mm above the base of corolla. Capsule ovoid, 4–8 mm long, narrowing at the apex, glabrous or sparsely pilose. Fig. 13, p. 35.

KENYA. Northern Frontier District: Southern Turkana, Loriu Plateau, Aug. 1968, *Mwangangi & Gwynne* 1072! & Dawa R., Murri, June 1951, *Kirika* 100! & Lotarr, Sept. 1944, *J.Adamson* 131 in *Bally* 3976!

DISTR. **K** 1; Eritrea, Ethiopia, Djibouti, Somalia; Egypt, Arabia, Pakistan, India, Bangladesh

HAB. Rocky desert, in crevices; 350–800 m

USES. None recorded on specimens from our area.

CONSERVATION NOTES. Least Concern (LC), but known from a few collections from Kenya

SYN. *Dodartia indica* L., Sp. Pl. 2: 633 (1753)

Bovea sinaica Decne. in Ann. Sci. Nat. 2nd sér. 253 (1834). Type: Sinai, *Bové* 64 (P, holo.; FI-WEBB, G-DC, K!, iso.)

Lindenbergia sinaica (Decne.) Benth., Scroph. Ind. 22 (1835) & in DC., Prodr. 10: 377 (1846); Hemsley & Skan in F.T.A. 4(2): 312 (1906); Cufodontis in B.J.B.B. 33: Suppl. 894 (1963)

L. abyssinica Benth. in DC., Prodr. 10: 377 (1846). Type: Ethiopia, *Schimper* 782 (G-DC, lecto.; BM!, BR, E, FT, UPS, W, isolecto.)

L. sinaica Decne. var. *abyssinica* (Benth.) Almagia in Ann. Inst. Roma 8: 140 (1904)

14. STEMODIA

L., Syst. Nat. ed. 10: 1091, 118, 1374 (1759), *nom. conserv.*

Stemodiacra P.Br., Hist. Jamaica: 261 (1756)

Herbs or undershrubs, glandular-pubescent, often aromatic. Leaves opposite or whorled. Flowers solitary in the axils of leaves or in a crowded leafy spike. Calyx deeply 5-lobed, lobes narrow, imbricate, equal or subequal. Corolla white or blue, 2-lipped, tube cylindrical. Stamens 4, didynamous, included, filaments filiform, anther-thecae stipitate, all fertile. Ovary with the style dilated at the apex, usually 2-lobed. Capsule narrow ovoid or globose, dehiscence loculicidal or sometimes septicidal, valves 2 or 4. Seeds numerous, minute.

About 30 species in the tropics of the Old and New Worlds; 2 species in Africa.

Stemodia serrata *Benth.* in DC., Prodr. 10: 381 (1846); Hemsley & Skan in F.T.A. 4(2): 314 (1906); Hepper in F.W.T.A. ed. 2, 2: 357 (1963); Fischer, F.A.C. Scrophulariaceae: 66, pl. 24 (1999) & in Fl. Ethiop. & Eritr. 5: 260 (2006). Types: Senegal [Senegambia], 1828, *Leprieur*, s.n.; *Perrottet*, 447, 448 & 575 (G, syn.)

Erect simple or usually much-branched herb, 12–32(–37) cm high; branches ± erect, 4-angled, glandular-pubescent, viscid with a foetid scent. Leaves narrowly elliptic or lanceolate, 0.5–5 cm long, 3–15 mm wide, narrowed to base or amplexicaul, acute at apex, margins serrate from the middle to the apex, finely glandular-pubescent. Flowers solitary in axils forming a leafy spike-like raceme;

FIG. 14. *STEMODIA SERRATA* — **1**, habit × $^{2}/_{3}$; **2**, flower × 6; **3**, flower dissected × 6; **4**, stamens × 12; **5**, capsule dehisced × 6. All from *Robinson* 1308. Drawn by Christine Grey-Wilson. Reproduced with permission from F.Z.

pedicels 1–3 mm long; bracteoles 2 at the base of the calyx, narrowly linear, ± 3 mm long, ciliate. Calyx ± 5 mm long in fruit, shorter in flower, lobes subulate-linear, acuminate, finely glandular-pubescent. Corolla white, ± 6 mm long, upper lip emarginate, lower lip broadly and shortly 3-lobed. Capsule 5–6 mm, narrow, acuminate, almost as long as the calyx, glabrous. Fig 14, p. 37.

KENYA. Tana River District: Tana R. National Primate Reserve, 17 Mar. 1990, *Luke et al.* 562!
TANZANIA. Dodoma District: Great North Road, 26 km N of Dodoma, Apr. 1962, *Polhill & Paulo* 2112!; Uzaramo District: Dar es Salaam, Nov. 1975, *Mwasumbi* 11353!; Kilwa District: Selous Game Reserve, July 1975, *Vollesen* 2535!
DISTR. **K** 7; **T** 5, 6, 8; Senegal, Mali, Ghana, Chad, Cameroon, Sudan, Ethiopia, Zambia, Malawi, Madagascar; India; widely distributed but probably overlooked in our area
HAB. Moist, grassy places, flood plains, drying out during dry season; 50–1100 m
USES. None recorded on specimens from our area
CONSERVATION NOTES. Least Concern (LC)

SYN. *Sutera serrata* Hochst. in Flora, 24, t. intell. 43 (1841), *nomen*

NOTE. This and other species of *Stemodia* are often mistaken for members of the Labiatae with which they have a superficial resemblance; the bilocular ovary is distinctive. In East Africa only *S. serrata* Benth. has been reported, yet *S. verticillata* (Mill.) Boldingh (syn. *S. parviflora* Ait.), an American species with blue flowers and globose capsules occurs as a creeping weed in West Africa, Mauritius and Java, and it may well be found in our region.

15. **STEMODIOPSIS**

Engl. in Ann. Ist. Bot. Roma 7: 25 (1897)

Perennial herbs, woody at the base; stems wiry, leafy. Leaves opposite, petiolate, toothed. Flowers axillary. Calyx deeply 5-lobed. Corolla tube broadly campanulate, not spurred, 2-lipped; upper lip rounded, lower lip with a palate and ± 3-lobed. Stamens 4 fertile + 1 staminode inserted in the tube; filaments of the middle longer pair twisted; anthers dehiscing longitudinally; staminode slender. Ovary elongated-conical, as long as the calyx, 2-locular. Capsule reflexed when mature, septicidal, many-seeded.

Ten species in Africa and Madagascar.

Corolla 4–5 mm long; capsule glabrous .... 1. *S. buchananii*
Corolla 3 mm long; capsule hairy .... 2. *S. rivae*

1. **Stemodiopsis buchananii** *Skan* in F.T.A. 4(2): 315 (1906); Philcox in F.Z.: 8(2): 40 (1990); U.K.W.F.: 255 (1994). Lectoype: Malawi, without locality, 1891, *Buchanan* 365 (K!, chosen by Philcox in F.Z.)

Small perennial herb, up to 40 cm, but often much smaller and tufted, rather woody at the base; stems often many arising from the base, slender, glabrous or with a few hairs on the younger parts. Leaves ovate, ovate-lanceolate or lanceolate, 5–40(–70) mm long and 4–22 mm wide, with 4–5 lateral veins on each side of the midrib, base cuneate, apex acute, margins coarsely 2–11-toothed on each side, ± glabrous; petiole nearly as long as the blade, narrowly 3-winged. Lower axillary cymes several-flowered, congested when internodes short, upper axils often with 1 flower; Pedicel filiform, 3–4 mm long, ascending in flower, reflexed in fruit. Calyx segments narrowly linear-triangular, 3–3.5 mm, ciliate. Corolla with yellow hood, lip white and purple towards the throat, 5–9 mm long. Capsule ellipsoid-conical, 3–4 mm, abruptly long-beaked, glabrous. Seeds cylindric.

var. **buchananii**

Plant glabrous except for a few scattered hairs on leaves and new shoots; plants generally small with leaves 5–40 mm long.

KENYA. Fort Hall District: Mabuloni Rock near Thika, 21 Dec. 1952, *Verdcourt* 845B!; Kitui District: Kitui, 2 km NE of Mwingi, 16 Dec. 1977, *Stannard & Gilbert* 1119–1138!; Teita District: Mzinga Hill, 8 Feb. 1953, *Bally* 8800!

TANZANIA. Dodoma District: 74 km N of Dodoma, 24 Jan. 1962, *Polhill & Paulo* 1250!; Ulanga District: near Kiberege, Mar. 1960, *Haerdi* 481/0!; Songea District: Unangwa Hill, 6 km E of Songea, 22 Mar. 1956, *Milne-Redhead & Taylor* 9278!

DISTR. **K** 4, 7; **T** 1, 5–8; Sudan, Malawi, Zambia

HAB. Crevices in boulders and rock cliffs; 450–1450 m

USES. None recorded on specimens from our area

CONSERVATION NOTES. Least Concern (LC)

var. **pubescens** Philcox in K.B. 40: 606 (1985) & in F.Z.: 8(2): 40 (1990). Type: Zimbabwe, Chimanimani Mts, *Grosvenor* 327 (K!, holo.; BR, LISC, PRE, iso.)

Stems and leaves pubescent; plant generally large (in our area), up to 40 cm with leaves up to 7 cm long.

KENYA. Meru District: Meru Game Reserve, Muguonga Hill, 10 Sept. 1963, *Verdcourt* 3743!; Teita District: Taita Hills, 11 June 1966, *Gillett & Burtt* 17191!

TANZANIA: Rufiji District: Stieglers Gorge, 7 Aug. 1976, *Vollesen* 3892!; Iringa District: Ruaha river area, Morogoro–Iringa, 18 July 1982, *Abdallah* 1144!; Mbeya District: Poroto Mts, May 1957, *Richards* s.n.!

DISTR. **K** 4, 7; **T** 4, 6, 7; Mozambique, Zimbabwe

HAB. Boulders and rocks; 150–800 m

USES. None recorded on specimens from our area

CONSERVATION NOTES. Least Concern (LC)

2. **Stemodiopsis rivae** *Engl.* in Ann. Ist. Bot. Roma 7: 25 (1897) & in E.J. 23: 497, t. 7, figs. A–F (1897); Skan in F.T.A. 4(2): 315 (1906); Cufodontis in B.B.J.B. 33, Suppl.: 895 (1963); Philcox in F.Z.: 8(2): 42, t. 17, figs 1–5 (1990); Fischer, F.A.C. Scrophulariaceae: 70, pl. 26 (1999) & in Fl. Ethiop. & Eritr. 5: 261 (2006). Type: Ethiopia, Robe Mt, at the source of Daua, *Riva* 162 (B†, holo.; FT!, lecto.)

Perennial herb or undershrub with short woody root-stock in older plants; stems tufted or with long pendent stems and several lateral branches; whole plant pubescent. Leaves ovate, 4–25 mm long and 2–15 mm wide, with 3–4 lateral veins on each side of the midrib, base cuneate, apex acute, margins coarsely 2–5-toothed on each side, usually shortly and densely pubescent; petioles about as long as the lamina. Flowers on short few-flowered cymes or solitary in axils; pedicels 2–5 mm long, erect in flower, reflexed in fruit. Calyx segments narrowly linear-triangular, 2–3 mm long. Corolla cream with a purple throat, ± 3 mm long. Capsule ellipsoid-conical, 4–5 mm, abruptly long-beaked, pubescent. Seeds cylindric. Fig. 15, p. 40.

UGANDA. West Nile District: Mt Otze, 7 June 1936, *A.S.Thomas* 1960!; Teso District: Soroti Rock, 9 May 1970, *Lye* 5378!

KENYA. Northern Frontier District: Sololo, Burroli Mt, 4 Sept. 1952, *Gillett* 13779!; Machakos/Masai District: Ngulia Hills, 30 June 1968, *Gilbert* 2751!; Teita District: Taita Hills, 3 km S of Manyani Prison, 11 Feb. 1966, *Gillett & Burtt* 17191!

TANZANIA. Lushoto District: Mkomazi Game Reserve, 1993, *Cox & Abdullah* 2353!; Mpwapwa District: Godegode Kopje, 8 Feb. 1933, *Burtt* 4562!; Arusha District: Meru Game Reserve, near Maua, 10 Sept. 1963, *Verdcourt* 3734!

DISTR. **U** 1, 3; **K** 1, 4, 6, 7; **T** 3, 5, 7; Nigeria, Sudan, Ethiopia, Zambia, Malawi, Mozambique, Zimbabwe

FIG. 15. *STEMODIOPSIS RIVAE* — **1**, flowering branch × $^{2}/_{3}$; **2**, flower × 4; **3**, flower dissected × 8; **4**, corolla transverse section showing hairs × 36; **5**, capsules × 4. All from *Ngoni* 177. Drawn by Christine Grey-Wilson. Reproduced with permission from F.Z.

HAB. Crevices of large boulders; 900–1600 m

USES. None recorded on specimens from our area

CONSERVATION NOTES. Least Concern (LC), but apparently uncommon in the Flora area as is known from only two collections from Uganda and three from Tanzania

SYN. *S. humilis* Skan in F.T.A. 4(1): 316 (1906); U.K.W.F.: 551 (1974). Type: Malawi, near Mt Chiradzulu, *Cameron* 182 (K!, holo.)

NOTE. It is remarkable how constant is the occurrence of these plants on exposed rock crevices that seem scarcely able to support any plant life.

16. LIMNOPHILA

R.Br., Prodr.: 442 (1810); Philcox in K.B. 24: 101–170 (1970), *nom. conserv.*

Ambulia Lam., Encycl. Meth. Bot. 1: 128 (1783)

Herbs, aquatic or in marshes, often aromatic on bruising, glabrous, pubescent or glandular; stems erect, simple or more usually branched, ± submerged, prostrate and rooting at nodes. Leaves of two types with one or both types on the same plant: submerged leaves finely divided, verticillate; aerial leaves undivided or variously laciniate, opposite or verticillate; ± punctate. Flowers sessile or pedicellate, solitary in the axils or in lax or compact, terminal or axillary spikes or racemes; bracteoles 0 or 2. Calyx tubular, 5-lobed, lobes unequal, tube with 0–5 prominent veins or striate with more than 10 prominent veins present at maturity. Corolla tubular or funnel-shaped, 5-lobed, bilabiate; upper (adaxial) lip outside in bud, entire or 2-lobed; lower (abaxial) lip 3-lobed. Stamens 4, didynamous, included, posterior pair shorter; anthers free, loculi stipitate. Ovary glabrous; stigma bilamellate; style filiform, deflexed at the apex. Capsule ellipsoid to globose, septicidally 4-valved. Seeds numerous.

35 species in the tropics and subtropics of the Old World with 7 occurring in Africa, 5 of which are represented in the Flora area.

All the species represented here are heterophyllous, bearing both aerial entire leaves and, in an aquatic state, divided filamentous submerged leaves; as such, all occur in section Limnophila (see Philcox in K.B. 24: 107 (1970)).

1. Most aerial leaves deeply lobed or divided; flowers solitary or in subsessile clusters 2
   All aerial leaves undivided; flowers mostly in spikes 4
2. Aerial stem without glands; corolla 4–6.5 mm long .... 3. *L. fluviatilis*
   Aerial stems glandular; corolla 6–12 mm long 3
3. Aerial stem stipitate-glandular becoming subglabrous; leaves not punctate; flowers on stalks up to 15 mm long or subsessile .... 1. *L. indica*
   Aerial stem sessile-glandular and hirsute, or subglabrous; leaves punctate; flowers sessile .... 2. *L. ceratophylloides*
4. Aerial stems and leaves glandular-hirsute; calyx 5–6 mm long; corolla 5–14 mm long .... 4. *L. bangweolensis*
   Aerial stems and leaves not glandular; calyx 2.5–3 mm long; corolla 3–4 mm long .... 5. *L. barteri*

1. **Limnophila indica** (*L.*) *Druce* in Rep. Bot. Excl. Club Brit. Isles 1913, 3: 420 (1914); F.P.S. 3: 138 (1956); Hepper in F.W.T.A. ed. 2, 2: 357 (1963); Philcox in K.B. 24: 115 (1970); Vollesen in Opera Bot. 59: 77 (1980); Philcox in F.Z.: (8) 2: 44 (1990); Koenders, Fl. Pemba 1: 47(1992); Fischer, F.A.C. Scrophulariaceae: 50, pl. 17 (1999) & in Fl. Ethiop. & Eritr. 5: 260 (2006). Lectotype: Herb. Linn. No. 204.2 (LINN) designated by Philcox in K.B. (1970)

Aquatic perennial herb; aerial stem 2.5–14 cm tall, simple to much-branched, slender with sessile or stipitate glands above becoming subglabrous; submerged stem up to 1 m long, much branched, glabrous. Leaves on aerial stem usually all verticillate and variously dissected, (2.5–)4–12(–22) mm long, sometimes 2–3 pairs of opposite, undissected, crenate-serrate to lacerate, 1–3-veined leaves towards the apex, up to 15 mm long, 4 mm wide, sessile-glandular to sub-glabrous, rarely all aerial leaves undissected; submerged leaves verticillate in whorls of 6–12, pinnatisect, up to 3 cm long with lobes flattened or capillary. Flowers solitary, axillary, slender pedicellate or very short-pedicellate, appearing sessile; pedicels (0.5–)3.5–10(–15) mm long, sessile-glandular to stipitate-glandular, usually longer than the subtending leaves; bracteoles 2, (1.5–)3–4 mm long, linear to linear-oblong to obovate-lanceolate, acute, entire to irregularly and remotely serrate-dentate to occasionally deeply incised, glandular to subglabrous. Calyx 3.5–6 mm long, sessile-glandular, rarely sparsely hirsute, not striate at maturity; lobes 2–3 mm long, broadly ovate to lanceolate, shortly acuminate, occasionally ciliate. Corolla white to pale yellow or yellow at base of tube, mauve-pink above, (6–)8–12 mm long, externally glabrous; lobes all entire. Stamens with anthers contiguous; posterior filaments 2 mm long, anterior 4 mm long, all glabrous; style up to 4.5 mm long with two lateral processes ± 0.2 mm wide at the apex and below the stigma. Capsule compressed ellipsoid to subglobose, ± 3.4 mm long, dark brown. Fig. 16, p. 43.

UGANDA. Lango District: Orumo, Sept. 1935, *Eggeling* 2212!; Teso District: Soroti, Omunyal swamp, 14 Sept. 1954, *Lind* 353!; Mengo District: 72 km on Kampala–Masindi road, *Lind* 2729!

TANZANIA. Tanga Ditrict: Magunga Estate, 28 Dec. 1053, *Faulkner* 1317!; Uzaramo District: 16 km SE of Dar, 1.2 km before Mnguvia R., 9 Sept. 1977, *Wingfield* 4149!; Tunduru District: 38 km E of Tunduru on Masasi road, 19 Nov. 1966, *Gillett* 17902!

DISTR. **U** 1, 3, 4; **T** 3, 4–6, 8; **Z**; **P**; Senegal, Mali, Togo, Nigeria, Cameroon, Gabon, Sudan, Angola, Zambia, Malawi, Mozambique, Zimbabwe and South Africa; throughout the Old World tropics

HAB. Submerged in shallow, fresh or stagnant, standing or slow-running water, or terrestrial on margins of rivers or lakes, or receding floodplains; 0–1250 m

USES. None recorded on specimens from our area

CONSERVATION NOTES. Least Concern (LC); widespread

SYN. *Hottonia indica* L., Sp. Pl., ed. 2, 1: 208 (1762)

Limnophila gratioloides R.Br., Prodr. Fl. Nov. Holl.: 442 (1810); Benth. & Hook.f. in Hook., Niger Fl.: 474 (1849); Skan in F.T.A. 4(2): 319 (1906); Fl. Sudan: 326 (1929); Hepper in F.W.T.A. 2: 223 (1931), *nom. illegit.* Type as for *L. indica*

Ambulia gratioloides (R.Br.) Wettst. in E.& P.Pf. IV, 3b: 73 (1891); P.O.A. C: 357 (1895); A.Chev., Expl. Bot. Afr. Occ. Fr. l: 470 (1921)

Limnophila gratioloides R.Br. var. *nana* Skan in F.T.A. 4(2): 319 (1906); Hepper in F.W.T.A. 2: 223 (1931). Type: Nigeria, Nupe, near Jebba, *Barter* 1709 (K!, holo.)

2. **Limnophila ceratophylloides** (*Hiern*) *Skan* in F.T.A. 4(2): 317 (1906); Peter, Wasserpfl. Deutsch Ostafr.: 127 (1928); Philcox in K.B. 24(1): 122 (1970) pro parte; A. Raynal & Philcox in Adansonia ser. 2, 15: 234, t. 1,2,5 (1975); Fischer, F.A.C. Scrophulariaceae: 55, pl. 19 (1999). Type: Angola, Huilla, between Lopollo and Humpata, *Welwitsch* 5778 (BM!, holo.; ?COI, K!, iso.)

Aquatic perennial; aerial stems up to 20 cm tall, simple or branching, glabrous to laxly white hirsute or lightly covered with small, sessile, yellow glands; submerged stems up to 60 cm long, simple or branching, glabrous. Leaves on aerial stems verticillate to opposite, irregularly pinnatisect to lacerate, or somewhat deeply divided, 5–8 mm long, 1–2 mm wide, densely punctate, glabrous, hirsute or yellow glandular; submerged leaves verticillate, pinnatisect-multifid, up to 2.5 cm long, segments capillary or, more usually, flattened, glabrous. Flowers solitary-axillary, sessile. Cleistogamous flowers present on submerged stems. Chasmogamous flowers: bracteoles 2, 1.5–4 mm long, narrowly linear, glabrous to very shortly hirsute; calyx

FIG. 16. *LIMNOPHILA INDICA* — **1**, habit × $^2/_3$; **2**, stem with capsules × 1; **3**, flower × 6; **4**, capsule × 6; **5**, seed × 32. 1 from *Greenway & Polhill* 1174; 2–5 from *Faulkner* 1317. Drawn by Juliet Williamson.

3–5 mm long, glabrous or hirsute, yellow-glandular; corolla 6–10 mm long, mauve to lilac with darker throat, externally glabrous, densely vinous within the tube mainly on the posterior side; stamens 4 with contiguous anthers at anthesis, filaments 0.5–3 mm long, anthers attached at one end to largely inflated connective, thecae divergent; stigma unequally bilobed with one lobe somewhat extended, narrowly deltoid and appearing hooked, perpendicular to the style; style 1–4 mm long. Cleistogamous flowers: bracteoles 2.5–4 mm long, narrowly linear, glabrous; calyx 3–3.5 mm long, glabrous; corolla 3.5–4 mm long, villous within the tube; stamens generally 2, filaments 0.5–1 mm long; style ± 1.5 mm long. Capsule broadly ovoid, ± 2.5 mm long, dark- to light brown, opaque.

TANZANIA. Songea District: Hanga Farm, 27 June 1956, *Milne-Redhead & Taylor* 10837A!
DISTR. **T** 8; Cameroon, Congo-Brazaville, Congo-Kinshasa, Angola, Burundi, Zambia, Zimbabwe, Botswana and Namibia
HAB. Submerged in shallow, fresh or stagnant, standing or slow to running water, or terrestrial on margins of rivers or lakes, or receding floodplains (similar to *L. indica*); ± 1100 m
USES. None recorded on specimens from our area
CONSERVATION NOTES. Least Concern (LC), but known from a single gathering in Tanzania

SYN. *Stemodiacra ceratophylloides* Hiern, Cat. Afr. P1. Welw. 1: 759 (1898)
[*S. sessiliflora* sensu Hiern, Cat. Afr. P1. Welw. 2: 758 (1898) quoad pl. afric., *non* (Vahl) Hiern sensu stricto]
Stemodia ceratophylloides (Hiern) K.Schum. in Just's Bot. Jahresb. 1898(1): 395 (1900)
Ambulia ceratophylloides (Hiern) Engl. & Gilg in Warb., Kunene-Sambesi-Exped.: 362 (1903)
A. baumii Engl. & Gilg in Warb., Kunene-Sambesi-Exped.: 361, t. 7 F–G (1903). Type: Angola, *Baum* 750 (B†, holo.; BM!, K!, M, S, W, iso.)

3. **Limnophila fluviatilis** *A.Chev.* in Bull. Mus. Hist. Nat. Paris, ser. 2, 4: 587 (1932); A. Raynal in Adansonia 7: 351 (1967); A. Raynal & Philcox in Adansonia ser. 2, 15: 236, t. 3, 4, 6 (1975); Fischer, F.A.C. Scrophulariaceae: 56, pl. 20 (1999). Type: Mali, Niger R., from Timbouctou to Gao, *Chevalier* 43079 (P!, holo.)

Aquatic perennial; aerial stems up to 10 cm tall, simple or rarely branched except at base, glabrous to sparsely hirsute, particularly above; submerged stems up to 35 cm long, branching, glabrous. Leaves on aerial stems verticillate, irregularly pinnatisect to lacerate or deeply divided, 4–18 mm long, 1–6 mm wide, glabrous, densely punctate, frequently with sessile yellow glands; submerged leaves verticillate, pinnatisect, multifid, up to 2.5 cm long, segments capillary or more usually flattened, glabrous. Flowers solitary, axillary, sessile on submerged stems, solitary or clustered on very short axillary few-flowered racemes on aerial stems. Cleistogamous flowers usually present on submerged stems. Chasmogamous flowers: bracteoles 2, 2–3.5 mm long, narrowly linear, glabrous to sparsely hirsute; calyx 3–4.5 mm long, glabrous, frequently with yellow glands; corolla 4–6.5 mm long, white to lilac, throat yellow, externally glabrous, occasionally very sparsely villous within the tube; stamens 4 with contiguous anthers at anthesis, filaments 0.5–2 mm long, anthers median attached to slightly swollen connective, thecae subparallel; stigma truncate to emarginate, style 1–2 mm long; capsule subflattened ovoid, 3–4 mm long, dark brown, subtruncate to emarginate. Cleistogamous flowers: bracteoles 2.5–6 mm long, narrowly linear, glabrous; calyx 3–6 mm long, glabrous; corolla 3.5–4.75 mm long, glabrous to sparsely villous within the tube; stamens 4 with contiguous anthers, filaments ± 0.3 mm long, anthers as above; style 0.8–1 mm long; capsule broadly ovoid, 2.3–4 mm long, light brown, translucent, emarginate to subspherical.

TANZANIA. Dodoma District: Manyoni, Kazikazi, 10 June 1932, *Burtt* 3704!; Songea District: Kwamponjore valley, ± 9.5 km SW of Songea, 19 June 1956, *Milne-Redhead & Taylor* 10837!
DISTR. **T** 5, 8; Senegal, Mali, Guinea, Nigeria, Cameroon, Central African Republic, Congo-Kinshasa, Zambia, Zimbabwe and Namibia

HAB. Submerged in shallow, fresh or stagnant, standing or slow-running water, or terrestrial on margins of rivers or lakes, or receeding floodplains; ± 1280 m
USES. None recorded on specimens from our area
CONSERVATION NOTES. Least Concern (LC); widespread

4. **Limnophila bangweolensis** (*R.E.Fries*) *Verdc.* in K.B. 5: 379 (1950); Philcox in F.Z.: 8 (2): 47, t. 18 (1990); Fischer, F.A.C. Scrophulariaceae: 54 (1999). Type: Zambia, Lake Bangweulu, *R.E. Fries* 895 (UPS, holo.)

Aquatic annual or perennial; stems up to 50 cm long, simple or branched towards apex, aerial part densely crisped white glandular-hirsute, submerged part glabrous. Leaves on aerial stem opposite becoming verticillate, sessile, broadly ovate, suborbicular or oblong, 6–13 mm long, 5–11 mm wide, densely glandular-hirsute, becoming glabrous below depending on degree of immersion, serrate-dentate with thickened margins, subamplexicaul, revolute, 3–5 parallel-veined; submerged leaves at base of aerial stem becoming laciniate to occasionally capillary-multifid; finely divided leaves with capillary segments, usually borne on sterile totally submerged stems, dying off before maturity of aerial stem. Flowers sessile to subsessile, bracteate, in densely flowered spikes, or occasionally solitary, axillary. Bracts reduced, leaflike; bracteoles 2, ± 3.5 mm long, linear, glandular-hirsute. Calyx 5–6 mm long, glandular-hirsute, not striate at maturity; lobes 3 mm long, narrowly lanceolate-acuminate. Corolla white, pale-pink, lilac to mauve-pink, 5–14 mm long, ?externally glabrous; adaxial lip broad emarginate; abaxial lip 3-lobed, lobes broadly ovate; tube with finely clavaculate papillae within. Stamens with posterior filaments 2 mm long, anterior 3.8–4 mm long, glabrous. Capsule compressed ovoid, ± 3 mm long, black.

TANZANIA. Mpanda District: Uruwira to Tabora road, Kambisama R., 30 Sept. 1970, *Richards & Arasululu* 26189!; Songea District: Kwamponjore Valley, 20 June 1956, *Milne-Redhead & Taylor* 10845!
DISTR. **T** 4, 8; Zambia
HAB. Wet boggy grasslands in shallow, standing or slow running water; 1000–1400 m
USES. None recorded on specimens from our area
CONSERVATION NOTES. Least Concern (LC)

SYN. *Ambulia bangweolensis* R.E.Fr., Wiss. Ergebn. Schwed. Rhod.-Kongo-Exped. 1: 288, t. 19/3 (1916)

5. **Limnophila barteri** *Skan* in F.T.A. 4(2): 317 (1906); A. Chev., Expl. Bot. Afr. Occ. Fr. 1: 470 (1921); Berhaut, Fl. Senegal: 68 (1954); Hepper in F.W.T.A. ed. 2, 2: 357 (1963), excl. *L. fluviatilis* Chev. et formae; Philcox in K.B. 24: 127 (1970); Vollesen in Opera Bot. 59: 77 (1980); Fischer, F.A.C. Scrophulariaceae: 52, pl. 18 (1999). Lectotype: Nigeria, without further locality, *Barter* 751 (K!, chosen by Philcox in K.B. (1970))

Semi-aquatic annual; aerial stems up to 40 cm tall, sparsely branched, densely short, white, patent hirsute above to glabrescent below; submerged stems up to 20 cm long, sparsely branched, glabrous, usually sterile. Leaves on aerial stem verticillate near base to opposite above, ovate-elliptic, 10–18 mm long, 6–9 mm long, sessile, subamplexicaul, margins crenate-serrate, shortly hirsute, densely and minutely punctate, 3–7 parallel-veined; submerged leaves capillary-pinnatifid, up to 18 mm long, glabrous. Flowers either sessile in axils of leaf-like bracts or in slender, distantly flowered spikes; bracteoles 2, ± 2 mm long, linear, subacute, densely pilose. Calyx 2.5–3 mm long, densely pilose with round yellowish glands, not striate at maturity; lobes ± 1.25 mm long, ovate-lanceolate, acuminate, divergent. Corolla tube pale-pink to white, limb white 3–4 mm long, totally glabrous; abaxial lip 3-lobed, lobes entire; adaxial lip erect, entire. Stamens with posterior filaments ± 0.4 mm long, anterior 1 mm long, glabrous. Capsule broadly ovoid, ± 2 mm long, pale brown.

TANZANIA. Songea District: S of Songea, by Likonde R., 26 June 1956, *Milne-Redhead & Taylor* 10899!

DISTR. **T** 8; Senegal, Mali, Guinea Bissau, Guinea, Sierra Leone, Ivory Coast, Ghana, Nigeria, Cameroon, Central African Republic and Zambia

HAB. Riversides and bogs in wet grasslands, surrounding rice fields; 750–1000 m

USES. None recorded on specimens from our area

CONSERVATION NOTES. Least Concern (LC), but known from a single gathering in Tanzania

17. MECARDONIA

Ruiz & Pav., Prodr.: 95 (1794); Rosso, Rev. Gen. Mecardonia, Candollea 42: 431 (1987)

Perennal herbs, glabrous; stems ascending to prostrate, angular. Leaves opposite, obovoid, obtuse, crenate, with ± revolute margins. Flowers purple to white, axillary, solitary, pedicellate; bracteoles 2. Calyx unequally deeply 5-lobed, outer 3 lobes broader than the inner 2. Corolla 2-lipped; corolla tube cylindrical. Stamens 4, didynamous, included; anthers with 2 separate thecae; connective 2-branched, each bearing an anther theca. Ovary 2-locular; stigma terminal, usually shortly 2-lobed. Capsule globose or ovoid, septicidal, 2-valved. Seeds numerous; reticulate.

15 species distributed in temperate and tropical Americas; one species introduced and naturalised in tropical Africa.

Mecardonia procumbens (*Mill.*) *Small*, Fl. SE U.S.: 1065, 1338 (1903); D'Arcy in Ann. Miss. Bot. Gard. 66: 240 (1979); Rosso in Candollea 42: 456 (1987). Type: Mexico, no locality cited, *Houston* s.n. (BM, holo.)

Annual herb; stems 4-angled, simple or branching from the base, decumbent to ascending to prostrate and creeping, glabrous. Leaves sessile to somewhat amplexicaul, ovate, (5–)8–20 mm long, (3–)5–11 mm wide, base tapering, obtuse to rounded at the apex, margins crenate, glabrous. Flowers solitary; pedicels slender, 12–20 mm long; bracteoles oblanceolate, ± 3 mm long. Calyx green, the 3 outer lobes ovate, 7–8 mm long, ± 4 mm wide, the 2 inner lobes linear-lanceolate, 6–7 mm long, ± 1 mm wide, somewhat transparent. Corolla yellow; corolla tube ± 6 mm long; upper lobe obovate, emarginate; lower lobe shallowly 3-lobed with rounded margins. Filaments 3–4 mm long, inserted about the middle of the corolla tube. Ovary ovoid. Capsule ovoid, 4–6 mm long, brown, glabrous, included in the calyx. Seeds ellipsoid, 0.3 mm long, reticulate. Fig. 17, p. 47.

TANZANIA. Muheza District: East Usambara Mts, Amani Medical Research Centre, May 1987, *Iversen et al.* 873451!

DISTR. **T** 3; Sierra Leone, Cameroon; tropical N & S America

HAB. Disturbed habitats, by roadsides, and in secondary vegetation; weedy; ± 900 m

USES. None recorded on specimens from our area.

CONSERVATION NOTES. Least Concern (LC), but known from a single gathering in Tanzania

SYN. *Erinus procumbens* Mill., Gard. Dict. ed. 8 (1768)
Monniera procumbens (Mill.) Kuntze, Rev. Gen. Plant. 2: 463 (1891)
Bacopa procumbens (Mill.) Green., Field Mus. Nac. Hist. Bot. Ser. 2: 261 (1907)
Herpestis procumbens (Mill.) Urb., Symb. Antill. 4: 558 (1911), *non* Spreng. (1819)

NOTE. Flower measurements are based on material fom Sierra Leone (*Deighton* 3432)

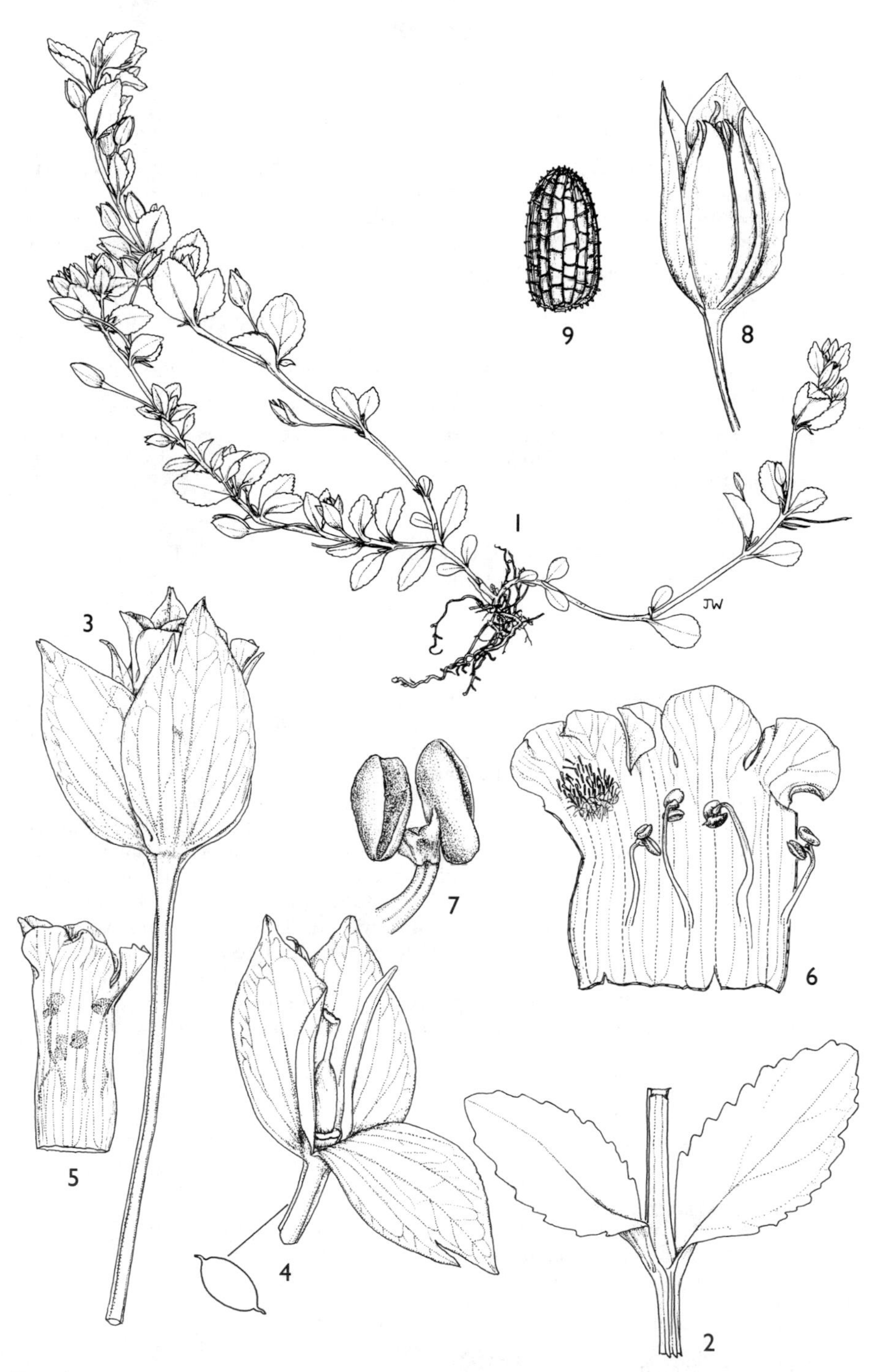

FIG. 17. *MECARDONIA PROCUMBENS* — **1**, habit × $^2/_3$; **2**, detail of leaf stalk × 3; **3**, flowering stalk × 6; **4**, flower dissected × 6; **5**, corolla tube × 6; **6**, corolla tube opened × 6; **7**, anther × 40; **8,** capsule × 5; **9**, seed × 50. 1, 3–7 from *Iversen, Persson & Petersson* 87345; 2 from *Borhidi, Iversen & Steiner* 86155; 8, 9 from *J. K. Morton* K947. Drawn by Juliet Williamson.

18. **BACOPA**

Aubl., Pl. Guian. 1: 128, t. 49 (1775), *nom. conserv.*

Moniera P.Browne, Hist. Jamaica: 269 (1756); Skan in F.T.A. 4(2): 319 (1906)

Perennial or annual herbs, erect, diffuse or submerged aquatics, usually glabrous. Leaves opposite, entire or dentate, variously dissected or multisect in submerged species, sometimes punctate. Flowers axillary, solitary or in racemes; pedicels with or without bracteoles. Calyx unequally 5-lobed, outer lobes broadest, inner narrow. Corolla almost regular, with all lobes equal or the upper 2 connate; corolla tube short. Stamens 4, didynamous, included. Stigma terminal, usually shortly 2-lobed. Capsule globose or ovoid, 2 or 4-celled, loculicidal or septicidal. Seeds numerous.

About 100 species distributed in the warmer parts of the world; usually moisture-loving or aquatic plants.

It is surprising that the widespread *Bacopa monnieri* (L.) Pennell has not been recorded from the Flora area, but is to be expected there. This is a creeping herb of moist and wet places with solitary, axillary white or pale blue flowers.

Erect annual; leaves entire; pedicels pubescent .... 1. *B. floribunda*
Stems usually creeping and rooting at the nodes; leaves toothed;
pedicels glabrous .... 2. *B. crenata*

1. **Bacopa floribunda** (*R.Br.*) *Wettst.* in E. & P. Pf. 4, 3B: 76 (1891); Engl. in P.O.A. C.: 357 (1895); Hepper in F.W.T.A. ed. 2, 2: 358 (1963); Philcox in F.Z.: 8(2): 51 (1990). Type: Australia, Queensland, Shoalwater Bay, *R. Brown* 2653 (BM!, holo.)

Erect annual 10–30(–55) cm high; stems 4-angled, simple or branched, finely puberulous to glabrous in the lower parts. Leaves slightly fleshy, aromatic, linear to linear-lanceolate, 12–30 mm long, 2–4 mm wide, sessile to somewhat amplexicaul at base, ± acute at the apex, margins entire, glandular-punctate. Flowers solitary, usually in the upper axils of leaves; pedicels 3–5 mm long, finely puberulous. Bracts setaceous, 1 mm long, inserted just below the calyx. Calyx finely puberulous and glandular, 3 outer lobes ovate, largest 6–8 mm long, 4–5 mm wide, in fruit prominently veined, 2 inner lobes linear-lanceolate, hyaline, 3–4 mm long. Corolla tube white, lobes bluish or white, yellow in throat, about as long as the largest calyx lobe; upper lip emarginate, lower lip equally 3-lobed; style and ovary glabrous; stigma lamellate. Capsule ovoid, 3–4 mm long, glabrous. Seeds shallowly reticulate, 0.3 mm long. Fig. 18, p. 49.

UGANDA. West Nile District: Madi, Dec. 1862, *Speke & Grant* s.n.!
KENYA. Kwale District: Mwasangombe forest, 27 Aug. 1953, *Drummond & Hemsley* 4010! & between Samburu and Mackinnon Road, 3 Sept. 1953, *Drummond & Hemsley* 4144!; Kilifi District: Mida, Arabuko-Sokoke Forest Reserve, 2 Dec. 1961, *Polhill & Paulo* 888!
TANZANIA. Lushoto District: Korogwe, 27 Sept. 1959, *Semsei* 2931!; Uzaramo District: Kinduchi, 1 Aug. 1970, *Harris & Mwasumbi* 4921!; Mbeya District: Ruaha National Park, May 1968, *Renvoize & Abdallah* 2290!
DISTR. **U** 1; **K** 7, **T** 3–8; **Z**, **P**; throughout much of tropical Africa, Zambia, Zimbabwe, and extending to Arabia, tropical Asia and Australia
HAB. Wet places and rice fields; 0–1400 m
USES. None recorded on specimens from our area.
CONSERVATION NOTES. Least Concern (LC); widespread

SYN. *Herpestis floribunda* R.Br., Prodr.: 442 (1810)
Monniera floribunda (R.Br.) T.Cooke, Fl. Pres. Bombay 2: 286 (1905); Skan in F.T.A. 4(2): 322 (1906)
M. pubescens Skan in F.T.A. 4(2): 322 (1906). Type: Sierra Leone, Falaba River, *Scott-Elliot* 5451 (K!, holo.)

Fig. 18. *BACOPA FLORIBUNDA* — **1**, habit × $^2/_3$; **2**, flower × 6; **3**, flower opened to show unequal calyx lobes × 6; **4**, flower dissected × 6; **5**, capsule dehisced × 6. All from *Robinson* 2838. Drawn by Christine Grey-Wilson. Reproduced with permission from F.Z.

NOTE. Small simple, annual plants from Mafia Island (*Greenway* 5118) are evidently a coastal ecotype; they match West African plants in similar situations in brackish habitats.

2. **Bacopa crenata** (*P.Beauv.*) *Hepper* in K.B. 14: 407 (1960) & in F.W.T.A. ed. 2, 2: 359 (1963); Philcox in F.Z.: 8(2): 49 (1990); Fischer, F.A.C. Scrophulariaceae: 28, pl. 7 (1999). Type: Nigeria, without locality, *P. Beauvois* s.n. (G!, holo.)

Ascending herb, sometimes a simple annual (see note), up to 40 cm high; stems fleshy, trailing, rooting at nodes, glabrous or nearly so. Leaves lanceolate to ovate, 1.5–4(–8) cm long, (0.2–)0.6–1.2(–2.3) cm wide, narrowed at the base into a short petiole or sessile or base amplexicaul, obtuse to acute at apex, margins crenate-serrate, glandular-punctate. Flowers solitary in most leaf axils; pedicels 1–3(–5) mm long. Bracts 2, 1.5 mm long, setaceous. Calyx glabrous, the outer lobes broadly ovate, posterior lobe largest, 8–9 mm long and 4–5 mm wide, in fruit reticulate-veined; other lobes successively smaller and narrower, lateral lobes somewhat keeled. Corolla pure white with yellow mark in the throat, slightly longer than the calyx; upper lip emarginate, lower lip equally 3-lobed; style and ovary glabrous. Capsule 3–4 mm, ovoid, glandular. Seeds ellipsoid.

KENYA. Kwale District: Mwasangombe Forest, 27 Aug. 1953, *Drummond & Hemsley* 4009! & Shimba Hills, 9 Feb. 1953, *Drummond & Hemsley* 1186! and Feb. 1968, *Magogo & Glover* 130!
TANZANIA. Lushoto District: Korogwe, by Lwengera River, 27 June 1953, *Drummond & Hemsley* 3036!; Uzaramo District: Dar es Salaam, 4 Sept. 1969, *Harris* 3204!; Singida District: Iramba Plateau, 30 Apr. 1962, *Polhill & Paulo* 2261!; Pemba, Wesha road, 30 Sept. 1929, *Vaughan* 686!
DISTR. **K** 7; **T** 3, 5, 6, 8; **Z**; **P**; Senegal eastwards to the Sudan and southwards to Mozambique and Angola; Madagascar
HAB. Grassy swamps and salt marshes; 0–1600 m
USES. None recorded on specimens from our area
CONSERVATION NOTES. Least Concern (LC); widespread

SYN. *Herpestis crenata* P.Beauv., Fl. Oware 2: 83, t. 112 (1819)
H. calycina Benth. in Hook., Comp. Bot. Mag. 2: 57 (1836). Types: Senegambia, *Guillemin* 12 (K!, syn.); Senegal, *Leprieur* 214 (K!, syn.)
Bacopa calycina (Benth.) De Wild. in Bull. Herb. Boiss. 1: 832 (1901)
[*B. monniera* sensu Engl. in P.O.A. C.: 357 (1895)]

NOTE. Typically this species has a stout ascending stem and rooting habit, but in brackish-marshy habitats it tends to be a small annual with a simple erect stem; these are ecotypes and cannot be considered as separate taxa.

19. DOPATRIUM

Benth. in Lindl., Bot. Reg. 21 ad t. 1770, f. 46 (1835); Fischer, Rev. *Dopatrium*, in Nord. J. Bot. 17(5): 527 (1997)

Annual herbs; stems often fleshy, simple or sparingly branched, mostly glabrous. Leaves opposite, usually small, somewhat closely arranged at the base of stem, larger ones at base decreasing in size and distantly placed upwards. Flowers in the axils of the upper leaves, solitary, forming short racemes. Calyx deeply 5-fid, lobes imbricate. Corolla bilabiate; corolla tube slender at the base, dilated at the throat; upper lip shortly 2-fid, lower lip spreading, broadly 3-lobed. Stamens 4, the 2 posterior fertile, included, filaments filiform, anther thecae parallel; anterior 2 reduced to staminodes, minute. Ovules numerous; style short, stigma 2-fid. Capsule small, globose or ovoid, loculicidally dehiscent. Seeds small, tuberculate.

About 20 species in the Old World tropics, most of them endemic to Africa.

Most species occur in shallow temporary pools on rocky outcrops in grasslands. They are ephemeral, flowering for a brief period with the seeds staying dormant in the dried up pans until the next rainy season. Hence, they are seldom collected, and their distribution is imperfectly known.

1. Lowermost flowers in inflorescence sessile and cleistogamous, upper flowers pedicellate .... 1. *D. junceum*
   All flowers pedicellate and with conspicuous corolla 2
2. Corolla tube campanulate, ± 3 mm long .... 2. *D. baoulense*
   Corolla tube cylindrical, > 3 mm long 3
3. Basal leaves up to 33 mm long; corolla tube cylindrical, not dilated at apex .... 3. *D. macranthum*
   Basal leaves up to 12 mm long; corolla tube ± dilated at apex 4
4. Stems fleshy; calyx teeth mucronate with hyaline margins .... 4. *D. dortmanna*
5. Stems not fleshy; calyx rounded .... 5. *D. stachytarphetoides*

1. **Dopatrium junceum** (*Roxb.*) *Benth.*, Scroph. Ind.: 31 (1835); Taylor in F.W.T.A., ed. 2, 2: 361 (1963); Hepper in Webbia 19: 609, figs. 8, 9 (maps) (1965); Philcox in F.Z.: 8(2): 51, t. 20/A1–3 (1990); U.K.W.F.: 255 (1994); Fischer in Nord. J. Bot. 17(5): 535, fig. 8, (1997) & F.A.C. Scrophulariaceae:: 39, pl. 12 (1999). Type: India, *Roxburgh* 1798 (K!, holo.)

Erect annual herb 4–28 cm high, usually with several stems from the base, simple or sparingly branched, fleshy. Lower leaves ovate-lanceolate, 8–30 mm long, 2–5 mm wide, sessile, apex obtuse, margins entire, with a strong marginal vein; cauline leaves ovate, decreasing in size and becoming scale-like bracts above, glabrous. Flowers solitary in axils, the lower flowers sessile and cleistogamous, upper flowers pedicellate and perfect; pedicels 6–9 mm long, filiform. Calyx ± 2 mm long, teeth 1 mm long, glabrous. Corolla purplish-blue or white, corolla tube campanulate, ± 3 mm long, limb 4–6 mm in diameter. Capsule globose, 2–3 mm in diameter, with remains of the persistent style. Seeds minute. Fig. 19: 1–3, p. 52.

KENYA. Fort Hall District: Chanya R., 1 July 1971, *Lye, Katende & Faden* 6366!; Machakos District: Thika Plateau, *Bogdan* in *Bally* 5195!
TANZANIA. Rufiji District: Selous Game Reserve, *Vollesen* MRC 4520!; Mbeya District: Trekimboga, 6 May 1970, *Greenway & Kanuri* 14464!; Songea District: Gumbiro, 10 May 1956, *Milne-Redhead & Taylor* 10036!
DISTR. **K** 4; **T** 6–8; widespread in tropical Africa and in Asia extending east to Japan and south to Australia
HAB. Temporary muddy clay pools; 250–1450 m
USES. None recorded on specimens from our area
CONSERVATION NOTES. Least Concern (LC)

SYN. *Gratiola juncea* Roxb., Pl. Corom. 2: 16, t. 129 (1799)

2. **Dopatrium baoulense** *Chev.* in Mem. Soc. Bot. Fr. 8: 184 (1912); Hepper in F.W.T.A., ed. 2, 2: 361 (1963); Fischer in Nord. J. Bot. 17(5): 538, fig. 9, (1997) & F.A.C. Scrophulariaceae:: 40, pl. 13 (1999) & in Fl. Ethiop. & Eritr. 5: 258 (2006). Type: Ivory Coast, valley between Langouassou and Mbayakro, *Chevalier* 22250 (P!, holo.; K!, iso.)

Erect annual herb 9–14 cm high; stems simple or sparingly branched, ± fleshy, glabrous. Lower leaves in a basal rosette, lanceolate-ovate, 7–11 mm long, 3–4 mm wide, sessile, base tapering, apex acute, margins entire, parallel-veined; cauline leaves

FIG. 19. *DOPATRIUM JUNCEUM* — **1**, habit × $^2/_3$; **2**, part of flowering stem × 4; **3**, carpel × 8. *DOPATRIUM STACHYTARPHETOIDES* — **4**, habit × $^2/_3$; **5**, flower × 3; **6**, dehiscing capsule × 6. 1–3 from *Richards* 10902; 4–6 from *Drummond & Rutherford-Smith* 7427. Drawn by Christine Grey-Wilson. Reproduced with permission from F.Z.

few and distantly placed, linear, ± 4 mm, decreasing in size, glabrous. Flowers in bracteate, few-flowered racemes; pedicels 6–10 mm long, filiform. Calyx ± 3 mm long, teeth acuminate, glabrous. Corolla purplish-blue; corolla tube campanulate, ± 3 mm long, upper limb 2–3 mm, entire, lower lip trilobed, 2–3 mm. Capsule ± 3 mm long.

TANZANIA. Kigoma District: Uvinza road, 26 April 1994, *Bidgood & Vollesen* 3216!
DISTR. **T** 4; Senegal, Ivory Coast, Ghana, Nigeria, Cameroon, Central African Republic, Burundi, Ethiopia
HAB. Temporary pools, wet seepages in grassland; ± 1000 m
USES. None recorded on specimens from our area
CONSERVATION NOTES. Least Concern (LC), widespread, but known from a single collection in the Flora area

3. **Dopatrium macranthum** *Oliv.* in Trans. Linn. Soc. 29: 120, t. 121A (1875); Skan in F.T.A. 4(2): 325 (1906); Taylor in F.W.T.A., ed. 2, 2: 36 fig. 287B (1963); Fischer in Nord. J. Bot. 17(5): 546, fig. 16, (1997) & F.A.C. Scrophulariaceae: 44, pl. 15 (1999) & in Fl. Ethiop. & Eritr. 5: 258 (2006). Type: Uganda, Acholi District: Koki, Dec. 1862, *Grant* s.n. (K!, holo., photo)

Erect annual herb 25–50 cm high; stems simple or sparingly branched, tufted, fleshy. Lower submerged leaves lanceolate, up to 33 mm long and 5 mm wide, upper stem leaves ovate, scale-like, 2–7 mm long, 2–3 mm wide, connate at the base, apex obtuse. Inflorescence a lax few-flowered panicle or raceme. Bracts minute; pedicels in fruit erect, up to 2 cm long. Calyx 3.5–5 mm long, teeth ± 1 mm long. Corolla deep violet, purple (yellow or white elsewhere in Africa) with a white spot on lower lip; corolla tube cylindric, 10–12 mm long, upper lip 3-lobed, lobes ovate, obtuse; lower lip twice as long, 2-lobed. Capsule ovoid, 4–5 mm long.

UGANDA. West Nile District: Midigo, Aringa country, Sept. 1937, *Eggeling* 3413!; Acholi District: Koki, Dec. 1862, *Grant* s.n.!; Mengo District: Mabira Forest, Kyagwe, Dec. 1904, *C.B. Fischer* 27!
DISTR. **U** 1, 3, 4; Sierra Leone, Mali, Ghana, Congo-Kinshasa, Sudan
HAB. Temporary pools on rock outcrops in grassland; ?900–1100 m
USES. None recorded on specimens from our area
CONSERVATION NOTES. Least Concern (LC)

SYN. *D. luteum* Engl. in E.J. 23: 498 (1897); Skan in F.T.A. 4(2): 324 (1906). Type: Sudan, Gir, *Schweinfurth* 2153 (K!, S, iso.)

4. **Dopatrium dortmanna** *S.Moore* in J.L.S. 37: 189 (1905); Skan in F.T.A. 4(2): 326 (1906); U.K.W.F.: 255 (1994); Fischer in Nord. J. Bot. 17(5): 551, fig. 19 (1997) & F.A.C. Scrophulariaceae: 42, pl. 14 (1999). Type: Uganda, Ankole District: Mulema, *Bagshawe* 315 (BM!, holo.)

Aquatic annual herb up to 34 cm high; stems erect, slender, simple or sparingly branched. Basal leaves oblong-ovate, ± 7 mm long (see note below), 3 mm wide, sessile, apex obtuse; cauline leaves ovate, scale-like, acute. Inflorescence a lax few-flowered raceme. Bracts minute; pedicels 4–5 mm long. Calyx campanulate, ± 3 mm long, minutely pubescent or scaly, teeth 1 mm long, mucronate with hyaline margins. Corolla deep mauve with a white to pale pink throat; corolla tube narrowly cylindric, 11–13 mm long, ± dilated and curved under the limb, upper lip shortly bilobed ± 6 mm long. Capsule ovoid, ± 4.5 mm long. Seeds oblong, 1 mm long, narrow, reticulate.

UGANDA. Acholi District: near Atiak, *Chorley* 2056!; Ankole District: Mulema, April 1903, *Bagshawe* 315!; Teso District: Kyere, 17 Mar. 1931, *Chandler* 808!
KENYA. Uasin Gishu District: Kipkarren, Aug. 1931, *Brodhurst-Hill* 33!

TANZANIA. Dodoma District: Manyoni, Kazikazi, 9 Apr. 1933, *Burtt* 4653!
DISTR. **U** 1–3; **K** 3; **T** 5; Rwanda, Burundi
HAB. In temporary pools on granite outcrops; 1100–1400 m
USES. None recorded on specimens from our area
CONSERVATION NOTES. Least Concern (LC)

NOTE. *Chandler* 808 bears the field note: leaves are only 1–1? inches (25–30? mm) long and scarce. No leaves are present on any herbarium specimens except on the type.

5. **Dopatrium stachytarphetoides** *Engl. & Gilg* in Warb., Kunene-Samb. Exped. Baum: 362 (1903); Skan in F.T.A. 4(2): 326 (1906); Philcox in F.Z.: 8(2): 53, t. 20/B1–3 (1990); Fischer in Nord. J. Bot. 17(5): 552, fig. 20, (1997) & F.A.C. Scrophulariaceae: 46, pl. 16 (1999). Type: Angola, between Kubano and Mundongo, *Baum* 923 (K!, lecto.; BM!, BR, S, isolecto.)

Annual herb up to 45 cm high; stems erect, fleshy, glabrous, simple or sparingly branched, and branching only within the inflorescence. Basal leaves oblong-ovate, 9–12 mm long, 4–7 mm wide, sessile, apex acute, margins entire; cauline leaves reduced and scale-like, ± 2 mm long, distantly placed, connate at the base. Inflorescence a lax, few-flowered terminal panicle. Bracts ± 1.5 mm; pedicels slender, 2–3 mm long. Calyx campanulate, 3.5–4.5 mm long, teeth ± 1 mm long, rounded. Corolla mauve to purple; corolla tube cylindrical, 11–13 mm long, ± dilated at apex under the limb, upper lip bilobed ± 3 mm long, lower-lip trilobed, 7–8 mm long. Capsule ellipsoid, ± 4 mm long. Fig. 19: 4–6, p. 52.

TANZANIA. Mpanda District: 3 km on Ugalla R. road from Mpanda–Uvinza road, 19 May 1997, *Bidgood et al.* 4048!
DISTR. **T** 4; Congo-Kinshasa, Angola, Zambia
HAB. In shallow pools on rocky outcrops; ± 1100 m
USES. None recorded on specimens from our area
CONSERVATION NOTES. Known from a single collection in the Flora area, elsewhere widespread and of Least Concern (LC)

20. ARTANEMA

D.Don in Sweet, Brit. Flow. Gard. ser. 2: 3, t. 234 (1834), *nom. conserv.*

Erect herbs with 4-angled stems. Leaves opposite, entire or serrate. Inflorescences terminal; flowers large, erect. Calyx 5-lobed, segments imbricate. Corolla broadly tubular, 5-lobed above; upper lip broad, emarginate; lower lip 3-fid. Stamens 4, didynamous; posterior pair included, filaments short, filiform; anterior pair inserted below the corolla throat, filaments long, arched, with a broad appendage at the base of each, connivent at the apex under the upper lip; anthers bithecal. Style long, filiform; stigma bilamellate. Fruit globose, septicidal. Seeds numerous, reticulate-rugose.

Small genus of 4 species of wet places extending from Africa through tropical Asia to Australia.

Artanema longifolium (*L.*) *Vatke* in Linnaea 43: 307 (1882); Engl. in P.O.A. C.: 357 (1895); Hepper in F.W.T.A. ed. 2, 2: 361 (1963); Fischer, F.A.C. Scrophulariaceae: 192, pl. 82 (1999). Lectotype: *"Bahel-tsjulli"* in Rheede, Hort. Malab., 9: 169, t. 87, 1689, designated by Cramer in Dassanayake & Fosberg (eds), Revised Handb. Fl. Ceylon 3: 404 (1981)

Erect perennial herb 30–120 cm high, sparingly branched; stems 4-angled, rather fleshy, glabrous. Leaves narrowly or broadly lanceolate, 4–14 cm long, 1–4.2 cm wide, base narrowing into a short petiole 2–5 mm, or base broad and ± amplexicaul, apex acute or acuminate, margins serrate, scabrid, lamina minutely pubescent above,

FIG. 20. *ARTANEMA LONGIFOLIUM* VAR. *LONGIFOLIUM* — **1**, habit × $^2/_3$; **2**, flower × 1; **3**, flower dissected × $1^1/_2$; **4**, stamens × 3; **5**, ovary & style × 3; **6**, stigma × 14. — *ARTANEMA LONGIFOLIUM* VAR. *AMPLEXICAULE* — **7,** habit × $^2/_3$. 1 from *Eggeling* 2064; 2–6 from *Kassner* 412; 7 from *Faulkner* 666. Drawn by Juliet Williamson.

glabrous beneath. Inflorescences leafless above with 2–5 flowers at each node. Bracts linear-lanceolate, 3–6 mm long; pedicels 8–10 mm long, glabrous. Calyx 5-lobed, segments lanceolate, 5–7 mm long, acuminate, glabrous or scabrid-hairy at the margins. Corolla purple-pink, ± 2.5 cm long, minutely pubescent on the outside; lobes 7–8 mm. Fruit a globose capsule, ± 1 cm in diameter.

var. **longifolium**

Leaves narrowed to the base, ± petiolate; lamina broadly lanceolate. Fig. 20: 1–6, p. 55.

UGANDA. Bunyoro District: Budongo, June 1935, *Eggeling* 2064 & without date, *Dawe* 1018; Kigezi District: Queen Elizabeth National Park, E of Kaizi R. Bridge, Mar. 1969, *Lock* 69/30!
DISTR. **U** 2; widespread in tropical Africa and extending to tropical Asia
HAB. Wet grassland and forest swamps; no altitude indicated; elsewhere 15–500 m
USES. None recorded on specimens from our area
CONSERVATION NOTES. Least Concern (LC)

SYN. *Columnea longifolia* L., Mant. P1.: 90 (1767)
Achimenes sesamoides Vahl, Symb. Bot. 2: 71 (1791)
Artanema sesamoides (Vahl) Benth., Scroph. Ind.: 39 (1835); Skan in F.T.A. 4(2): 347 (1906)

var. **amplexicaule** *Vatke* in Linnaea 43: 307 (1882). Type: Kenya, near Mombasa, *Hildebrandt* 2001b (K!, iso.)

Leaves broad at the base, ± dilated and amplexicaul; lamina narrowly lanceolate. Fig. 20: 7, p. 55.

KENYA. Mombasa District: near Mombasa, June 1876, *Hildebrandt* 2001b!; Kwale District: Majoreni, Nov. 1929, *Graham* 2217! & Mwena, May 1990, *Luke & Robertson* 2334!
TANZANIA. Pangani District: Bushiri, 19 Sept. 1950, *Faulkner* 6661!; Bagamoyo District: Kikoka Forest Reserve, Apr. 1964, *Semsei* 3825!; Kilwa District: Tundus, Apr. 1968, *Rodgers* 216!
DISTR. **K** 7; **T** 3, 6, 8; apparently endemic to East Africa
HAB. Wet grassland, forest swamps, plantations; 0–800 m
USES. None recorded on specimens from our area
CONSERVATION NOTES. Least Concern (LC)

SYN. *Artanema sesamoides* (Vatke) Benth. var. *amplexicaule* (Vatke) Skan in F.T.A. 4(2): 328 (1906)

21. **CRATEROSTIGMA**

Hochst. in Flora 1841: 668 (1841)

Rosetted perennial herbs; stem very reduced and often encased in old leaf-bases, hirsute to ± glabrous; roots usually coloured yellow, orange or red, fibrous. Leaves linear-lanceolate to broadly ovate, sessile or with short petiole. Inflorescences capitate or shortly racemose, usually pedunculate. Flowers with a short or long pedicel. Calyx tubular, 5-ribbed or -winged, teeth 5, subequal. Corolla blue, purple or white, upper lip hooded, entire or emarginate, lower lip with 3 broad lobes. Stamens 4; upper pair with broad straight filaments, included under the hood, lower pair inserted in the throat; filaments with a yellow appendage at the base and sharply angled about the middle; anthers bithecal, thecae divaricate; style filiform, dilated and very shortly bilobed. Fruit an oblong to globose capsule. Seeds numerous.

About 15 species mainly in tropical and South Africa, extending to Madagascar, Arabia and India. In the concept of the genus accepted here the species with slender erect stems included earlier in *Craterostigma* are now placed in *Torenia*.

1. Leaves ± pubescent only on veins beneath, glabrous above; lamina lanceolate 2
   Leaves pilose or pubescent beneath, glabrous, pubescent or scabrid above; lamina ovate to lanceolate 3
2. Calyx prominently 5-winged; flowers long-pedicellate, in corymbs .... 1. *C. alatum*
   Calyx not winged; flowers shortly pedicellate, in racemes 2. *C. lanceolatum*
3. Pedicels 1.5–3 cm long 4
   Pedicels 1 cm long or less, much shorter than peduncle 5
4. Capsule long-exserted from calyx: leaves above minutely pubescent or scabrid; inflorescence very lax .... 4. *C. longicarpum*
   Capsule included within or as long as calyx; leaves glabrous above or appressed-pubescent at least near the tip .... 5. *C. pumilum*
5. Leaves hairy above, whole plant pubescent or pilose 6
   Leaves ± glabrous above; corolla purple, blue or pink with white margins 7
6. Corolla white with blue markings .... 3. *C. hirsutum*
   Corolla purple with white markings .... 7. *C. plantagineum* (see note)
7. Plants clustered, forming mats; inflorescence ± capitate, few-flowered; bracts always longer than pedicels; corolla deep blue and white .... 6. *C. smithii*
   Plants often solitary; inflorescence shortly racemose; bracts ± shorter than pedicels; corolla purple and white ...... 7. *C. plantagineum*

1. **Craterostigma alatum** *Hepper* in K.B. 42: 943 (1987); Fischer, Syst. Afrik. Lindernieae, Trop. und subtrop. Plflanz.: 81: 113, t. 58, 59 (1992); U.K.W.F.: 256 (1994). Type: Kenya, Tsavo National Park, *R.B. & A.J. Faden & Kingston* 74/541 (K!, holo.; EA, MO, iso.)

Rosette perennial herb up to 15 cm; rhizome very reduced with many orange-brown roots. Leaves linear-lanceolate, 3.5–10.5 cm long, 6–18 mm wide, base narrowed to a slender petiole, apex obtuse, margins entire, ciliate, lamina glabrous. Inflorescence a pedunculate corymb; peduncle 3–7 cm. Bracts several pairs, lanceolate, 3–14 mm long, 2–5 mm wide. Pedicels up to 3 cm long. Calyx strongly 5-winged, 6–9 mm long, teeth acute, ciliate. Corolla bright blue or purple (with yellow stamen appendage), up to 2 cm long, lower lip up to 1 cm wide. Stigma white. Capsule linear-oblong, included in calyx.

KENYA. Northern Frontier District: Ura River, Kinna area, Dec. 1959, *J. Adamson* EA11745!; Kitui District: Sasoma, Dec. 1953, *Edwards* 203!; Teita District: Tsavo East, Sala, 4 Dec. 1969, *Hucks* 1208!

DISTR. **K** 1, 4, 7; not found elsewhere

HAB. Shallow laterite soil by rock pools; 550–900 m

USES. None recorded on specimens from our area

CONSERVATION NOTES. Known only from a few specimens, but where found not at threat as far as is known; assessed here as of Least Concern (LC)

2. **Craterostigma lanceolatum** (*Engl.*) *Skan* in F.T.A. 4(2): 331 (1906); Philcox in F.Z.: 8(2): 54 (1990); U.K.W.F.: 256 (1994); Fischer, Syst. Afrik. Lindernieae, Trop. und subtrop. Plflanz.: 81: 116, t. 60–62 (1992) & F.A.C. Scrophulariaceae: 82, pl. 32 (1999). Type: Malawi, *Buchanan* 796 (B†, holo.; K!, iso.)

Rosette perennial herb; rhizome stout, horizontal, with numerous roots. Leaves held erect, lanceolate, 3–7 cm long, 1–2.5 cm wide, gradually narrowing at base to form a broad petiole up to half as long as lamina, entire, apex acute to subacute, 3–5-veined, veins prominent beneath, glabrous and shining above, pubescent on the veins beneath. Inflorescence 3–12 cm long, shortly racemose in upper third or half, sparsely pilose, usually with several pairs of ovate flowerless bracts well below inflorescence; flowers shortly pedicellate with pedicel 1–2 mm. Bracts ovate, 12–14 mm long. Calyx ± 8 mm long, shortly and acutely toothed, sparsely hispid on veins. Corolla white with purple markings, ± 1.5 cm long, upper lip nearly as long as lower, lobes very obtuse. Capsule about as long as calyx.

UGANDA. Mengo District: Bulumaji, *Dummer* 3977!

KENYA. Machakos District: without locality, 20 Dec. 1931, *van Someren* 1591!; Kiambu District: Lukenya, Mua Hills, 21 May 1961, *Lucas & Williams* EA12345! & Kamiti forest, 3 Dec. 1960, *Archer* 211!

TANZANIA. Tabora District: Urambo railway ststion, March 1964, *Msagamosi* 5699 & between Zanzibar and Uyui, *Taylor* 1886!

DISTR. **U** 4; **K** 4; **T** 4; Congo-Kinshasa, Rwanda, Zambia, Malawi, Zimbabwe

HAB. Seasonally wet shallow soils over rock; 1700–1850 m

USES. None recorded on specimens from our area

CONSERVATION NOTES. Least Concern (LC)

SYN. *C. nanum* Engl. var. *lanceolatum* Engl. in P.O.A. C: 357 (1895)

NOTE. See comments on variation after *C. hirsutum*.

3. **Craterostigma hirsutum** *S.Moore* in J.B. 38: 461 (1900); Skan in F.T.A. 4(2): 330 (1906); F.P.U.: 133 (1962); Philcox in F.Z.: 8(2): 54, t. 21 (1990); U.K.W.F.: 256, pl. 110 (1994); Fischer, Syst. Afrik. Lindernieae, Trop. und subtrop. Plflanz.: 81: 95, t. 49, 50 (1992) & F.A.C. Scrophulariaceae: 78, pl. 30 (1999) & in Fl. Ethiop. & Eritr. 5: 268 (2006). Type: Kenya, Masai District: Kapiti [Kapte] Plains, 1893, *Gregory* s.n. (BM!, holo.)

Perennial, hairy herb with a small rosette of leaves; stems very short, covered by fibrous remains of persistent leaf veins; roots fibrous. Leaves oblong-spatulate, 1.5–3(–4.5) cm long, 0.5–1(–2) cm wide, gradually tapering at base, apex obtuse, margins entire or undulate, densely pilose on both sides, parallel 5-veined. Inflorescence scape 3–5(–9) cm long, pilose; flowers sub-sessile or with pedicels up to 8 mm long, in dense apical racemes ± 1 cm long, sometimes lax. Bracts ovate to ovate-lanceolate, 4–8 mm long. Calyx 4–5 mm long in flower, 6 mm in fruit, shortly 5-toothed, pilose, especially on the veins. Corolla white with violet patches on each of three lower lobes, 2 yellow protuberances in throat, ± 9 cm long, upper lip hooded, shortly 2-lobed. Ovary ovoid-oblong, apiculate, little longer than the calyx, glabrous. Fig. 21, p. 59.

UGANDA. Karamoja District: Mt Debasien, Jan. 1936, *Eggeling* 2772!; Toro District: Kyegegwa, 9 Dec. 1933, *A.S. Thomas* 1020!; Masaka District: 18 km SE of Ntusi, 19 Oct. 1969, *Lye & Rwaburindore* 4507!

KENYA. Meru District: Timau, June 1974, *Faden* 74/865!; Machakos District: 30 km SE of Nairobi, 17 May 1975, *Hepper & Field* 5533!; Kwale District: Taru, 3 Sept. 1953, *Drummond & Hemsley* 4149!

TANZANIA. Ngara District: Kabogo, Shanga, 6 Mar. 1961, Ngara, *Tanner* 5861!; Masai District: Mt Longido foot, 12 Dec. 1959, *Verdcourt* 2527!; Kondoa District: Kinyassi Scarp, 2 Jan. 1928, *Burtt* 946!

DISTR. **U** 1, 2, 4; **K** 1–7; **T** 1, 2, 5; Congo-Kinshasa, Rwanda, Zambia, Malawi, Zimbabwe

HAB. Shallow soil over rock where water runs during rains; 350–2000 m

USES. None recorded on specimens from our area

CONSERVATION NOTES. Least Concern (LC)

SYN. *C. ndassekerense* Engl. in E.J. 57: 611 (1922). Type: Tanzania, Masai District: Ndassekera, *Jaeger* 361 (B†, holo.)

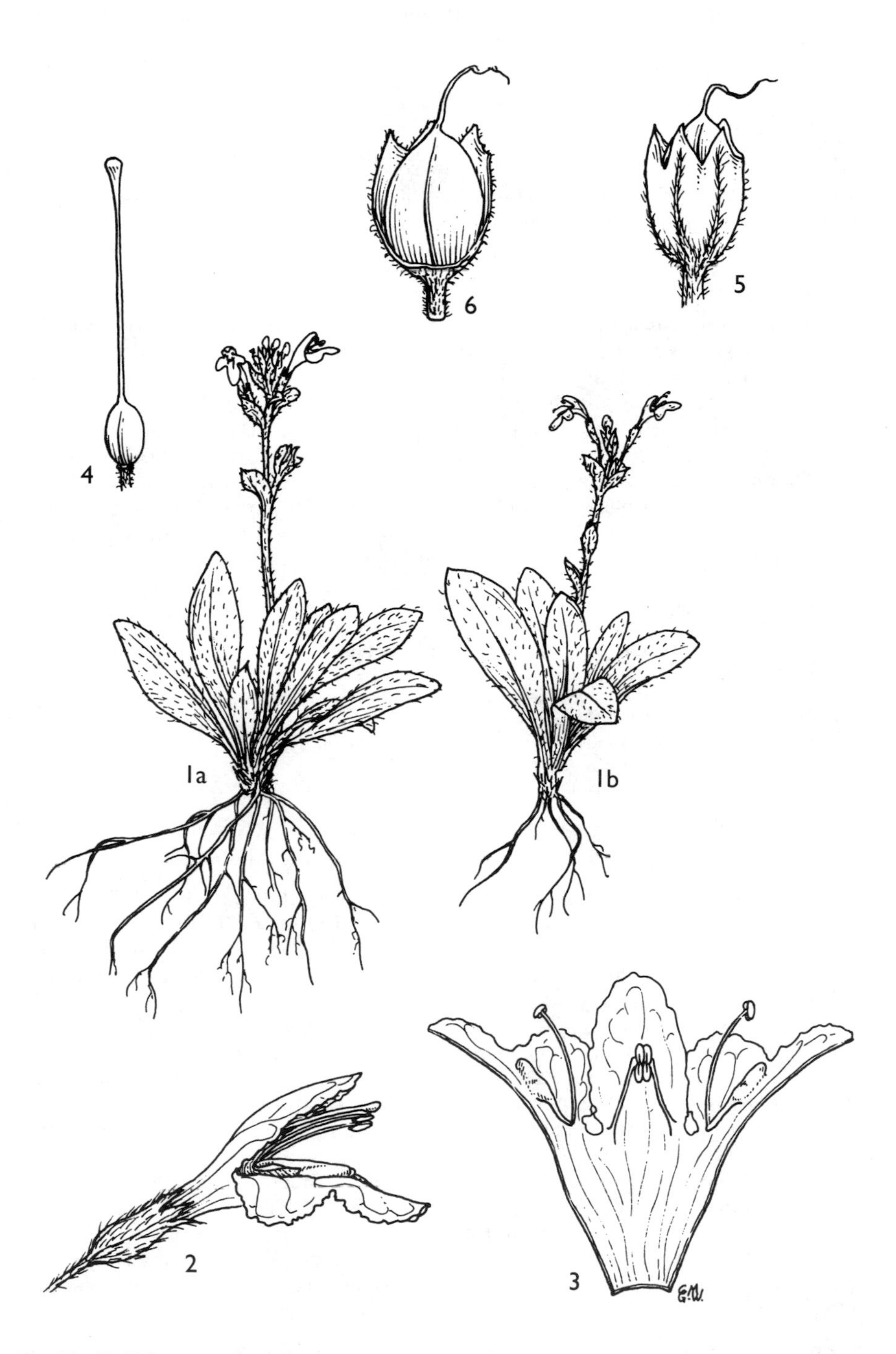

FIG. 21. *CRATEROSTIGMA HIRSUTUM* – **1a & b**, habit × $^2/_3$; **2**, flower × 3; **3**, flower dissected × 3; **4**, carpel × 3; **5**, capsule with calyx × 3; **6**, capsule with front calyx lobes removed × 3. 1 from *Mavi* 143; 2–4 from *Hanger* 175–32; 5, 6 from *Hill* s.n. Drawn by Christine Grey-Wilson. Reproduced with permission from F.Z.

NOTE. Some plants may be sparsely pilose on the upper surface of the leaves and grade into the almost glabrous *C. lanceolatum*. Plants from parts of Tanzania (e.g. West Lake Prov., *Tanner* 4624) have broader leaves and a longer more discontinuous inflorescence than is usual elsewhere. Some long-leaved specimens may be *C. ndassekerense* Engl. which cannot satisfactorily be maintained as distinct from *C. hirsutum*.

The plants remain dormant during dry season and flower rapidly after rain.

4. **Craterostigma longicarpum** *Hepper* in K.B. 42: 943 (1987); Fischer, Syst. Afrik. Lindernieae, Trop. und subtrop. Plflanz.: 81: 99, t. 51, 52 (1992) & in Fl. Ethiop. & Eritr. 5: 269 (2006). Type: Kenya, Northern Frontier District: Ndoto Mts, Ngoronet, *Hepper & Jaeger* 7241 (K!, holo.; EA, P!, iso.)

Perennial rosette herb, up to 15 cm; rhizome shortly horizontal with numerous red fibrous roots. Leaves ovate to broadly ovate, 20–30 mm long, 8–21 mm wide, abruptly narrowing at base to petiole, apex acute to obtuse, margins entire, upper surface uniformly minutely scabrid-pubescent, lower surface densely pubescent, 5-veined. Inflorescence lax, with flowers usually in opposite pairs. Peduncle 3–9 cm long, pubescent. Bracts lanceolate up to 10 mm long, much shorter than pedicels; pedicels up to 20 mm long in fruit. Calyx ± 5 mm long, up to 11 mm in fruit, acute, 5-veined, sparsely pubescent. Corolla white with pale blue on violet patches, yellow in throat, 10–12 mm long, upper lip hooded. Capsule up to 3.5 long, 15 mm wide, long-exserted from calyx. Seeds prismatic, 0.5 mm long, brown.

UGANDA. Karamoja District: N of Kacheliba, 7 May 1953, *Padwa* 74!
KENYA. Northern Frontier District: Ndoto Mts, Ngoronet, 3 Dec. 1978, *Hepper & Jaeger* 7241!; Turkana District: Kacheliba, 7 Oct. 1964, *Leippert* 5045!; North Nyeri District: Ngare Ndare Farm, 19 April 1981, *Gilbert* 6071!
TANZANIA. Mbulu/Singida District: Yaida valley, 25 Jan. 1970, *Richards* 25304!
DISTR. **U** 1; **K** 1–3, 4/7; **T** 2; Ethiopia
HAB. Open eroded gravel slopes over rock outcrops; 600–1500 m
USES. None recorded on specimens from our area
CONSERVATION NOTES. Least Concern (LC)

NOTE. In the fruiting stage this species is characterised by its long, exserted capsules.

5. **Craterostigma pumilum** *Hochst.* in Flora 1841: 67 (1841); Skan in F.T.A. 4(2): 330 (1906); Philcox in F.Z.: 8(2): 54 (1990); U.K.W.F.: 256, pl. 110 (1994); Fischer, Syst. Afrik. Lindernieae, Trop. und subtrop. Plflanz.: 81: 102, t. 53, 54 (1992) & in Fl. Ethiop. & Eritr. 5: 270 (2006). Type: Ethiopia, Semien, near Enjedcap, *Schimper* 986 (K!, holo; P!, iso.)

Perennial rosette herb up to 7 cm; rhizome short with many reddish orange roots. Leaves lying flat on the ground, obovate, 2–4 cm long, 1–2 cm wide, base narrowing into a short broad petiole, apex obtuse to acute or shortly acuminate, margins entire, densely pubescent beneath, closely ciliate along the entire margins, glabrous and smooth above or at most with a few cilia towards the apex. Flowers arising from the centre of the rosette on a ± sessile inflorescence. Bracts lanceolate, up to 1 cm long, ciliate; pedicels 2–3 cm long, pilose. Calyx 6 mm long, up to 9 mm long in fruit, teeth acuminate, pubescent along the median veins. Corolla variable from purple or blue to pink with white and yellowish throat, ± 15 mm long, upper lip entire, lower lip 3-lobed, 10 mm long. Capsule as long as or slightly longer than the calyx, broadly ovoid.

UGANDA. Mbale District: Kapchorwa, Sept. 1954, *Lind* 453! & Mt Elgon, Sebei, 20 July 1924, *Snowden* 934! & 5 May 1955, *Norman* 261!
KENYA. Elgeyo District: Cherangani Hills, near Kapsowar, 16 April 1975, *Hepper & Field* 5026!; Naivasha District: Chania stream, 22 Nov. 1953, *Verdcourt* 1042!; Masai District: foot of Mau escarpment, 7 Dec. 1969, *Greenway & Kanuri* 13888!

TANZANIA. Arusha District: left bank of Tululuise R., Ngurdoto National Park, 7 Nov. 1965, *Greenway & Kanuri* 12327!; Dodoma District: Maboro village, 16 Dec. 1931, *Burtt* 3506!; Iringa District: Ruaha National Park, Magangwe air strip, 16 Dec. 1972, *Bjørnstad* AB2129!
DISTR. **U** 3; **K** 1–6; **T** 2, 3, 5–7; Ethiopia, Somalia, Zambia, Zimbabwe; SW Arabia (Yemen, Oman)
HAB. Open montane grassland, on moist shallow soil overlying rock; 2000–2600 m
USES. None recorded on specimens from our area
CONSERVATION NOTES. Least Concern (LC); widespread

NOTE. A variant (*C.* sp. *A* of U.K.W.F.: 256) with a distinct peduncle, large leaves and long bracts is recorded from Uganda (**U** 1: Karamoja, *Padwa* 75!), Kenya (**K** 1–4: Kanyuere, *Gatheri, Mungai* & *Kanuri* 79/10! and many others) and a more pubescent, pedunculate form (**K** 4: Kenya, Lukenya, *Gilbert* 6135!), and Tanzania (**T** 2: Mbulu, *Greenway & Kanuri* 11078!; **T** 2: Ketumbeine, *Greenway* 4315!; **T** 2: Kilimanjaro, *Beesley* 293!; Biruki Hill, *Richards* 20071!; **T** 4: Ufipa Plateau, ± 6 km from Namamyere, *Moyer & Sanane* 51!). I (F.N.H.) have seen several specimens showing other variations and am therefore reluctant to describe these as a new taxon unless new evidence convinces me that it is truly distinct and not a variant of *C. pumilum.*

6. **Craterostigma smithii** *S.Moore* in J.B. 1900: 461 (1900); Skan in F.T.A. 4(2): 331 (1906); Fischer, Syst. Afrik. Lindernieae, Trop. und subtrop. Plflanz.: 81: 109, t. 56, 57 (1992) & in Fl. Ethiop. & Eritr. 5: 270 (2006). Type: Somalia, Jara, Oct. 1899, *Smith* s.n. (BM!, holo.)

Small rosette perennial herb up to 10 cm, often forming mats; rhizome horizontal with orange roots. Leaves ovate, 2–4 cm long, 1–2 cm wide, base gradually tapering, apex acute to obtuse, margins entire, strongly 5–7-veined, densely pubescent beneath, glabrous above. Inflorescence capitate or nearly so. Peduncle (2–)3–6(–8) cm long, pubescent. Bracts broadly ovate, ± overlapping, 5–10 mm long, 3–5 mm wide. Flowers almost sessile or pedicels up to 1 cm long. Calyx 4–5 mm long, lobes acute, 5-angled, pubescent on angles (veins). Corolla edged white with deep blue patch on each of lower lobes, yellow in throat, ± 13 mm long. Capsule included in calyx.

KENYA. Moyale District: Moyale, 17 April 1952, *Gillett* 12835! & 3 May 1958, *Everard* EA11446!; Laikipia District: Lower Narok R., 50 km N of Rumuruti, 20 April 1975, *Hepper & Field* 5080!
TANZANIA. Musoma/Maswa District: Serengeti National Park, 9 Dec. 1970, *Harris* 5468!
DISTR. **K** 1, 3, 4; **T** 1; Rwanda, Ethiopia, Somalia, Angola
HAB. Sandy soil over rock, saturated in the rains; 1100–2000 m
USES. None recorded on specimens from our area
CONSERVATION NOTES. Least Concern (LC)

7. **Craterostigma plantagineum** *Hochst.* in Flora 24: 669 (1841); Skan in F.T.A. 4(2): 329 (1906); F.P.U.: 133 (1962); Philcox in F.Z.: 8(2): 56 (1990); U.K.W.F.: 256, pl. 110 (1994); Fischer, Syst. Afrik. Lindernieae, Trop. und subtrop. Plflanz.: 81: 87, t. 47, 48 (1992) & in Fl. Ethiop. & Eritr. 5: 268 (2006). Type: Ethiopia, Mt Scholada, *Schimper* 310 (B†, holo.; K!, P!, iso.)

Rosette perennial herb; rhizome short and stout with numerous orange-yellow roots. Leaves leaves often purple beneath, broadly ovate to lanceolate, 1.5–5 cm long, 1–3 cm wide, broadly cuneate at base, apex acute to obtuse, margins entire (or crenate in Kenya), glabrous above, densely pubescent beneath, numerous parallel veins obscure or indented above, prominent beneath; petiole short and broad. Inflorescence 4–8 cm long, with flowers racemose in upper third, pubescent. Bracts opposite, lanceolate, 4–10 mm long, shorter than the pedicels at least in fruit. Calyx ± 5 mm long, up to 9 mm in fruit, shortly toothed, pubescent mainly on the veins. Corolla purple and white, up to 18 mm long, upper lip shortly bilobed, lower lip obtusely 3-lobed, the central lobe very broad. Capsule ovoid-oblong, glabrous, slightly longer than calyx.

UGANDA. Karamoja District: Kacheliba, 20 May 1940, *A.S. Thomas* 3386!; Ankole District: Ruizi R., 3 Nov. 1950, *Jarrett* 229!; Kigezi District: Queen Elizabeth Park, Ishasha R. camp, 8 May 1961, *Symes* 662!

KENYA. Elgeyo-Marakwet District: Tambach, 18 April 1975, *Hepper & Field* 5063!; Mt Elgon, Feb. 1936, *Tweedie* 308!; Masai District: Loita Plains near Goregore, 20 April 1961, *Glover, Gwynne & Samuel* 743!

TANZANIA. Musoma District: Seronera, Serengeti, 10 April 1961, *Greenway* 10009!; Masai District: Naibardad, Nov. 1962, *Newbould* 6309!; Buha District: near Murungu, 8 Dec. 1949, *Shabani* 74!

DISTR. **U** 1–4; **K** 3–7; **T** 1, 2, 4; Niger, Congo-Kinshasa, Rwanda, Burundi, Sudan, Ethiopia, Somalia, Angola, Zambia, Zimbabwe, Botswana, Namibia, South Africa; SW Arabia

HAB. Shallow soil over rock; 900–2200 m

USES. None recorded on specimens from our area

CONSERVATION NOTES. Least Concern (LC); widespread

SYN. *Torenia plantagineum* (Hochst.) Benth. in DC., Prodr. 10: 411 (1846) (excl. syn. *Dunalia acaulis*)

NOTE. A very hairy form with leaves pubescent on both sides, but blue flowers, occurs in Kenya (**K** 3, 4, 5, e.g. Timau to Isiolo road, *Gilbert* 6081) which is referable to *C. plantagineum*.

22. TORENIA

L., Sp. P1.: 619 (1753)

Annual or perennial herbs; stems erect, straggling, creeping and rooting. Leaves opposite, toothed or entire, petiolate or sessile. Flowers solitary, paired or subumbellate, seldom racemose in elongated racemes and panicles terminating the stems; usually long-pedicellate; bracteoles linear at base of petioles. Calyx tubular, 3–5-winged, rarely only ridged, 3–5-toothed or 2-lipped. Corolla tubular, tube cylindric, 2-lipped with upper lip emarginate or entire, lower lip larger with 3 ± equal lobes. Stamens 4, didynamous; posterior pair inserted in the tube; anterior pair inserted in the corolla throat; filaments arched, each with a tooth-like appendage at the base; anthers bithecal; style shortly divided. Capsule oblong, included in the persistent calyx, septicidal. Seeds numerous.

About 60 species in tropical and subtropical countries.

1. Flowers with pedicels at least 1 cm long; plants ± procumbent with long stems .... 1. *T. thouarsii*
   Flowers shortly pedicellate (pedicel < 1 cm long) or ± subsessile 2
2. Inflorescence spicate; calyx conspicuously winged 3
   Inflorescences ± capitate; calyx inconspicuously winged or not winged 4
3. Erect annual 4–17 cm high; leaves lanceolate; inflorescence leafy .... 2. *T. spicata*
   Perennial, with creeping and ascending stems to 50 cm high; leaves ovate; inflorescence not leafy .... 3. *T. goetzei*
4. Broad bracts enclosing inflorescence .... 4. *T. latibracteata*
   Inflorescence without broad bracts 5
5. Inflorescence capitate; stems seldom branched .... 5. *T. schweinfurthii*
   Inflorescence subcapitate or interrupted; stems several-branched 6
6. Stem very slender; leaves narrow linear up to 1 mm wide; calyx lobes longer than the calyx tube .... 6. *T. tenuifolia*
   Stem slender, but not as above; leaves linear to lanceolate, 2–7 mm wide; calyx lobes shorter than the calyx tube .. 7. *T. ledermannii*

1. **Torenia thouarsii** (*Cham. & Schlecht.*) *Kuntze*, Rev. Gen. Pl. 2: 468 (1891); Hepper in F.W.T.A. ed. 2, 2: 363 (1963); Philcox in F.Z.: 8(2): 57, t. 22/A1–6 (1990); U.K.W.F.: 256 (1994); Fischer, F.A.C. Scrophulariaceae: 188, pl. 80 (1999); Hepper in Fl. Masc. 129: 17 (2000). Lectotype: Madagascar, *Thouars* (B–W, holo., K!, photo), presumably chosen by W.G. D'Arcy in Ann. Miss. Bot. Gard. 66: 269 (1979)

Slender branched, procumbent or weakly ascending herb, up to 40 cm, rooting at the nodes and sometimes along internodes; stems quadrangular, glabrous to ± pilose. Leaves lanceolate to ovate-lanceolate, 10–35 mm long, 7–16 mm broad, rounded at base, ± acute at apex, margins crenate-serrate, sparsely pubescent, at least on the veins; petioles 4–9 mm long. Flowers axillary, solitary or a few together on a short leafless peduncle; bracteoles at the base of the pedicels linear, 2–5 mm long; pedicels 1.5–3 cm long, usually reflexing in fruit. Calyx slightly winged, at least in fruit, 5–6 mm long in flower, up to 10 mm long in fruit, with 2 short acute teeth on the upper lip and 3 on the lower. Corolla 9–10 mm long, white, variably suffused mauve or deep blue, sometimes with red spots in throat. Capsule oblong-ellipsoid, shorter than calyx, 7.5–9 mm long, ± 3 mm wide. Seeds pitted. Fig. 22: 1–6, p. 64.

UGANDA. West Nile District: Maracha, 27 July 1953, *Chancellor* 64!; Busoga District: Kibibi, 6 Feb. 1953, *Wood* 625!; Mengo District: Kisubi, 9 Jan. 1969, *Lye* 116!

KENYA. Trans-Nzoia District: Kitale, 12 Sept. 1956, *Bogdan* 4283!; Kilifi District: Arabuko, no date, *Graham* 1704!; Kwale District: SW of Kwale, 27 Aug. 1953, *Drummond & Hemsley* 4012!

TANZANIA. Tanga District: Tanga, Jan. 1893, *Volkens* 21!; Mpanda District: Lubulungu, 23 Sept. 1958, *Newbould & Jefford* 2589!; Songea District: Kigonsera, 26 Dec. 1973, *Mhoro* 1794!; Zanzibar, Kidichi, 9 July 1960, *Faulkner* 2632!

DISTR. **U** 1–4; **K** 3, 7; **T** 1, 3, 4, 6–8; **Z**; **P**; throughout much of lowland tropical Africa, Madagascar, Asia and America

HAB. Wet sandy soil, often shady, and by streams or pools; 0–1400 m

USES. None recorded on specimens from our area

CONSERVATION NOTES. Least Concern (LC); widespread

SYN. *Nortenia thouarsii* Cham. & Schlecht. in Linnaea 3: 18 (1828)

Torenia parviflora Benth., Scroph. Ind.: 39 (1835); Hook. f., Fl. Brit. Ind. 4: 278 (1884); Skan in F.T.A. 4(2): 35 (1906); Li in Taiwania 9: 4 (1963), *nom. illegit. superfl.* based on *Nortenia thouarsii*

T. ramosissima Vatke in Oest. Bot. Zeitschr. 25: 10 (1875). Type: Tanzania, Zanzibar, *Hildebrandt* 986 (K!, iso.)

2. **Torenia spicata** *Engl.* in E.J. 23: 502, t. 7, figs. G–M (1897); Skan in F.T.A. 4(2): 334 (1906); Hepper in F.W.T.A. ed. 2, 2: 363 (1963); Philcox in F.Z.: 8(2): 59, t. 22/B1–5 (1990). Type: Sudan, Djur, *Schweinfurth* 4296 (B†, holo; K!, iso.)

Slender erect annual, 4–17 cm high; stem simple or branched, quadrangular, narrowly winged, glabrous or with a few glands at or near the nodes. Leaves sessile, lanceolate, (5–)8–21 mm long, 1.5–3 mm wide, apex ± acute, entire or slightly toothed, glabrous. Inflorescence laxly spicate; flowers solitary, axillary. Lower bracts leaf-like, upper much shorter than calyx. Pedicel 0.5–1 mm, glabrous or occasionally minutely glandular. Calyx 3.5–4 mm long, elongating to ± 6 mm in fruit, 5-toothed and 2-lipped, conspicuously winged on each vein. Corolla blue and white, 5–8 mm long, upper lip hooded, lower lip 3-lobed, rounded; anthers ± exserted. Capsule oblong-ovoid, 3–4 mm long, 2–2.5 mm wide, acute, dehiscing into 2 valves. Fig. 22: 7–11, p. 64.

TANZANIA. Mwanza District: Mbarika, 6 May 1952, *Tanner* 742!; Singida District: Iramba Plateau, 28 April 1962, *Polhill & Paulo* 2226!; Iringa District: Msembi to Trekimboga, 11 May 1970, *Greenway & Kanuri* 14496!

DISTR. **T** 1, 4, 5, 7, 8; Mali, Guinea, Ghana, Nigeria, Cameroon, Congo-Kinshasa, Sudan, Angola, Zambia, Malawi, Zimbabwe, Botswana, South Africa

HAB. In wet sandy places in woodland, and in cultivations; 900–1350 m

FIG. 22. *TORENIA THOUARSII* — **1**, habit × ²⁄₃; **2**, flower × 4; **3**, flower dissected × 4; **4**, carpel × 4; **5**, fruiting calyx × 4; **6**, capsule with calyx removed × 4. *TORENIA SPICATA* — **7**, habit × ²⁄₃; **8**, flower × 4; **9**, corolla opened showing androecium and gynoecium × 4; **10**, fruiting calyx × 4; **11**, capsule with calyx removed × 4. 1, 5, 6 from *Goldsmith* 82/69; 2–4 from *Smith* 1462; 7–11 from *Philcox & Drummond* 9051. Drawn by Christine Grey-Wilson. Reproduced with permission from F.Z.

USES. None recorded on specimens from our area
CONSERVATION NOTES. Least concern (LC); widespread

SYN. *Canscora ramossisima* Baker in K.B. 1898: 158 (1898). Type: Malawi, Fort Hill, *Whyte* s.n. (K!, holo.)
Crepidorhopalon spicatus (Engl.) Fischer in F.R. 100: 444 (1989) & F.A.C. Scrophulariaceae: 122, pl. 50 (1999)

3. **Torenia goetzei** (*Engl.*) *Hepper* in Bol. Soc. Brot. 60: 271 (1978); Philcox in F.Z.: 8(2): 59 (1990). Type: Tanzania, Iringa District: Udzungwa [Utschungwe], Mt near Kisinga [Kissinga], *Goetze* 581 (B†, holo.). Neotype: Tanzania, Iringa District: Mufindi, Ngwazi marsh, *Goyder et al.* 3911 (K!, chosen here)

Creeping and ascending herb 15–50 cm high, branched from the base; stems quadrangular, glabrous above, ± tomentose below. Lower internodes ± 1 cm long, upper ± 10 cm long. Leaves ovate, purplish beneath, 8–14 mm long, 5–11 mm broad, subcordate at base, rounded at apex, margins obscurely toothed, pellucid-punctate, 3–5-veined. Bracts 1–2 mm, narrow-triangular, acute, hispid-ciliate. Flowers in lax racemes, clustered towards apex, shortly pedicellate. Calyx often purple, 5–6 mm long, prominently 5-winged producing an orbicular outline; teeth triangular, acuminate. Corolla rich purple, 9–12 mm long, minutely pubescent outside, upper lip ovate, subentire, half as long as 3-lobed lower lip, throat densely pubescent. Capsule ovoid, 5 mm long 2–2.5 mm wide, slightly shorter than the calyx, acute at apex. Seeds ovoid, longitudinally grooved and minutely tuberculate.

KENYA. Kwale District: near Lunguma, 20 Aug. 1994, *Luke & Gray* 4061!
TANZANIA. Ufipa District: Sumbawanga, Nkundi–Kamwanga road, 16 June 1996, *Faden et al.* 96/241!; Mbeya District: near Sao Hill, May 1975, *Hepper, Field & Mhoro* 5253!; Songea District: 12 km W of Songea, Jan. 1956, *Milne-Redhead & Taylor* 8070!
DISTR. **K** 7; **T** 4, 7, 8; Congo-Kinshasa, Burundi, Zambia, Malawi
HAB. Boggy places beside streams; 140–2200 m
USES. None recorded on specimens from our area
CONSERVATION NOTES. Least Concern (LC); common

SYN. *Craterostigma goetzei* Engl. in E.J. 28: 477 (1900); Skan in F.T.A. 4(2): 332 (1906); Cribb & Leedal, Mount. Fl. S Tanz.: 118, pl. 28F (1982)
Torenia brevifolia Engl. & Pilger in E.J. 45: 214 (1910). Type: Tanzania, Songea District: Songea, *Busse* 801 (K!, fragment)
Crepidorhopalon goetzei (Engl.) Fischer in F.R. 100: 444 (1989) & F.A.C. Scrophulariaceae: 108, pl. 43 (1999)

4. **Torenia latibracteata** (*Skan*) *Hepper* in Bol. Soc. Brot., sér. 2, 60: 271 (1987). Type: Congo-Kinshasa, Dolo, *Schlechter* 12440 (K!, holo.)

Annual herb 18–35 cm high; stems often purple, simple or branched, quadrangular, glabrous to ± ciliate on the angles. Leaves distantly placed, lanceolate, 11–28 mm long, 5–11 mm wide, sessile, apex obtuse or acute, margins shallowly dentate to subentire, 3-veined, ciliate on the veins and margins. Inflorescence capitate, terminal; flowers few to many, shortly pedicellate, with involucral bracts. Outer bracts 2, leafy, broadly lanceolate, up to 1 cm long, acute; inner bracts 2, broadly ovate, obtuse, toothed, cilate. Calyx ± 6 mm long, densely to sparsely pilose, lobes ± 2 mm, triangular, acute, ciliate. Corolla bright purple, ± 16 mm long, upper lip entire, lower lip 3-lobed with club-shaped hairs in the throat. Capsule cylindric-ellipsoid, ± 3 mm long, 1.5 mm wide.

subsp. **parviflora** *Philcox* in Bol. Soc. Brot., sér. 2, 60: 267 (1987); Philcox in F.Z.: 8(2): 60 (1990). Type: Zambia, Kawimbe, *Richards* 5792 (K!, holo.)

Stems cilate on the angles; flowers ± 11 mm long.

KENYA. Trans-Nzoia District: Maboonde, Dec. 1971, *Tweedie* 4204!
TANZANIA. Buha District: 35 km on Kasulu–Kibondo road, 13 June 1997, *Bidgood et al.* 4342!; Iringa District: Iringa, 3 Sept. 1932, *Geilinger* BXIV.13!; Songea District: Kwamponjore valley, 20 June 1956, *Milne-Redhead & Taylor* 10851!
DISTR. **K** 3; **T** 4, 7, 8; Congo-Brazzaville, Congo-Kinshasa, Zambia, Malawi, Mozambique
HAB. Grassy swamps; 1000–1800 m
USES. None recorded on specimens from our area
CONSERVATION NOTES. Least Concern

SYN. *Craterostigma latibracteatum* Skan in F.T.A. 4(2): 333 (1906)
[*Lindernia latibracteata* Engl. in Schlecht., Westafr. Kautsch.-Exped. 313 (1900), nom. nud.]
Crepidorhopalon parviflorus (Philcox) Eb.Fisch. in B.J.B.B. 67: 378 (1999) & F.A.C. Scrophulariaceae: 102, pl. 40 (1999). Type as for *T. latibracteata* subsp. *parviflora* Philcox
Crepidorhopalon latibracteatus (Skan) Eb.Fisch. in F.R. 100: 444 (1989) & Trop. & subtrop. Pflanzenwelt, 81: 149 (1992) & F.A.C. Scrophulariaceae: 100, pl. 39 (1999)

5. **Torenia schweinfurthii** *Oliv.* in Hook., Ic. Pl. t. 1251 (1878); Philcox in F.Z.: 8(2): 60 (1990). Type: Sudan, Bongo, R. Lesi, *Schweinfurth* 4009 (K!, holo.)

Spindly, erect perennial, 15–60 cm; stems simple or sparingly branched, ± creeping, rooting at the base, quadrangular, glabrous. Leaves linear-lanceolate, 6–12(–30) mm long, 2–8(–10) mm wide, apex acute, margins entire or minutely toothed, scabrid-ciliate; lower leaves purple beneath, broader than the smaller and distant upper leaves, 3-veined. Inflorescences capitate, ± 1 cm in diameter, up to 2.5 cm in fruit. Bracts subulate. Flowers subsessile. Calyx 5 mm long, purple, prominently 5-angled, teeth triangular with setaceous apex. Corolla 9–15 mm long, rich blue-purple, brown-tomentellous outside, upper lip slightly emarginate, lower lip deeply 3-lobed. Capsule as long as the calyx, opening 1/3 of its length. Seeds yellowish, angular, minute, nearly as broad as long.

UGANDA. West Nile District: Kobboko, Feb. 1934, *Eggeling* 1488! & Mar. 1935, *Eggeling* 1850! & June 1938, *Hazel* 545!
TANZANIA. Mpanda District: Uruwira, 30 Sept. 1970, *Richards & Arasululu* 26201!
DISTR. **U** 1; **T** 4, 8; Ghana, Benin, Nigeria, Cameroon, Sudan, Angola, Zambia
HAB. Swamps; 850–1300 m
USES. None recorded on specimens from our area
CONSERVATION NOTES. Least Concern (LC)

SYN. *Craterostigma schweinfurthii* (Oliv.) Engl. in E.J. 23: 501 (1897); Skan in F.T.A. 4(2): 332 (1906); Hepper in F.W.T.A. ed. 2, 2: 361 (1963); Fischer, F.A.C. Scrophulariaceae: 88, pl. 33 (1999)
Crepidorhopalon tanzanicus Fischer in B.J.B.B. 67: 378 (1999). Type: Tanzania, Songea District: ± 1.5 km of R. Mtandasi, *Milne-Redhead & Taylor* 10030 (B, holo.; BR, K!, S, iso.)

NOTE. *Milne-Redhead & Taylor* 10030 from Tanzania comes closest to *T. schweinfurthii* but has a more leafy stem and the leaves are on average broader (up to 1 cm). Fischer in B.J.B.B. 67: 378 (1999) has described this specimen as a new species (placed in synonymy here), but I (S.A.G.) feel that the material is insufficient to give it an independent status.

6. **Torenia tenuifolia** *Philcox* in Bol. Soc. Brot. ser. 2, 60: 267 (1987) & in F.Z. 8(2): 61 (1990). Type: Zambia, Chilongowelo, *Richards* 1690 (K!, holo.)

Erect annual herb, up to 40 cm high; stems simple or usually branched in upper part, quadrangular, glabrous. Leaves linear, 10–30 mm long, 0.5–1 mm wide, sessile, apex acute, margins entire or with one or two minute teeth, glabrous or with a few hairs at the base, 1-veined. Flowers terminating each branch, in the axils of bract-like reduced leaves appearing capitate, pedicellate. Bracts linear to lanceolate, up to 3 mm long, glabrous. Calyx 3.5–4 mm long, minutely glandular, lobes 2–3 mm, narrowly lanceolate, longer than the calyx-tube, 1-veined, scabrid-cilate. Corolla blue, 4–5 mm long, lower lip 3-lobed, whitish with hairs on the palate, upper lip blue to mauve. Anterior filaments with appendage at base. Capsule subrotund, compressed, 3–4 mm in diameter.

TANZANIA. Mpanda District: Uzundo Plateau, Mpanda–Uvinza road, 29 May 2000, *Bidgood et al.* 4530!
DISTR. **T** 4; Zambia
HAB. Seasonally inundated grassland, with *Kotschya bullockii*, *Protea* and *Combretum*, on grey sandy-peaty soil; ± 1670 m
USES. None recorded on specimens from our area
CONSERVATION NOTES. Least Concern (LC), common in Zambia, but in the Flora area known from a single gathering; possibly undercollected

SYN. *Crepidorhopalon tenuifolius* (Philcox) Eb.Fisch. in F.R. 100: 443 (1989) & F.A.C. Scrophulariaceae: 92, pl. 35 (1999)

7. **Torenia ledermannii** *Hepper* in K.B. 35: (1981). Type: Cameroon, Garoua near Tchambutu, *Ledermann* 5047 (B†, holo.; K!, photo.)

Erect annual herb, (6–)25–40 cm high; stems suffused purple, simple or usually branched in upper part, quadrangular, glabrous. Leaves linear to lanceolate, 3–19 mm long, 2–7 mm wide, sessile, apex acute, margins ± serrate, scabrid-ciliate on the margins, 1–3-veined. Flowers terminating each branch, few together or shortly spicate, almost sessile. Bracts lanceolate. Calyx 5 mm long with 5 wide purple longitudinal bands, teeth triangular, acute, scabrid. Corolla 5–7 mm long, lower lip 3-lobed, violet with a whitish spot in middle and two yellow marks in throat, upper lip deep mauve; anthers bluish, pollen white. Capsule oblong, 5 mm long, acute. Seeds suborbicular, grooved down one side.

TANZANIA. Songea District: by R. Luhira, near Mshangano, 15 June 1956, *Milne-Redhead & Taylor* 10817!
DISTR. **T** 8; Cameroon, Zambia
HAB. Moist soil on rock outcrops; ± 1000 m
USES. None recorded on specimens from our area
CONSERVATION NOTES. Least Concern (LC), but in the Flora area known from a single gathering that was collected in 1956

SYN. *Craterostigma gracile* Pilger in E.J. 45: 213 (1910); Raynal in Adansonia ser. 2, 6: 431, pl. (1966), *non Torenia gracilis* Benth. in Wall. Cat. 3952
Crepidorhopalon gracilis (Pilger) Eb.Fisch. in F.R. 100: 443 (1989) & F.A.C. Scrophulariaceae: 90, pl. 34 (1999)

NOTE. The Zambia specimens have pubescent angles of the stems, and the flowers are conspicuously pubescent in a capitate inflorescence. If Raynal (l.c.) is correct in including *C. guineese* Hepper (K.B. 14: 407 (1960)) in the synonymy, the distribution would extend to Guinea Republic, Guinea Bissau and Senegal, but I still consider it to be a distinct species.

23. **LINDERNIA**

Allioni, Misc. Taur. 3: 178, t. 5, fig. 1 (1762)

Vandellia L., Mant. 1: 12 (1767)
Bonnaya Link & Otto, Ic. P1. Select.: 25, t. 11 (1820)
Ilysanthes Rafin., Ann. Nat.: 13 (1820)

Annual or perennial herbs; stems slender, obscurely to distinctly quadrangular. Leaves opposite to almost opposite, simple, sessile or petiolate, pinnately or 3–5-veined with veins arising at base of lamina. Flowers small, solitary, axillary or terminal, or in terminal or axillary racemes or clusters, pedicellate or occasionally subsessile. Calyx 5-lobed, shallowly lobed with lobes spreading or somewhat connivent when mature, or lobed almost to base with lobes lanceolate to linear lanceolate; tube 5-veined, each vein with obscure or distinct rib or minute wing. Corolla tubular, bilabiate; anterior lip trilobed, lobes usually spreading; posterior lip either entire or emarginate to bilobed, suberect. Stamens either 4, all antheriferous and fertile, or 2, with posterior pair fertile and anterior pair reduced to staminodes, posterior pair affixed to corolla tube, anterior pair or staminodes arising in corolla throat; frequently anterior filaments each with distinct spur arising at or near base; anthers free or contiguous, thecae divaricate; style slender, erect; stigma bilamellate. Capsule globose to cylindric to narrowly so, bivalved. Seeds numerous, smooth to alveolate, variously shaped.

A genus of about 120 species from the tropics, subtropics and warm temperate regions.

1. Stamens 4 2
   Stamens 2; staminodes 2 18
2. Stamens all fertile or apparently so 3
   Posterior 2 stamens with normal anthers, fertile; anterior 2 stamens reduced in size, frequently sterile 30
3. Plant few-leaved, with all leaves ± 2 mm long, bract-like .... 1. *L. stuhlmannii*
   Plant not as above, leaves more than 2 mm long, not all bract-like 4
4. Anterior filaments distinctly clavate at or near base 5
   Anterior filaments bent or geniculate, not clavate 10
5. Plant decumbent, ascending or straggling; stems up to 60 cm long, glabrous; capsule to 9.5 mm or more long, narrowly oblong-ellipsoid .... 2. *L. whytei*
   Plant with erect to spreading stems up to 20 cm tall, glandular, glandular-hispid to -pubescent, hirsute or glabrous; capsule 2.5–4 mm long (8 mm in *L. hepperi*), not narrowly cylindric-ellipsoid 6
6. Stems much branched 7
   Stems rarely branched more than once, more usually simple .... 7. *L. bifolia*
7. Leaves 8–25(–30) mm long, 0.1–0.3 mm wide; calyx divided below midway, tube not strongly ridged or shallowly winged 8
   Cauline leaves 0.6–1.5 mm long, 0.3–1.75 mm wide; calyx not divided to midway or below, tube strongly 5-veined with veins pronounced into shallow wings .... 6. *L. tenuis*
8. Plant procumbent, straggling; leaves subreniform, circular or broadly ovate; pedicels 7–16 mm long .... 3. *L. humilis*
   Plant erect to spreading; leaves not as above; pedicels 0.5–3 mm long 9
9. Plant strictly erect; calyx ± 3 mm long, glabrous, lobes not widely spreading; corolla pale mauve or blue; capsule 7–8 mm long, lanceolate-ellipsoid . ... 4. *L. hepperi*
   Plant erect to spreading; calyx 4–5 mm long, glandular-pubescent, lobes widely spreading; corolla yellow, orange or white; capsule to 4 mm long, globose to globose-obovoid .... 5. *L. rupestris*
10. Plant narrow-leaved; leaves linear to linear-lanceolate 11
    Plant wider-leaved; leaves ovate, not linear or linear-lanceolate 13

11. Plant erect, up to 35 cm tall; leaves 10–20 mm long, 0.3–4 mm wide; pedicels 5–17 mm long, spreading to reflexed in fruit; capsule 8–15 mm long 12
    Plant prostrate to suberect, rarely more than 10 cm tall; leaves 1.5–11 mm long, 0.2–5 mm wide; pedicels 0.5–1 mm long, not reflexed in fruit; fruit 2–3 mm long . . . . . 10. *L. debilis*
12. Plant up to 35 cm tall; leaves 7–15(–40) mm long, 0.5–4 mm wide; pedicels straight, spreading, reflexed in fruit; corolla blue or mauve; capsule 8.5–15 mm long . . . . . 8. *L. oliveriana*
    Plant up to 12 cm tall; leaves 10–20 mm long, to 1 mm wide; pedicels upwardly curved, spreading-reflexed in fruit; corolla red or bronze; capsule 7–9 mm long . . . . . 9. *L. lindernioides*
13. Stems 20–40 cm long, often creeping and rooting at nodes 14
    Stems not exceeding 15 cm long, erect or suberect 16
14. Ripe fruit twice or more longer than calyx 15
    Ripe fruit equal to or 1/3 longer than calyx . . . . . 13. *L. diffusa*
15. Stems ± glabrous; pedicels 3–8(–11) mm long . . . 11. *L. brevidens*
    Stems hispid on angles; pedicels 2–5 mm long . . . 12. *L. subracemosa*
16. Fruit less than, or rarely more than 1.5 times length of calyx 17
    Fruit 2.5–3 times length of calyx . . . . . 16. *L. longicarpa*
17. Leaves ovate to subcircular; calyx 2.5–3.5 mm long 14. *L. nummulariifolia*
    Leaves ovate to obovate; calyx 3.5–6.5 mm long . . . 15. *L. abyssinica*
18. Calyx deeply 5-lobed almost to base 19
    Calyx shallowly lobed, never beyond midway 22
19. Stem erect or decumbent or ascending, never creeping 20
    Stem creeping, rooting at nodes; leaves mostly as long as wide, orbicular or broadly ovate . . . . . 22. *L. rotundifolia*
20. Flowers variously arranged along stem, but not congested 21
    Flowers congested in glomerules . . . . . 21. *L. congesta*
21. Leaves broadly ovate, ovate-lanceolate or elliptic, entire to rarely shallowly dentate; fruit 1.5–4.5 (–6) mm long, cylindric-ellipsoid to subglobose, equal to or slightly longer than calyx . . . . . 19. *L. parviflora*
    Leaves narrowly oblanceolate to subelliptic, remotely crenate-dentate; fruit up to 9 mm long, narrowly cylindric, up to twice as long as calyx . . 20. *L. zanzibarica*
22. Flowers markedly pedicellate with pedicels up to 20 mm long 23
    Flowers subsessile or very shortly pedicellate with pedicels not exceeding 1.5 mm long . . . . . 31. *L. nana*
23. Stems short-hispid or hispid-pubescent 24
    Stems not hispid 25
24. Leaves 2.5–4 mm long, 1 mm wide; pedicels 3–7 mm long; calyx 2–2.5 mm long; corolla 6 mm long . . 23. *L. ugandensis*
    Leaves 5–14 mm long, 1.5–4 mm wide; pedicels 8–20 mm long; calyx 3.5–5.5 mm long; corolla 8–10(–14) mm long . . . . . 24. *L. pulchella*
25. Woody-based perennial, much branched 26
    Non-woody, herbaceous annual 27

26. Plant 10–20 cm tall, erect, somewhat tufted; pedicels 5–7.5 mm long, widely spreading or at times reflexed .... 25. *L. niamniamensis*
    Plant suberect, 4–10 cm tall from a creeping rhizome; pedicels 3.5–7.5 mm long, erect, not spreading or reflexed .... 26. *L. acicularis*
27. Pedicels 4–6 mm long, widely spreading or deflexed in fruit; calyx 1.5–3 mm long, lobes 0.4–0.5 mm long; capsule ± 4 mm long .... 27. *L. schweinfurthii*
    Plant not as above 28
28. Basal leaves appearing rosetted or clustered; calyx lobes acute, ciliolate, these and pedicels minutely glandular-pubescent .... 28. *L. wilmsii*
    Basal leaves absent or occurring as 1 or 2 pairs; calyx lobes acute to obtuse, these and pedicels glabrous. 29
29. Plant 3–10 cm with creeping to ascending stems rooting at nodes; calyx glabrous .... 29. *L. serpens*
    Plant 3 cm; stems erect, not rooting at nodes; calyx pilose at apex .... 30. *L. hartlii*
30. Plant 2.5–5(–10 cm) tall; basal leaves rosetted; cauline leaves narrowed into petiole-like base .... 17. *L. bolusii*
    Plant to 35 cm tall; basal leaves not rosetted; cauline leaves sessile, not narrowed at base .... 18. *L.* sp. '*alpha*'

1. **Lindernia stuhlmannii** *Engl.*, P.O.A. C: 357 (1895); Skan in F.T.A. 4(2): 344 (1906). Type: Tanzania, Mwanza/ Biharamulo District: Uzinza, *Stuhlmann* 3550 (B†, holo.; K!, photo.)

Erect, ± 14 cm tall; stems branched from base, quadrangular, glabrous. Leaves few, 2–2.5 mm long, 0.5–0.7 mm wide, glabrous, bract-like throughout; basal leaves linear, stem leaves short, lanceolate, channelled. Flowers laxly racemose in axils of upper leaves, appearing pedicellate, if so then pedicels exceeding subtending leaves, ± equalling flowers and said to become deflexed at flowering. Calyx 2 mm long, glabrous, lobes 1 mm long, acutely triangular. Corolla with upper lip shortly bilobed, equalling lobes of lower lip. Stamens 4, fertile, anterior filaments geniculate. Capsule elongate fusiform, twice as long as calyx.

TANZANIA. Mwanza/Biharamulo District: Uzinza, Mar. 1892, *Stuhlmann* 3550!
DISTR. **T** 2; known only from the type collection
HAB. Not known; ?± 1600 m
USES. None recorded on specimens from our area
CONSERVATION NOTES. Data Deficient (DD) and possibly extinct as it has not been collected for more than 100 years

NOTE. The above description was made solely from Engler's original description in conjunction with a photograph of the holotype and a life-size drawing made from that specimen. From the evidence thus provided, I (D.P.) have no reason to consider that these two sources represent anything other than a good species. Unlike Fischer (Syst. Afrik. Lindernieae, Trop. und subtrop. Plflanz. 81: 165 (1992)), I cannot subscribe to it belonging to *L. oliverana* Dandy.

2. **Lindernia whytei** *Skan* in F.T.A. 4(2): 340 (1906); Philcox in F.Z.: 8 (2): 62 (1990); U.K.W.F.: 257 (1994). Type: Uganda, first and second days' march from Mumias, 6–7 Dec. 1898, *Whyte* s.n. (K!, holo.)

Decumbent or ascending annual; stems up to 60 cm long, simple or branched, quadrangular, glabrous. Leaves sessile, broadly ovate, subcircular to ovate-lanceolate, 8–24(–32) mm long, 8–20(–28) mm wide, rounded to cordate at base, obtuse to shortly apiculate, dentate or glabrous with few hairs on margins or major veins beneath. Flowers terminally racemose, lax. Bracts smaller than leaves, ovate below becoming linear to subulate above; pedicels 1.7–3(–6) mm long. Calyx 6–7 mm long, 5-ribbed, 5-lobed, glabrous, or ciliate on ribs; lobes 3.5–6.5 mm long, linear, acuminate, spreading, keeled, glabrous or ciliate. Corolla blue, lilac, purple or yellow, 8–12(–20) mm long; tube 5–6(–9) mm long, funnel-shaped, minutely glandular-pubescent outside; upper lip ovate, slightly emarginate, ciliate, pubescent outside; lower lip trilobed. Stamens 4, all fertile; anterior filaments 4–5(–7) mm long, arched with broadly clavate, glandular-papillose spur about 1 mm or more long produced above base; base of filament extended into pronounced antero-lateral ridge. Capsule to 9.5 mm or more long, 3 mm broad, narrowly oblong-ellipsoid.

UGANDA. West Nile District: Logiri, Mar. 1935, *Eggeling* 1874!; Ankole District: Igara, Mar. 1939, *Purseglove* 616!; Teso District: Soroti, Sept. 1954, *Lind* 394!

KENYA. Uasin Gishu District: near Kipkarren, 27 Mar. 1952, *Cooke* 16!; South Kavirondo District: Mumias, 11 May 1979, *Bridson* 80!; Kwale District: Mwasangombe Forest, 24 km SW of Kwale, 27 Aug. 1953, *Drummond & Hemsley* 4007!

TANZANIA. Bukaba District: 18 Jan. 1959, *Lind* 2373!; Mpanda District: Uruwira–Mpanda road, 19 km from Uruwira, 25 Sept. 1970, *Richards & Arasululu* 26150!; Iringa District: Iringa, 29 Feb. 1999, *Price* WK350!

DISTR. **U** 1, 3, 4; **K** 3–7; **T** 1, 3–6; Rwanda, Burundi, Sudan, Ethiopia, Angola, Mozambique, Zimbabwe

HAB. Stream and riversides, marshy ground surrounding ponds and waterholes and in montane forests; 200–2250 m

USES. None recorded on specimens from our area

CONSERVATION NOTES. Least Concern (LC); widespread

SYN. *L. flava* S.Moore in J.L.S. Bot. 40: 153 (1911). Types: Mozambique, Mt Maruma, *Swynnerton* 1922, (BM!, K!, syn.); Zimbabwe, Chirinda, *Swynnerton* 19 & 6 (BM!, K!, syn.)

Torenia mildbraedii Pilger, Z.A.E., 2: 285 (I914). Type: Rwanda, Luhondo Valley, *Mildbraed* 701 (B†, holo.; K!, iso. fragment)

Crepidorhopalon whytei (Skan) Fischer in F.R. 100: 444 (1989) & Trop. and subtrop. Pflanzenw. 81: 165, t. 85, 86 (1992) & F.A.C. 110, pl. 44 (1999) & in Fl. Ethiop. & Eritr. 5: 270 (2006)

NOTE. The flower colour of all the material I (D.P.) have studied for this treatment ranges through various shades of blue, with no record of any yellow coloured flowers being present in the area. This latter colour mentioned in the description, relates to that seen in material from the Flora Zambesiaca area under the name *L. flava* S.Moore.

3. **Lindernia humilis** *Bonati* in Bull. Soc. Bot. Genève, Ser. 2, 15: 100 (1924). Type: Madagascar, Mangoky R. Basin, *Perrier de la Bâthie* 8502 (P!, holo.; iso.)

Creeping or straggling, procumbent annual, laxly branched throughout; stems up to 28 cm long, slender, leafy throughout, acutely quadrangular, minutely glandular-pubescent, sparsely above to subglabrous, denser below. Leaves sessile, subreniform, circular or broadly ovate, (4–)9–14 mm long, (4–)8–14 mm wide, base subcordate or truncate, apex obtuse, margins dentate or broadly somewhat serrate, rarely subentire, glandular-pubescent especially on major veins beneath. Flowers solitary in leaf axils throughout most of stem; pedicels 7–16 mm long, very slender, densely to sparsely glandular-pubescent with slender gland-tipped hairs. Calyx 2–4 mm long, densely glandular-pubescent, 5-lobed almost to base, lobes 2–3 mm long, narrowly lanceolate, acute, shortly ciliate, erect, not spreading. Corolla white with pink, lilac to purple markings, or yellow, 4–5 mm long; upper lip ± cucculate, rounded to shallowly emarginate, lower lip trilobed with rounded lobes. Stamens 4, all fertile;

anthers 0.5–0.6 mm long; posterior filaments 0.5–1 mm long, anterior filaments 1–2 mm long, very slender with clavate appendage (0.1–)0.5–0.6 mm long at base, antero-lateral ridge prominent. Capsule subglobose, ± 2.5 mm in diameter, shorter than calyx.

KENYA. Kwale District: Mwasangombe Forest, 24 km SW of Kwale, 27 Aug. 1953, *Drummond & Hemsley* 4008!; Shimba Hills, 16 Mar. 1991, *Luke & Robertson* 2714!; Lamu District: Gongoni Forest, 5 Jan. 1992, *Luke* 3044!

TANZANIA. Uzaromo District: Dar University campus, 3 Nov. 1974, *Wingfield* 2820! & 17 km WSW of Dar, 2 km on Gongolamboto–Pagu road, 8 May 1971, *Wingfield* 1545!; Zanzibar: Mkokotoni, 27 June 1960, *Faulkner* 262IV!

DISTR. **K** 7; **T** 6, 8; **Z**; Mozambique; Madagascar

HAB. Margins of pools and water holes in open country and rice fields, stream sides in lowland forests; 0–400 m

USES. None recorded on specimens from our area

CONSERVATION NOTES. Least Concern (LC)

SYN. *Lindernia subreniformis* Philcox in Bol. Soc. Brot. ser. 2, 60: 268 (1987) & in F.Z.: 8, 2: 65 (1990). Type: Tanzania, Kilwa District: Miombo Valley, about 19 km SSW of Kingupira, *Vollesen* in MRC 3931 (K! holo.; EA, WAG, iso.)

4. **Lindernia hepperi** (*Eb.Fisch.*) *Philcox*, **comb. nov**. Type: Tanzania, Iringa, just N of township, *Milne-Redhead & Taylor* 11072 (K!, holo.)

Erect annual, 12–20 cm tall; stems branched, quadrangular, subglabrous above, patent hirsute below. Lower and median leaves ovate-lanceolate to broadly ovate-elliptic, 10–25(–30) mm long, 5–15(–18) mm wide, margins coarsely dentate to subentire, sparsely hirsute on both surfaces to subglabrous, (3–)5-veined from base, veins prominent beneath; upper leaves lanceolate, bract-like, ± 1.5 mm long, shortly ciliate. Flowers solitary in axils of leaves, erect, somewhat appressed to stem especially when fruiting; pedicels 0.5–1(–2) mm long, glabrous, rarely spreading. Calyx ± 3 mm long, extending to 4.5 mm long at fruiting, lobed to about midway; lobes lanceolate, 1.5–2.5 mm long, acuminate, glabrous, hispid-ciliate. Corolla pale mauve or blue, ± 12 mm long, tube 4–7 mm long; upper lip ± 2.5 mm long, shallowly emarginate; lower lip 4–5 mm long, 3-lobed, lobes rounded, 1.5–1.8 mm wide, throat densely orange papillose or stipitate-glandular. Stamens 4, fertile; anterior filaments orange, 4–4.5 mm long, curved, not geniculate, clavate at base, with spur ± 0.8 mm long, swollen at apex ± 0.5 mm in diameter. Capsule 7–8 mm long, 2 mm broad, narrowly lanceolate, ellipsoid, style somewhat persistent.

KENYA. Kwale District: 2 km N of Gazi village, 12 Oct. 1991, *Luke* 2899! & Majoremi area, 18 Aug. 1993, *Luke* 3796! & Lunguma, 20 Aug. 1994, *Luke & Gray* 4061!

TANZANIA. Kigoma District: Tubira [?Tubilal Forest], 1 April 1994, *Bidgood & Vollesen* 3010!; Mpanda District: 20 km from Mwese from Mpanda–Uvinza road, 6 June 2000, *Bidgood et al.* 4603! & 4564!; Iringa District: N of Iringa township, 13 July 1956, *Milne-Redhead & Taylor* 11072!

DISTR. **K** 7; **T** 4, 7; known only from the above collections; not known elsewhere

HAB. Secondary grassland and wet forests; 0–1650 m

USES. None recorded on specimens from our area

CONSERVATION NOTES. Least Concern (LC) but may qualify as Near Threatened (NT) due to continued degradation of habitat (if no new material is collected)

SYN. *Crepidorhopalon hepperi* Eb.Fisch. in B.J.B.B. 60; 410 (1990) & Trop. and subtrop. Pflanzenw. 81: 178, t. 90 (1992)

5. **Lindernia rupestris** *Engl.* in E.J. 30: 402 (1901). Type: Tanzania, Njombe District: Livingstone Mt, *Goetze* 811 (B†, holo.). Neotype: Tanzania, Rungwe District: Kyimbila, N of Lake Nyassa, *Stolz* 791 (K!, chosen here)

Erect to spreading annual, (3–)6–14(–18) cm tall, simple or branched; stems leafy below or towards base, quadrangular, minutely glandular to subglabrous; branches opposite or less frequently alternate. Leaves sessile, or lowermost with short, petiole-like base, broadly ovate to elliptic-oblong, 8–18(–25) mm long, 3.5–13(–18) mm wide, rounded at base or lowermost subcuneate, apex obtuse or rounded, margins shortly dentate or lower subentire, sparsely glandular-pubescent to subglabrous, palmately 5-veined, veins ± prominent beneath. Flowers in lax, terminal or lateral leafless racemes. Bracts similar to leaves but smaller, becoming more so above, each member of pair of different size with smaller subtending flower; pedicels 0.5–3 mm long. Calyx 4–5 mm long, minutely glandular-pubescent; lobes narrowly linear, 3–4 mm long, acuminate, widely spreading. Corolla yellow, orange-yellow or white, up to 7(–10) mm long, upper lip oblong-ovate, lobes of lower lip with large, yellow clavate hairs ± 0.15 mm long at throat, rounded. Stamens 4, all fertile; anterior filaments 4–7 mm long with papillose appendage ± 1 mm long present at base, clavate at base, antero-lateral ridge not evident. Capsule (2.5–)3–4 mm long, 3–4 mm broad, globose or globose-obovoid, shorter than calyx, glabrous.

UGANDA. Masaka District: Lake Victoria, Sese, Bugalla Is., 4 Oct. 1958, *Symes* 424! & 5 June 1932, *A.S. Thomas* 120!

KENYA. North Kavirondo District: Bungoma, Sept. 1967, *Tweedie* 3486!

TANZANIA. Tanga District: Kange, 14 Aug. 1958, *Faulkner* 2173!; Kigoma District: Gombe Valley, 27 Jan. 1964, *Pirozynski* 315!; Songea District: 2.5 km NE of Kigonsera, 13 Apr. 1956, *Milne Redhead & Taylor* 9632!

DISTR. **U** 3, 4; **K** 5; **T** 1, 3, 4, 7, 8; Burundi, Zambia, Malawi, Mozambique, Zimbabwe

HAB. Open grassland and *Brachystegia* and *Julbernardia* and other woodlands, on wet sandy and stony or peaty soils; 1000–1750 m

USES. None recorded on specimens from our area

CONSERVATION NOTES. Least Concern (LC)

SYN. *Lindernia insularis* Skan in F.T.A. 4(2): 342 (1906); Philcox in F.Z.: 8 (2): 63 (1990). Type: Uganda, Masaka District: Lake Victoria, Sesse Is., *Brown* 117 (K!, holo.)

L. subscaposa Mildbr. in N.B.G.B. 8: 233 (1922). Type: Uganda, Elgon, *Lindblom* s.n. (B†, holo.; K!, icon.)

Crepidorhapalon insularis (Skan) Eb.Fisch., Trop. and subtrop. Pflanzenw. 81: 172, t. 88 (1992) & F.A.C. Scrophulariaceae: 114, pl. 46 (1999), *nom. illegit.*

6. **Lindernia tenuis** *S.Moore* in Journ. Bot. 56: 10 (1918). Type: Congo-Kinshasa, Lubumbashi [Elizabethville], *Rogers* 10886 (BM, holo.; K!, iso.)

Erect annual up to 18 cm tall; stems very slender, much branched especially towards base, quadrangular, glabrous above, at times sparsely short hispid below. Lower leaves opposite, narrowly ovate-lanceolate, 4.5–10 mm long, 1–4 mm wide, apex obtuse, margins subentire or sparsely bluntly dentate, midvein not markedly pronounced beneath, subglabrous; upper leaves much reduced, oblong, 0.6–1.5 long, 0.1–0.3 mm wide, glabrous. Flowers solitary, alternate, in axils of small bract-like leaves; pedicels 1–1.5(–2.5) mm long, glabrous or sparsely, minutely pubescent. Calyx 2.5–4.5 mm long, glabrous, mid vein pronounced into a narrow wing running full length of tube; lobes lanceolate, 1–1.5 mm long, ciliolate or not but often inrolled at maturity, appearing subulate and at times spreading. Corolla lilac, bluish, upper lip shortly bifid, lobes rounded, lower lip 3–4-lobed, lobes rounded, palate with large orange yellow clavate hairs. Stamens 4, fertile; anterior filaments 3–4 mm long, slender, curved but not geniculate, long extruded with short clavate spur ± 1 mm long at base. Capsule pale fawn, 3–3.5 mm long, 1–1.5 mm wide, broadly ovoid, obscurely longitudinally shallowly ridged.

TANZANIA. Dodoma District: Lake Chaya, 24 April 1964, *Greenway & Polhill* 11749!

DISTR. **T** 5; Congo-Kinshasa, Zambia

HAB. Wet sandy grassland, and amongst rocky outcrops, and in wet pools on rocks; 1600–2200 m
USES. None recorded on specimens from our area
CONSERVATION NOTES. Data Deficient (DD); known only from a few collections in its range of distribution

SYN. *Lindernia damblonii* Duvignaud in Bull. Soc. Roy. Bot. Belg. 90: 256 (1958); Philcox in F.Z.: 8 (2): 67 (1990). Type: Congo-Kinshasa, L'Etoile, Lubumbashi [Elizabethville], *Duvigneaud & Damblon* 2870 (K!, holo.)
Crepidorhopalon tenuis (S.Moore) Eb.Fischer in B.J.B.B. 59: 457 (1989) & in Trop. & subtrop. Pflanzenw. 81: 191 (1992) & F.A.C. Scrophulariaceae: 128, pl. 53 (1999)

7. **Lindernia bifolia** *Skan* in F.T.A. 4(2): 343 (1906); Philcox in F.Z.: 8 (2): 63 (1990). Type: Zambia, Kambole, SW of Lake Tanganyika, ± 1525 m [5000 ft], 1896, *Nutt* s.n. (K!, holo.)

Erect, slender herb, (2.5–)6–12(–18) cm tall, branched from base or at times above; stems quadrangular, very sparsely short spreading hispid, frequently mixed with short stipitate glands. Leaves sessile, basal in one or two pairs, broadly ovate to elliptic-ovate, 5–20 mm long, 3–9 mm wide, thin, base rounded, apex acute or obtuse, margins obscurely few toothed, short pubescent to subglabous; one or two pairs of cauline leaves occasionally present, similar to basal but smaller. Flowers few, mostly laxly alternate in terminal racemes. Bracts subulate or linear-lanceolate, 0.5–1(–2.5) mm long; pedicels 1–2.5 mm long, slender. Calyx 3–5(–6) mm long, shortly hispid to glandular or subglabrous; lobes narrowly lanceolate, 2–4(–5) mm long. Corolla pale yellow to white, upper lip oblong-ovate, lateral lobes of lower lip elliptic, median lobe subcircular. Stamens 4, all fertile; anterior filaments 4–5.5 mm long with clavate appendages 0.8–1 mm long produced at base; antero-lateral ridge not evident. Capsule broadly ellipsoid to ellipsoid globose, ± 3.5 mm long, 1.5–2.5 mm broad, apiculate.

TANZANIA. Ufipa District: Kalambo Falls, 10 Apr. 1950, *Bullock* 2851!
DISTR. **T** 4; Zambia
HAB. Wet areas; ± 1550 m
USES. None recorded on specimens from our area
CONSERVATION NOTES. Data Deficient (DD); known only from a few collections in its range of distribution

SYN. *Crepidorhopalon bifolius* (Skan) Eb.Fisch. in Trop. and subtrop. Pflanzenw. 81: 176, t. 89 (1992) & F.A.C. Scrophulariaceae: 116, pl. 47 (1999)

NOTE. The collection included here from Kalambo Falls was made by Bullock but no further information on the locality is available from the label with the specimen. As the Falls are on the border between Zambia and Tanzania the actual locality must remain uncertain, but as the usual access is by road within Tanzania, the probability is that this collection was made within that country and as such has been included here. There is a strong possibility that, at a later date, it may be found occurring elsewhere within the Flora area.

8. **Lindernia oliveriana** *Dandy* in F.P.S. 3: 139 (1956); Philcox in F.Z.: 8 (2): 64 (1990); Fischer in Trop. and subtrop. Pflanzenw. 82: 267, t. 128, 129 (1992) & F.A.C. Scrophulariaceae: 154, pl. 65 (1999), excl. *Craterostigma lindernioides*; U.K.W.F.: 257 (1994). Type: Uganda, West Nile District: Madi, Dec. 1862, *Grant* s.n. (K!, holo.)

Erect or ascending annual, (6–)15–35(–45) cm tall, simple or branched usually at or towards base, subglabrous. Leaves linear to linear-lanceolate, (7–)12–15(–40) mm long, (0.5–)1–2.5(–4) mm wide, sessile, apex acute, margins entire or rarely obscurely, shallowly crenate towards apex, glabrous or ± ciliate when young. Flowers solitary, axillary; pedicels 5–17 mm long in flower, slender, suberect, up to 40 mm long in fruit, stouter, spreading to reflexed. Calyx (3–)5–17 mm long, 5-lobed, slightly ridged; lobes

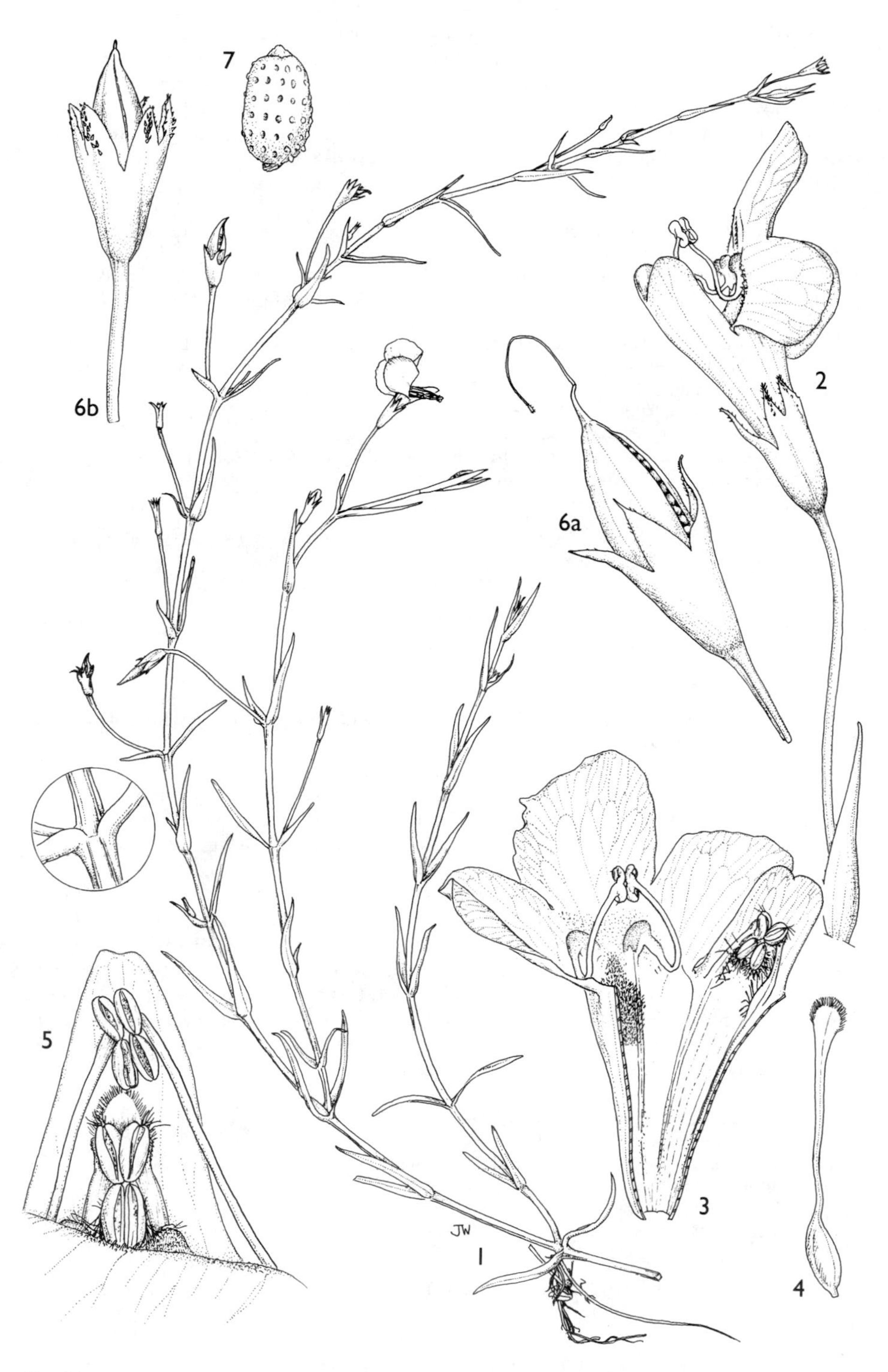

FIG. 23. *LINDERNIA OLIVERIANA* — **1**, habit × $^{2}/_{3}$; **2**, flower × 5; **3**, flower opened × 5; **4**, ovary & style × 5; **5**, stamens × 12; **6a & b,** capsule × 4; **7,** seed × 32. 1, 5 from *Major Lugard* 148; 2 from *Chandler* 610; 3, 4, 6–8 from *Napper* 2138. Drawn by Juliet Williamson.

(1–)1.5–2.5 mm, lanceolate, acute, glabrous or shortly hispid on margins or central area. Corolla bright or violet blue to mauve or brownish, (6–)10–13(–18.5) mm long, tube yellow or white, to 8 mm long; upper lip to 8 mm long, ± 3 mm wide, broadly oblong; lower lip trilobed, middle lobe to 10 mm long, 7 mm wide, all lobes rounded. Stamens 4, all fertile; anterior filaments 6–7 mm long, sharply geniculate above, gibbous, minutely densely papillose at base. Capsule linear- oblong in outline, 8.5–9(–15) mm long, (1–)3 mm broad, acute or acuminate. Fig. 23, p. 75.

UGANDA. West Nile District: Logiri, 18 Mar. 1945, *Greenway & Eggeling* 7220!; Teso District: Serere, Mar. 1932, *Chandler* 610!

KENYA. Uasin Gishu District: Kapsaret Farm, Jan. 1948, *Bickford* in *Bally* 6260!; Nyanza Basin, 29 July 1913, *Battiscombe* 682!; Kilifi District: Arabuko Sokoke Forest Reserve, 24 km S of Malindi, 3 Dec. 1961, *Polhill & Paulo* 893!

TANZANIA. Shinyanga District: Msalala, *Hannington* s.n.!; Buha District: about 115 km on Kibondo–Kasulu road, 15 July 1960, *Verdcourt* 2856!; Songea District: 12 km E of Songea by Nonganonga stream, 28 Dec. 1955, *Milne-Redhead & Taylor* 7786!

DISTR. **U** 1, 3; **K** 3, 5, 7; **T** 1, 4, 5, 7, 8; **Z**; Togo, Nigeria, Congo-Kinshasa, Sudan, Ethiopia, Angola, Zambia, Malawi, Zimbabwe

HAB. Wet grassland, marshes, banks of streams and coasts; 0–1500(–2200) m

USES. None recorded on specimens from our area

CONSERVATION NOTES. Least Concern (LC)

SYN. *Vandellia lobelioides* Oliv. in Trans. Linn. Soc., Bot. 29: 120, t. 121B (1875), *non* F.Muell. (1858). Type as for *L. oliveriana*

Lindernia lobelioides (Oliv.) Wettst. in P.O.A. C: 357 (1895); Skan in F.T. A. 4, 2: 340 (1906)

NOTE. The coastal specimens from Kenya and Zanzibar (*Vaughan* 2265 & *Mrs Taylor* 5, both at EA) have broader leaves (up to 5 mm)

9. **Lindernia lindernioides** (*E.A.Bruce*) *Philcox*, **comb. nov**. Type: Tanzania, Dodoma District: Manyoni, Kazikazi, *Burtt* 3612 (K!, holo.)

Erect, slender annual, up to 12 cm tall; stems ± branched, somewhat flattened, furrowed, glabrous. Leaves linear, 1–2 cm long, up to 1 mm wide, sessile, acute, margins entire. Flowers few, solitary, axillary. Pedicel ± 1 cm long, slender, erect in flower, spreading-reflexed, markedly upward curved in fruit. Calyx 3–3.5 mm long, lobes 1–1.5 mm long, lanceolate, ciliate. Corolla ± 5 mm long, tube short, ± 2 mm long, glabrous; upper lip purple to purple-blue including throat, oblong-ovate, 3 mm long, up to 2 mm wide, emarginate; lower lip 3-lobed, reddish, bronze or orange, two bright yellow patches or honey-guides at throat on middle lobe, lobes rounded, spreading to flower width of 7–8 mm. Stamens 4, fertile; anterior filaments ± 2 mm long, gently geniculate above midway, gibbous at base; posterior filaments 1–2 mm long, inserted at mouth of tube. Capsule narrowly lanceolate-ovoid to ellipsoid, 7–9 mm long, up to 2 mm broad, slightly arcuate.

UGANDA. Mt Elgon, Kapchorwa, 7 Sept. 1954, *Lind* 238!

TANZANIA. Singida District: ± 40 km from Sekenke on Singida road, 1 May 1962, *Polhill & Paulo* 2265!; Dodoma District: Manyani, Kazikazi, 17 May 1932, *Burtt* 3612!; Iringa District: Nyangolo scarp, about 48 km N of Iringa, 17 Apr. 1962, *Polhill & Paulo* 2043!

DISTR. **U** 3; **T** 5, 7; not known elsewhere

HAB. Wet seasonal ponds, marshes and damp sandy areas in open grassland; 1300–1600 m

CONSERVATION NOTES. Least Concern (LC)

SYN. *Craterostigma lindernioides* E.A.Bruce in K.B. 1933: 474 (1933)

10. **Lindernia debilis** *Skan* in F.T.A. 4(2): 344 (1906); Fischer in Trop. and subtrop. Pflanzenw. 81: 306, t. 146 (1992). Type: Sudan, Bongo, Gir, *Schweinfurth* 2516 (K!, holo.)

Annual, prostrate to suberect to 5(–8) cm tall, much branched; branches up to 9 cm long, weak, quadrangular, glabrous. Leaves sessile, very thin, unequal, lanceolate to linear lanceolate, (1.5–)3–8(–11) mm long, 0.5–3(–5.5) mm wide, slightly narrowed at base, apex obtuse or acute, margins entire or denticulate, glabrous. Flowers solitary in axils of smaller leaf of upper pairs; pedicels 0.5–1 mm long. Calyx 2.5–5 mm long, glabrous, or sparsely glandular-pubescent especially when young, somewhat prominently 5-ribbed; lobes subulate, 1–2.5 mm long, minutely ciliate, unequal. Corolla yellow or white, tinged with purple, 3–4 mm long; upper lip emarginate; lower lip with purple lines, deeply 3-lobed. Stamens 4, fertile; anterior filaments geniculate about or below midway, thickened but not clavate at base. Capsule oblong, 2–3 mm long, 1–1.5 mm broad, slightly to somewhat shorter than calyx, obtuse.

UGANDA. Lango/Karamoja District: Orumo [Oruma], Sept. 1935, *Eggeling* 2208!; Masaka District: near Lake Nabugabo, Aug. 1935, *Chandler* 1290!

TANZANIA. Njombe District: Iyayai, about 230 km S of Iringa, 15 Apr. 1962, *Polhill & Paulo* 2012!; Masasi District: near Masasi, 8 Mar. 1991, *Bidgood et al.* 1841!

DISTR. **U** 1, 4; **T** 7, 8; West Africa, Sudan

HAB. Damp places in sandy soil, scrubland and marshy areas surrounding rocky outcrops; 450–1400 m

USES. None recorded on specimens from our area

CONSERVATION NOTES. Least Concern (LC)

SYN. *Crepidorhopalon debilis* (Skan) Eb.Fisch. in F.R. 106: 10 (1995) & F.A.C. Scrophulariaceae: 132, pl. 55 (1999)

11. **Lindernia brevidens** *Skan* in F.T.A. 4(2): 339 (1906); Fischer in Trop. and subtrop. Pflanzenw. 18: 239, t. 114 (1992). Type: Tanzania, Lushoto District: Usambara, *Buchwald* 295 (K!, holo.)

Annual, rarely exceeding 5 cm in height; stems up to 30 cm long, diffusely branched, frequently rooting at lower nodes, sharply quadrangular, glabrous or occasionally with few minute white hairs at nodes. Leaves very shortly petiolate to subsessile, broadly ovate to ovate-suborbicular, (5–)3–18(–22) mm, (3.5–)11–18 mm wide, broadly cuneate to rounded at base, apex obtuse or rounded, margins coarsely serrate, glabrous or almost so, occasionally minutely ciliolate especially towards base, 3(–5)-veined from base, veins prominent beneath. Flowers throughout most of stem, solitary, axillary, alternate; pedicels 3–8(–11) mm long in flower, extending to 15–20 mm in fruit, very slender, glabrous, widely spreading in fruit to at times becoming reflexed. Calyx 3.5–4 mm long, accrescent to 4–5 mm long in fruit, glabrous, shallowly 5-ribbed; lobes unequal, ovate, 1–1.5 mm long, acute, remaining close to capsule in fruit, not spreading. Corolla white or with various amounts of blue, mauve or purple markings, 6–9.5 mm long, tube 3.5– 4.5 mm long, externally minutely stipitate-glandular, internally with antero-lateral ridge extending mostly length of tube, densely stipitate-glandular; upper lip broadly ovate to subrectangular, ± 3 mm long, apex rounded to subtruncate; lower lip 3-lobed. Stamens 4, fertile; anterior filaments very slender, geniculate about midway, gibbous at base, glabrous throughout. Capsule (7–)10–13 mm long, 1.5–2.5 mm broad, narrowly oblong-ellipsoid, acute to acuminate.

KENYA. Teita District: summit of Mwanda [Vuria], 2 May 1975, *Friis & Hansen* 2676!; Ngaongao Forest, 15 May 1985, *Faden et al.* 494!

TANZANIA. Lushoto District: Mgamba, 24 Jan. 1950, *Faulkner* 501! & W Usambara Mts, Soni, 1 Dec.1970, *Faulkner* 4503!

DISTR. **K** 7; **T** 3; not known elsewhere

HAB. Rocky hillsides, pools on rock outcrops, and wet seepage and swampy areas, forest clearings and surrounds, weed of cultivation; 1200–2500 m

USES. None recorded on specimens from our area

CONSERVATION NOTES. Least Concern (LC)

12. **Lindernia subracemosa** *De Wild.*, P1. Bequaert. 5: 425 (1932); Fischer in Trop. and subtrop. Pflanzenw. 81: 243, t. 116 (1992) & F.A.C. Scrophulariaceae: 148, pl. 62 (1999) & in Fl. Ethiop. & Eritr. 5: 272 (2006). Type: Congo-Kinshasa, Kabango, *Bequaert* 6143 (BR, holo.; K!, photo)

Annual or perennial, up to 30 cm tall, branched from base; stems up to 40 cm long, slender, trailing, subquadrangular, patent hispid, especially on angles, becoming subglabrous. Leaves ovate or subelliptic, 12–35(–40) mm long, 8–30 mm wide, subsesssile or shortly petiolate, base cordate, apex obtuse or rounded, margins finely serrate or shallowly crenate, generally appressed finely hispid above to rarely subglabrous, hispid beneath mainly on major veins, ciliolate, 3–5-veined from base, veins prominent; petioles (0–)2–3(–7) mm long. Flowers in axillary inflorescence of 3–5(–7) flowers or solitary in axils of leaf-like bracts, shortly pedicellate. Bracts similar to leaves but much smaller, ± 5–7 mm long, (1.5–)3–4.5 mm wide. Peduncle where present 10–35(–48) mm long; pedicels 2–5 mm long, erect. Calyx funnel-shaped, 4.5–6 mm long, accrescent to 8 mm in fruit, strongly 5-ribbed, glabrous except shortly hispid on ribs; teeth subtriangular, 1.5–2 mm long, 2–3.5 mm long in fruit, ciliolate. Corolla pink through pinkish blue to blue or blue purple; tube 5–6 mm long, upper lip ± 3 mm long, emarginate. Stamens 4, fertile; anterior filaments geniculate towards base. Capsule ellipsoid, 13–17 mm long, acute.

UGANDA. Kigezi District: Mushumgero, Bufumbira, Oct. 1947, *Purseglove* 2497! & Bukimbiri, May 1950, *Purseglove* 3374! Masaka District: Sesse, Bugalla Is, Lake Victoria, 12 Oct. 1958, *Symes* 506!
TANZANIA. Bukoba District: near Bukoba, Aug. 1931, *Haarer* 2139!; Morogoro District: Uluguru Mts, Mkambaku [Mkumbaku], 14 June 1978, *Thulin & Mhoro* 3221!; Iringa District: Dabaga Highlands, Udzungwa Scarp Forest Reserve, 90 km SE of Iringa, 29 Jan. 1971, *Mabberley* 617!
DISTR. **U** 2, 4; **T** 1, 6, 7; Congo-Kinshasa, Rwanda, Burundi, Ethiopia
HAB. In open country on grassy banks, paths and roadsides, edges of swamps, streamsides in wet forests; 1200– 2400 m
USES. None recorded on specimens from our area
CONSERVATION NOTES. Least Concern (LC)

NOTE. It may be of interest to note, that of all the material of this species seen for this work, that from Uganda, without exception, is represented by the smallest measurements given above and being a much more compact plant.

13. **Lindernia diffusa** (*L.*) *Wettst.* in E. & P. Pf. 4, 3B: 80 (1891); U.K.W.F.: 257 (1994); Fischer, F.A.C. Scrophulariaceae: 144 (1999). Lectotype: Virgin Is, St. Thomas, *Browne* s.n., Herb. Linn. No. 795.3 (LINN), designated by Edmondson in Davis (ed.), Fl. Turkey 6 : 680 (1978)

Annual, much branched, prostrate, 2.5–5 cm tall; stems up to 20 cm long, often rooting at lower nodes, quadrangular but not sharply so, short, white, patent hirsute throughout, mainly on angles. Leaves broadly ovate to subcircular, 12–19 (–25) mm long, 9–15(–19) mm wide, rounded to broadly cuneate at base, apex obtuse, margins finely crenate-serrate, glabrous above, shortly spreading hirsute on major veins beneath, somewhat densely ciliolate, 3-veined from base with mid vein once or twice clearly pinnately branched. Petioles 1–2 mm long, densely hairy. Flowers solitary in axils mostly of upper leaves; pedicels 2–3.5 mm long in flower, extending to 4–8 mm in fruit, rather stout, densely hairy, ascending in fruit, never reflexed. Calyx prominently 5-ribbed, 3.5–4.5 mm long in flower, patent white-hirsute; lobes narrowly lanceolate to subulate, 1.8–2(–2.5) mm long, acuminate, ciliate; lobes accrescent to 3–5 mm long, widely spreading and distorted after dehiscence of capsule. Corolla white or cream ("mauve blue", *Thomas* 853), 5–6 mm long, tube ± 4.5 mm long; upper lip shallowly emarginate or minutely irregularly toothed; lower lip somewhat deeply 3-lobed with median lobe rounded to subtruncate, minutely dentate. Stamens 4, fertile; anterior filaments geniculate, gibbous at base below geniculation, minutely yellow-glandular. Capsule oblong-ellipsoid, 7–7.5(–12) mm long, 2–3.5 mm broad. Fig. 24, p. 79.

FIG. 24. *LINDERNIA DIFFUSA* — **1**, habit × 2/3; **2**, part of fruiting stem × 2/3; **3**, fruit and leaf detail × 2; **4**, flower × 8; **5**, flower opened × 8; **6**, anterior stamens × 24; **7**, posterior stamens × 24; **8**, style and ovary × 8; **9**, capsule × 4; **10**, seed × 32. 1 from *K.A. Lye* 1768; 2,3 from *Jackson* U122; 4, 7, 9, 10 from *Norman* 17; 5, 6, 8 from *Lock* 69/164. Drawn by Juliet Williamson.

UGANDA. Busoga District: Lake Victoria, Lalui I., 19 May 1964, *Jackson* 122!; Masaka District: Lake Nabugabo, July 1937, *Chandler* 1769!
TANZANIA. Bukoba District: near Kitwe, Oct. 1931, *Haarer* 2205!
DISTR. **U** 3, 4; **T** 1, 3 (fide Iverson, Usambara rain forest project, Uppsala, 1988); West Africa from Sierra Leone to Cameroon, Gabon, Congo (Brazzaville), Central African Republic, Congo-Kinshasa, Burundi; Madagascar; Central and South America
HAB. Forest borders, damp grassland surrounding lakes and other standing water, path and roadsides; 1100–1250 m
USES. None recorded on specimens from our area
CONSERVATION NOTES. Least Concern (LC)

SYN. *Vandellia diffusa* L., Mant. 1: 89 (1767)
Pyxidaria diffusa (L.) Ktze., Rev. Gen. 2: 464 (1891)

14. **Lindernia nummulariifolia** (*D.Don*) *Wettst.* in E. & P. Pf, 3B: 79 (1891); Engl., P.O.A. C: 357 (1895); Hiern, Cat. Afr. Pl. Welw. 1: 763 (1898); Skan in F.T.A. 4(2): 341 (1906); F.P.U.: 136 (1962); Philcox in F.Z.: 8 (2): 65 (1990); Fischer in Trop. and subtrop. Pflanzenw. 81: 225, t. 110 (1992) & F.A.C. Scrophulariaceae: 142, pl. 59 (1999) & in Fl. Ethiop. & Eritr. 5: 272 (2006); U.K.W.F.: 257 (1994). Type: Nepal, 1821, *Wallich* s.n. ex Herb. Hook. (K!, lecto.)

Erect annual, 3–14 cm tall; stems simple or branched from base or above, shortly hairy to subglabrous; branches ascending, 3–12 cm long. Leaves ovate to subcircular, 9–15(–22) mm long, 5–15 mm wide, subsessile, apex acute or obtuse, margins serrate or crenate-serrate, short hairy on margins and major veins beneath. Flowers solitary in axils of upper leaves; pedicels (1–)3–10(–25) mm long, very slender, often becoming stouter in fruit. Calyx 2.5–3.5 mm long, sparsely pubescent to subglabrous; lobes unequal, 1–1.5 mm long, acute. Corolla blue, lilac, violet or purple, to about 6 mm long; upper lip entire, oblong; lobes of lower lip rounded. Stamens 4, all fertile; anterior filaments ± 3 mm long, swollen geniculate at base. Capsule ovoid-cylindric, 4.5–6 mm long, 1.5–3 mm broad, shortly beaked.

UGANDA. Ankole District: Bunyaruguru, Feb. 1939, *Purseglove* 561!; Mbale District: Bududa, 5 Sept. 1954, *Norman* 232!; Masaka District: Lake Victoria, Sese, Bugala Is., 12 Oct. 1958, *Symes* 494!
KENYA. Nandi District: Kaimosi, 800 m N of Yala Bridge, 4 June 1933, *Gilbert Rogers* 714!
TANZANIA. Lushoto District: Usambara Mts, Oaklands, 17 Mar. 1970, *Batty* 986!; Kigoma District: Kasye Forest, 19 Mar. 1994, *Bidgood et al.* 2821!; Songea District: Matagoro Hills, S of Songea, 22 Feb. 1956, *Milne-Redhead & Taylor* 8858!
DISTR. **U** 2–4; **K** 3; **T** 2–4, 7, 8; West and Central Africa, Sudan, Ethiopia, Angola, Zambia, Malawi, Mozambique, Zimbabwe; Madagascar; India, Nepal, Sri Lanka, Myanmar, China, Vietnam, Thailand
HAB. Weed of cultivation, but also found by streamsides in both open country and wet, mossy, evergreen forests, and in thickets on boulder strewn hillsides; 900–1950 m
USES. None recorded on specimens from our area
CONSERVATION NOTES. Least Concern (LC)

SYN. *Vandellia nummularifolia* D.Don, Prodr. Fl. Nep.: 86 (1825); Benth. in Prodr. 10: 416 (1846)
Torenia sessiliflora Benth. in Wall. Cat. 3959 (1831), *nomen*
Vandellia sessiliflora Benth., Scroph. Ind.: 37 (1835). Type: Myanmar, Tavoy, 1826, *Wallich* 3959 (K!, holo.)
Lindernia sessiliflora (Benth.) Wettst., in E.& P. Pf. 4, 3B: 79 (1891)
Pyxidaria nummularifolia (D.Don) Kuntze, Rev. Gen. 2: 464 (1891)
Lindernia nummulariifolia (D.Don) var. *sessiliflora* (Benth.) Hiern, Cat. Afr. P1. Welw. 1: 763 (1898)

NOTE. D. Don when describing *Vandellia nummularifolia* typified his name by a collection made by Wallich from Nepal. There are no specimens in the Wallich Herbarium (K-W) corresponding to this citation, and I have chosen a specimen representing that name, collected by Wallich in 1821, and once housed in Hooker's herbarium now at Kew, as the lectotype.

15. **Lindernia abyssinica** *Engl.* in E.J. 23: 503, t. 9, fig. A–E (1897); Skan in F.T.A. 4(2): 343 (1906); Fischer in Trop. and subtrop. Pflanzenw. 81: 250, t. 119 (1992); U.K.W.F.: 257 (1994); Fischer in Fl. Ethiop. & Eritr. 5: 273 (2006). Type: Ethiopia [Abyssinia], Begemeder, *Schimper* 1164 (K!, iso.)

Erect perennial, 4–6(–10) cm tall, stoloniferous, much branched from base, with many fibrous roots; stems sparsely leafy below, denser above, quadrangular, glabrous or at times sparingly minutely pubescent below nodes and on angles of stem. Leaves ovate to obovate, 5–12(–20) mm long, 3–5.5(–10) mm wide, narrowly cuneate at base, apex obtuse, margins laxly serrulate to subentire, glabrous above, minutely pubescent beneath especially on midrib, shortly ciliate. Flowers solitary, usually in alternate leaf axils; pedicels 5–7 mm long, lengthening to about 12 mm long in fruit, glabrous. Calyx 3.5–5(–6.5) mm long, glabrous to minutely hispid pubescent, strongly 5-ribbed; lobes narrowly lanceolate, 1.5–2.5 mm, long, acute. Corolla pink, or various shades of blue through lilac to purple [*Lind* 432 reports colour as deep yellow], 9–12 mm long, tube ± 5 mm long; lower lip deeply 3-lobed, lobes rounded. Stamens 4, fertile; anterior filaments 5–5.5 mm long, slender, geniculate about midway, glabrous, gibbous below. Capsule oblong-ellipsoid, 5–6.5 mm long, 1.5–2.5 mm broad, acute.

UGANDA. Karamoja District: Napak, 28 May 1940, *A.S. Thomas* 3624!; Mbale District: Bugisu [Bugishu], Bulago, 28 Aug. 1932, *A.S. Thomas* 346! & Kapchorwa, 9 Sept. 1954, *Lind* 295!
KENYA. Trans-Nzoia District: Cherangani, E of Kitale, 22 May 1949, *Maas Geesteranus* 4744!; Mt Elgon, S slopes, Feb. 1930, *Gardner* 2258! & Jan. 1936, *Tweedie* 295!
TANZANIA. Ufipa District: Nsanga Highlands, Rukwa Escarpment, *Richards* 15865!; Iringa District: Ndundulu, Udzungwa Mts, Kilombera Forest Reserve, 14 Feb. 2000, *Marshall* WK387!
DISTR. **U** 1, 3; **K** 2, 3, ?5, 6; **T** 4, 7; Nigeria, Cameroon, Sudan, Ethiopia
HAB. In crevices and in short grass on rocks in damp montane grassland and riverine forests; 1800–2500 m
USES. None recorded on specimens from our area
CONSERVATION NOTES. Least Concern (LC)

16. **Lindernia longicarpa** *Eb.Fisch. & Hepper* in K.B. 46, 3: 534, fig. 4 (1991); Fischer in Trop. and subtrop. Pflanzenw. 81: 252, t. 120 (1992). Type: Tanzania, Handeni District, Kideleko, *Faulkner* 1399 (K!, holo.)

Erect perennial, 6–20 cm tall; stems much branched, markedly quadrangular, densely minutely pubescent on the angles throughout; branches up to 15 cm or more long, spreading. Leaves sessile, broadly ovate, 6–14 mm long, 3–7 mm wide, margins dentate, minutely pubescent on both surfaces. Flowers solitary, in axils of most leaves in upper half of branch; pedicels 4–10(–15) mm long, slender, glabrous. Calyx 4–4.5 mm long, lengthening up to 8.5 mm in fruit, shallowly 5-ribbed, glabrous; lobes lanceolate, 1.2–2 mm long. Corolla white with purple or violet markings, 7–8 mm long, glabrous; upper lip bilobed, lower lip 3-lobed, 4(–5) mm long, lobes 3–4 mm long. Stamens 4, fertile; anterior filaments ± 2.5 mm long, geniculate just below midway, gibbous base fused with lower lip; posterior filaments 2 mm long. Capsule 15–20 mm long.

TANZANIA. Handeni District: Kwamkono, 25 June 1966, *Archbold* 721! & 30 June 1966, *Archbold* 804!
DISTR. **T** 3; only recorded from Handeni District, Tanzania
HAB. Damp pockets of soil on rocky outcrops; 500–600 m
USES. None recorded on specimens from our area
CONSERVATION NOTES. Endemic to Tanzania with not much material. Here assessed as Endangered EN B1& B2 ab, with the extent of occurrence estimated to be less than 100 km^2 and area of occupancy less than 10 km^2 and continuing decline in the quality of habitat.

NOTE. This species was described in general terms as being 'glabrous' as, more especially, were the leaves. I have studied all the material cited in the original publication and find both the stems and leaves to be densely pubescent.

17. **Lindernia bolusii** (*Hiern*) *Eb.Fisch.* in Trop. and subtrop. Pflanzenw. 81: 261, t. 125 (1992) & in Fl. Ethiop. & Eritr. 5: 273 (2006). Type: South Africa, Transvaal, Ramakoja, *Schlechter* 4503 (K!, holo.)

Erect annual or short-lived perennial, 2.5–5(–10) cm tall; stems simple or branched, leafy at base, sparingly so above, quadrangular, minutely glandular-pubescent. Radical leaves rosetted, obovate or narrowly elliptic, (3–)5–8(–12) mm long, (1–)2–3.5 mm wide, base tapering into a short petiole, apex obtuse, margins entire or repand to sinuate, densely, minutely glandular-pubescent especially beneath; lower stem leaves opposite, similar to radical leaves; upper stem leaves opposite, oblong, 7–18 mm long, 1–1.5(–5.5) mm wide, base lightly narrowing into the petiole, at times appearing subsessile, apex obtuse or subacute, margins entire or minutely, sparsely shallowly dentate. Flowers solitary, axillary; pedicels 2–12(–20) mm long, filiform, quadrangular, subglabrous to lightly glandular-pubescent. Calyx 3–5 mm long, lightly pubescent, mid vein barely prominent; lobes lanceolate, 1.5–2 mm long, acute. Corolla white flushed with pink or purple, pink or mauve, 6–6.5(–9) mm long, tube 3–3.5(–6.5) mm long; upper lip rounded, appearing crenate; lower lip 3-lobed, lobes rounded. Stamens 4; anterior pair with much reduced polleniferous anther thecae, at times appearing sterile. Capsule narrowly oblong-ellipsoid, 7–13 mm long, acute.

UGANDA. Karamoja District: Kacheliba, 20 May 1940, *A.S. Thomas* 3389!
KENYA. Northern Frontier District: Ngorinit [Nguronit] Mission Station, near Ndoto Mts, 31 Oct. 1978, *Gilbert & Gachathi* 5347!; Kilifi District: Mariakani, 17 Sept. 1961, *Polhill & Paulo* 486!; Kwale District: near Taru, between Samburu and Mackinnon Road, 3 Sept. 1953, *Drummond & Hemsley* 4150!
TANZANIA. Tunduru District: 96 km from Masasi, 21 Mar. 1963, *Richards* 18012!
DISTR. **U** 1; **K** 1, 7; **T** 8; Ethiopia, South Africa
HAB. Wet, shallow, sandy soil surrounding granite outcrops, and in depressions and crevices in rocks; 250–1600 m
USES. None recorded on specimens from our area
CONSERVATION NOTES. Least Concern (LC)

SYN. *Ilysanthes bolusii* Hiern, Fl. Cap. 4, 2: 367 (1904)

18. **Lindernia** sp. **'alpha'**

Annual, slender herb, up to 35 cm tall; stems erect or ascending and rooting at lower nodes, simple or more rarely 1- or 2-branched above, apparently glabrous but densely minutely glandular, markedly spongy at base. Leaves narrowly linear, 7–20 mm long, to 0.6 mm wide, sessile, apex acute, margins minutely serrate, appearing glabrous; basal leaves not rosetted. Flowers solitary, axillary, bluish purple with white tube, slender; pedicels to ± 7 mm long, minutely glandular. Calyx 3–3.5 mm long, minutely, densely glandular; lobes narrowly lanceolate, 1–1.5 mm long, acute. Corolla 7–7.5 mm long; tube 3.5–4 mm long, externally glabrous; upper lip erect ± 3 mm long, somewhat deeply emarginate; lower lip 3-lobed. Stamens 4, the 2 posterior fertile and 2 anterior sterile with filaments ± 2 mm long, shallowly geniculate below midway with bilobed cupular protrusion openly sheathing filament at point of junction with gibbous base, terminated with minute rudimentary anther ± 0.2 mm long, 0.1 mm wide. Capsule not seen, material immature.

TANZANIA. Kigoma District: 35 km on Uvinza road from Kigoma–Kasulu, 24 April 1994, *Bidgood & Vollesen* 3214!
DISTR. **T** 4; only known from the above collection
HAB. Wet seepage grassland on grey sandy soil; ± 1000 m
USES. None recorded on specimens from our area
CONSERVATION NOTES. Data Deficient (DD) as is known from a single collection; but possibly qualifies as Vulnerable, VU D1 as it is known from roadside.

NOTE. This species comes very close to *L. oliveriana* Dandy but due to the lack of more mature material it is not given a specific name at this stage.

19. **Lindernia parviflora** (*Roxb.*) *Haines*, Bot. Bihar & Orissa 4: 635 (1922); Philcox in F.Z.: 8(2): 69 (1990); Fischer in Trop. and subtrop. Pflanzenw. 81: 312, t. 148 (1992) & F.A.C. Scrophulariaceae: 174, pl. 74 (1999); U.K.W.F.: 257 (1994); Fischer in Thulin (ed.) Fl. Somal. 281, fig. 180 (2005) & in Fl. Ethiop. & Eritr. 5: 277 (2006). Type: India, Roxb., Corom. P1. 3: 3, t. 204 (1819)

Erect, decumbent or ascending annual, 5–15(–25) cm tall; stems usually diffusely branched, glabrous to rarely slightly pubescent. Leaves sessile or shortly petiolate, ovate, ovate-lanceolate or elliptic, 2–5(–16) mm long, 1–5(–8) mm wide, apex acute or obtuse, margins entire, or larger leaves occasionally minutely, shallowly dentate, 1–3(–5)-veined; petiole 0–2 mm long. Flowers axillary or in terminal racemes, slender; pedicels (5–)7–12 mm long in flower, extending in fruit to 9–17(–25) mm long. Calyx (1–)2–4 mm long, glabrous or with few glandular hairs (e.g. *Kibuwa* 2450 from Kenya), divided almost to base; lobes lanceolate to linear-lanceolate, up to 0.5 mm wide, acute. Corolla white or pale blue, with darker blue or purple markings especially towards base, 4–7(–12) mm long, upper lip emarginate to shallowly bifid. Fertile stamens 2; filaments 0.4–1 mm long; staminodes two, 0.2–0.5 mm long, slender, clavate at apex occasionally with short spur ± 0.7 mm long just below. Capsule obovoid or broadly cylindric-ellipsoid to subglobose, 1.5–4.5(–6) mm long, 1–2 mm broad, usually slightly longer than calyx. Fig. 25, p. 84.

UGANDA. Toro District: Queen Elizabeth National Park, 16 Oct. 1967, *Lock* 67/164!; Teso District: Serere, Dec. 1931, *Chandler* 349!; Masaka District: Lake Victoria, Sese, Bugala I., 11 Oct. 1958, *Symes* 481!

KENYA. Northern Frontier District: Lolokwe, opposite Subata Repeater Station, 15 Apr. 1979, *Gilbert* 5384!; Machakos District: S end of Mua Hills, 2 Feb. 1969, *Napper* 1861!; Kilifi District: Malindi, Oct. 1951, *Tweedie* 1039!

TANZANIA. Moshi District: Gwari [Kware], Aug. 1928, *Haarer* 1451!; Mpanda District: near Tumba, 26 Jan. 1951, *Bullock* 3628!; Songea District: Kwamponjore Valley, 20 June 1956, *Milne-Redhead & Taylor* 10854! Zanzibar: Kizimbani, 31 Mar. 1961, *Faulkner* 2796!

DISTR. **U** 2–4; **K** 1, 3, 4, 7; **T** 2–8; **Z**; **P**; West Africa, Sudan, Ethiopia, Zambia, Malawi, Mozambique, Zimbabwe, Botswana, Namibia; Madagascar; India, Sri Lanka, Vietnam

HAB. Sandy or muddy river or streambanks in flood plain areas, wet areas around and in ephemeral pools on rocky outcrops, muddy margins of seasonal or permanent pools or pans, rice fields; 1500–2000 m

USES. None recorded on specimens from our area

CONSERVATION NOTES. Least Concern (LC)

SYN. *Gratiola parviflora* Roxb., P1. Corom. 3: 3, t. 203 (1811)
Bonnaya parviflora (Roxb.) Benth. in Wall.Cat. 3867 (I831) & Scroph. Ind.: 34 (1835)
Ilysanthes parviflora (Roxb.) Benth. in DC., Prodr. 10: 419 (1846); Skan in F.T.A. 4(2): 346 (1906); Hepper in F.W.T.A. ed. 2, 2: 365, f. 289 (1963)

20. **Lindernia zanzibarica** *Eb.Fisch. & Hepper* in K.B. 46, 3: 529, fig. 1 (1991) & in Trop. and subtrop. Pflanzenw. 81: 248, t. 118 (1992). Type: Tanzania, 27 km NNW of Dar es Salaam, 1 km S of Log Cabins, *Wingfield* 2083 (K!, holo.)

Erect annual, 15–20 cm tall, glabrous, loosely branched; branches spreading, up to 12 cm long. Leaves sessile, narrowly oblanceolate or subelliptic, 15–25(–40) mm long, 3–6 mm wide, margins remotely crenate-dentate, pinnately veined, glabrous. Flowers usually throughout length of branch with mostly, but not always, only one flower per node, subtending bract narrowly linear-lanceolate, 2–3 mm long, up to 0.5 mm wide, opposite bract leaf-like but smaller, often with rudimentary flower bud present; pedicels 3–5 mm long, extending by maturity to 7 mm long. Calyx deeply

FIG. 25. *LINDERNIA PARVIFLORA* — **1**, habit × $^2/_3$; **2**, part of shoot with flower × 1; **3**, flower × 10; **4**, young flower with glabrous calyx × 6; **5**, flower opened × 14; **6**, stigma detail × 36; **7**, immature capsule with calyx × 6; **8**, mature capsule with seeds × 12; **9**, seed × 60. 1, 3, 5, 6, from *Kibuwa* 2450; 2, 4, 7–9 from *Hepper & Jaeger* 6686. Drawn by Juliet Williamson.

divided almost to base, 3.5–4.5 mm long; lobes 3–4 mm long, sparsely pubescent on main vein only or not. Corolla blue to violet, 4.5–5 mm long; tube 2.5–3 mm long, sparsely glandular-pubescent outside; upper lip 2 mm long; lower lip 2 mm long, tripartite. Stamens 2; filaments 10–12 mm long, fused with lower lip of corolla; anthers ± 0.8 mm long; staminodes 2, minute, clavate, 0.2–0.3 mm long, without geniculations at base, glandular-pubescent. Ovary 1.5 mm long; style 3 mm long. Capsule ± 9 mm long.

TANZANIA. Uzaramo District: 27 km NNW of Dar es Salaam, 23 July 1972, *Wingfield* 2083! & July 1891, *Sacleux* 1612!, 1613! photos!; Zanzibar: Mwera Swamp, 20 Aug. 1964, *Faulkner* 3409!
DISTR. **T** 6; **Z**; Mozambique; Madagascar
HAB. Damp grasslands, coastal swamps and rice fields; sea level
USES. None recorded on specimens from our area
CONSERVATION NOTES. Least Concern (LC)

21. **Lindernia congesta** (*A.Raynal*) *Eb.Fisch.* in Trop. and subtrop. Pflanzenw. 81: 317, t. 149 (1992). Type: Senegal, Kayar, 4 km NE of village, *Raynal* 7170 (P!, holo. & iso.)

Annual, erect, up to 9 cm tall; stems simple, becoming sparsely branched, quadrangular, glabrous. Leaves linear, 5–7 mm long, ± 1 mm wide, sessile, apex acute or subacute, margins entire, pale, slightly thickened, glabrous, obscurely 3-veined from base. Flowers aggregated into compact glomerules up to 5 mm wide, minutely glandular, with up to 5 flowers in each and subtended by 2 leaf-like bracts, lanceolate, 2.5–3.5 mm long, 0.7 mm wide, acute; each flower shortly pedicellate with pedicels ± 0.5 mm long, subtended by 2 bracteoles up to 1.5 mm long. Calyx 1–2.5 mm long, divided almost to base, glabrous; lobes ± 0.15 mm wide. Corolla white, 1.5–2 mm long, glabrous; anterior lip 3-lobed; posterior lip suberect. Stamens 2 fertile; anthers ± 0.2 mm long; filament less than 0.1 mm long; staminodes 2, minute, present towards mouth of corolla; style persistent. Capsule subglobose or broadly ovoid, 1–1.5 mm long.

TANZANIA. Uzaramo District: 28 km NNW of Dar es Salaam, 222 m S of Log Cabins, 15 July 1972, *Wingfield* 2037!
DISTR. **T** 6; Senegal
HAB. Depressions in sand dunes which are frequently inundated with water, usually of low salinity; sea level
CONSERVATION NOTES. Least Concern (LC) but in the Flora area only recorded from a single location which is close to a main city

SYN. *Ilysanthes congesta* A.Raynal in Adansonia n.s. 7: 348 (1967)

22. **Lindernia rotundifolia** (*L.*) *Alston* in Trimen, Handb. Fl. Ceylon 6, Suppl.: 214 (1931); Fischer in Trop. and subtrop. Pflanzenw. 81: 309, t. 147 (1992); U.K.W.F.: 257 (1994). Lectotype: "Habitat in Malabariae arenosis", Herb. Linn. No. 30.4 (LINN), designated by Cramer in Dassanayake & Fosberg (ed.), Revised Handb. Fl. Ceylon 3: 417 (1981)

Annual or perennial, decumbent and rooting at lower nodes; branches weak, ascending up to ± 20 cm high, glabrous or slightly pubescent. Leaves sessile, orbicular to broadly ovate, 5–13 mm long, 4–12 mm wide, truncate at base to semi-amplexicaul, apex obtuse, margins entire to distantly serrulate especially towards apex, 4–5-veined, glabrous, minutely glandular-punctate above. Flowers solitary, axillary; pedicels 5–15 mm long, slender, glabrous, spreading, extending in fruit to ± 20 mm, somewhat reflexing. Calyx 4–4.5 mm long, divided almost to base; lobes subequal, lanceolate-oblong, 0.8–1 mm wide, acute, glabrous. Corolla bluish to mauve and white, 8–10 mm long; tube 5–6 mm long, externally glabrous; lower lip

3-lobed with lobes rounded, upper lip deeply emarginate. Fertile stamens 2; filaments very short, inserted adaxially below mouth of tube; staminodes 2, linear, abaxial, 1–5 × 0.2 mm, slightly arcuate towards apex. Capsule ellipsoid, 2.5–3 mm long, (?immature).

UGANDA. Toro District: Ruwenzori, May 1894, *Scott Elliot* 7793!
KENYA. Kiambu District: Limuru, 16 Feb. 1965, *Agnew* EA7081!
TANZANIA. Rungwe District: Mwakaleli, near Mwatesi R., 9 May 1975, *Hepper et al.* 5456!
DISTR. **U** 2; **K** 4; **T** 7; Nigeria, Cameroon, Congo-Kinshasa, Rwanda, Ethiopia, Malawi; Madagascar; India, Sri Lanka, China
HAB. Mostly a weed of cultivation in wet areas; 1600–2000 m
USES. None recorded on specimens from our area
CONSERVATION NOTES. Least Concern (LC)

SYN. *Gratiola rotundifolia* L., Mant.: 174 (1767); Roxb., Pl. Corom. 3: 3, t. 204 (1811)
Ilysanthes rotundifolia (L.) Benth. in DC., Prodr. 10: 420 (1846); Skan in F.T.A. 4(2): 346 (1906)
Ilysanthes rotundata Pilger in E.J. 45: 214 (1910). Type: Cameroon, Babadju, *Ledermann* 1849 (B†, holo.; K!, iso.)
Lindernia rotundata (Pilg.) Eb.Fisch. in B.J.B.B. 67: 366 (1999) & F.A.C.: 172 (1999) & in Fl. Ethiop. & Eritr. 5: 277 (2006)

23. **Lindernia ugandensis** (*Skan*) *Philcox*, **comb. nov**. Type: Uganda, Ankole District: near Rufuha R., *Bagshawe* 519A (BM!, holo.)

Erect annual, 3.5–4 cm tall, branched; stems quadrangular, glabrous above, sparsely minutely hispid-pubescent particularly on angles. Leaves sessile, narrowly ovate to ovate-oblong, 2.5–4 mm long, 1 mm wide, apex obtuse, margins entire, glabrous. Flowers solitary in axils of upper leaves; pedicels 3–7 mm long, slender, suberect, glabrous. Calyx 2–2.5 mm long, sparsely minutely hispid pubescent; lobes shortly ovate-deltoid, 0.5–0.7 mm long, subobtuse. Corolla colour not known, ± 6 mm long; tube 3.5–4 mm long; upper lip very shallowly emarginate to subtruncate, lower lip 3-lobed, middle lobe broadly oblong, truncate. Stamens 2 fertile, included in throat; staminodes 2, very small. Capsule narrowly linear oblong, ± 5 mm long, 1 mm broad, acute.

UGANDA. Ankole District: near Ibanda, 2 Oct. 1906, *Bagshawe* 1252!
DISTR. **U** 2; known only from the above collections; not known elsewhere
HAB. Not known, as no details are given on specimens
USES. None recorded on specimens from our area
CONSERVATION NOTES. Endangered, probably extinct, as all collections have been made at the same time

SYN. *Ilysanthes andongensis* sensu S.Moore in J.L.S. 37: 190 (1905), *non* Hiern
Ilysanthes ugandensis Skan in F.T.A. 4(2): 348 (1906)

24. **Lindernia pulchella** (*Skan*) *Philcox* in Bol. Soc. Brot. ser. 2, 60: 268 (1987) & in Philcox in F.Z.: 8(2): 69 (1990); Fischer in Trop. and subtrop. Pflanzenw. 81: 300, t. 143,144 (1992) & F.A.C. Scrophulariaceae: 170, pl. 72 (1999) & in Fl. Ethiop. & Eritr. 5: 277 (2006); U.K.W.F.: 257 (1994). Type: Malawi, Zomba Plateau, *Whyte* s.n. (K!, holo.)

Erect or procumbent annual or perennial, 4–8(–18) cm tall, branched from or towards base; stems quadrangular, variously covered with short, patent to somewhat retrorse, stiff white hairs. Leaves sessile or lower shortly petiolate, elliptic-oblong, 5–10(–14) mm long, 1.5–2(–4) mm wide, apex obtuse or acute, margins entire to sparsely crenate or crenate-dentate, glabrous to somewhat densely, short white-hispid, especially beneath. Petioles, where present, up to 4 mm long. Flowers solitary, axillary; pedicels 8–20 mm long, slender. Calyx 3.5–5.5 mm long, subglabrous to minutely pubescent; lobes 0.7–1.5 mm, ovate-lanceolate, obtuse, glabrous on margins or shortly ciliate, unequal to subequal. Corolla blue to blue-mauve or violet, lavender, pink or white, 8–10(–14) mm long; upper lip deeply bifid; lower lip trilobed, lobes rounded, wavy margined, ciliate. Stamens 2, fertile; filaments 0.5–1(–1.5) mm long; anthers 1–1.5 mm long; staminodes 2, up to 1.8 mm long, somewhat recurved, slightly clavate, geniculate above base but appearing gibbous due to tissue forming a web in angle of geniculation, tissue produced laterally forming short, slender or blunt projection 0.2–0.4 mm long; antero-lateral ridge barely evident. Capsule cylindric ellipsoid, 5–7(–14) mm long, 1.5–2(–2.5) mm broad, acute.

UGANDA. Mbale District: Kapchorwa, Mt Elgon, *Lind* 296!
KENYA. Northern Frontier District: Ndoto Mts, Ngorinit [Ngoronet], *Hepper & Jaeger* 7213!; West Suk District: 16 km S of Kongelai, *Tweedie* 3810!; Narok District: Mara–Morijo, 12 Feb. 2001, *Luke & Luke* 7337!
TANZANIA. Songea District: Matengo Hills, Liwiri Kitesa [Luwiri Kitesa], *Milne-Redhead & Taylor* 9038!
DISTR. **U** 3; **K** 1, 3, 4, 6, 7; **T** 8; Sudan, Ethiopia, Angola, Zambia, Malawi, Mozambique, Zimbabwe, South Africa
HAB. On damp rocks and in wet pools on outcrops, seepage areas at base of rocks and generally wet, sandy soils; 400–2300 m
USES. None recorded on specimens from our area
CONSERVATION NOTES. Least Concern (LC)

SYN. *Ilysanthes pulchella* Skan in F.T.A. 4(2): 348 (1906)

25. **Lindernia niamniamensis** *Eb.Fisch. & Hepper* in K.B. 46(3): 534, fig. 3 (1991); Fischer in Trop. and subtrop. Pflanzenw. 81: 291, t. 139 (1992). Type: Sudan, Niamniam, Gumango R., *Schweinfurth* 3931 (K!, holo.; B, iso.)

Woody-based perennial, 10–20 cm tall, erect, somewhat tufted, glabrous. Leaves linear, 1.5–10.5 mm long, 0.8–1 mm wide, slightly terete, glabrous. Flowers throughout upper $^1/_3$ of most major branches, with one flower at each node subtended by bract or small leaf; pedicels 5–7.5 mm long, very slender, widely spreading to reflexed at maturity. Calyx (2–)2.5–4 mm long, glabrous, usually with somewhat prominent ribs; lobes lanceolate, 1–1.5 mm long, acute or subacute. Corolla blue, up to 8 mm long, glabrous with tube 3–4 mm long: upper lip bilobed, 3–4 mm long; lower lip trilobed, 5 mm long, each lobe ± 3 mm long. Stamens 2, fertile; filaments 2 mm long; anthers 1.5 mm long; staminodes 2, 1 mm long, geniculate; geniculations papillose and yellow glandular, fused with lower lip. Ovary 1 mm; style 4 mm long. Capsule ± 7 mm long.

UGANDA. Bunyoro District: Bunyoro, near Lukohe Hill, Dec. 1933, *Eggeling* 1470!; Teso District: Kyere, Feb. 1933, *Chandler* 1079!
DISTR. **U** 2, 3; Sudan
HAB. Depressions on boulders and rocky outcrops with seasonal standing water; 1100–1200 m
USES. None recorded on specimens from our area
CONSERVATION NOTES. Least Concern (LC)

26. **Lindernia acicularis** *Eb.Fisch.* in B.J.B.B. 59: 446, fig. 1 (1989) & Trop. and subtrop. Pflanzenw. 81: 293, t. 140 (1992) & F.A.C. Scrophulariaceae: 164, pl. 69 (1999). Type: Burundi, Ruyigi, Muremera, *Reekmans* 5965 (BR, holo.; K!, iso.)

Woody herb, 4–10 cm tall, erect, much branched, glabrous; rhizome creeping. Leaves narrowly to linear-lanceolate, 3.5–9.5 mm long, 0.5–1 mm wide, sessile, apex acute, margins entire, glabrous, 3-veined from base for about 2/3 length, veins prominent, deeply channelled between. Flowers solitary in leaf axils throughout most of stem length; pedicel 3.5–7.5 mm long, strongly quadrangular, glabrous. Calyx 5–6 mm long, tube 5-ribbed glabrous; lobes lanceolate, 1–1.3 mm long, to ± 1 mm wide at base, acute, margins scarious, shortly ciliate. Corolla mauve or white, 7–10 mm long; tube 6–7.5 mm long; upper lip 3–3.5 mm long, 2-lobed, glabrous inside and out; lower lip 3-lobed, lobes 3 mm long. Stamens 2, fertile, on upper lip; filaments 2 mm long; staminodes 2, ± 2 mm long, inserted on lower lip, filiform, with rounded apex, covered with glandular hairs and papillae. Ovary ovoid, 2 mm long. Capsule oblong-ellipsoid, 5–6 mm long, 1–1.5 mm broad, about equalling or slightly exceeding calyx.

TANZANIA. Bukoba District: Ihanjiro, 23 Feb. 1892, *Stuhlmann* 3367!; Ngara District: Ngara, 12 Dec. 1959, *Tanner* 4619!

DISTR. **T** 1; Burundi

HAB. In water filled pools on rocky outcrops; 1100–1600 m

USES. None recorded on specimens from our area

CONSERVATION NOTES. Data Deficient (DD); known from a few specimens, and in the Flora area from two collections

SYN. *Strigina pusilla* Engl. in E.J. 23: 516 (1897). Type: Tanzania, Ihanjiro, *Stuhlmann* 3367 (B†, holo.; K!, iso.)

NOTE. The epithet 'pusilla' cannot be used here as it is *nom. illegit.* See note under *L. serpens* Philcox

27. **Lindernia schweinfurthii** (*Engl.*) *Dandy* in F.P.S. 3: 139 (1956); Philcox in F.Z.: 8 (2): 72 (1990); Fischer in Trop. and subtrop. Pflanzenw. 81: 272, t. 130 (1992) & F.A.C. Scrophulariaceae: 158, pl. 66 (1999) & in Fl. Ethiop. & Eritr. 5: 275 (2006); U.K.W.F.: 257 (1994). Lectotype: Sudan, Niamniam country, Nganje on Ibba R., *Schweinfurth* 3991 (B†, holo.; K!, iso.)

Annual herb, (3–)6–20 cm tall, slender; stems erect, simple or branched at or towards base, quadrangular, glabrous or slightly pubescent. Leaves oblong-ovate to linear-ovate, 2–5 mm long, 0.5–1 mm wide, apex obtuse, margins entire; leaves smaller above. Flowers few, distantly racemose, in axils of bract-like reduced leaves; pedicels 4–6 mm long, slender, somewhat deflexed in fruit. Calyx 1.5–3 mm long, glabrous, prominently 5-veined; lobes lanceolate, 0.4–0.5 mm long, acute, ciliate. Corolla blue or mauve and white with yellow throat, 4–4.5(–9) mm long; upper lip oblong, ± 2.5 mm long, 0.5 mm wide, bifid at apex with lobes narrowly acute to acuminate; lower lip 3-lobed, lobes oblong, rounded. Stamens 2, fertile; posterior filaments ± 1 mm long; staminodes 2, oblong, 0.2(–1) mm long, 0.1(–0.5) mm wide, obtuse, papillose. Capsule linear-ellipsoid, ± 4 mm long, 1.5 mm broad, acute.

UGANDA. West Nile District: near Arua, Sept. 1940, *Purseglove* 1019!; Teso District: Mt Abela, SW of Katakwi, 10 May 1970, *Lye* 5411!; Mengo District: near Bugombe, Kome Island, 27 Oct. 1968, *Lye* 111!

KENYA. West Suk District: Kongelai [Kongoli], June 1961, *Lucas* 174!; North Kavirondo District: Broderick Falls, above Nzoia Falls, Aug. 1965, *Tweedie* 3082!

TANZANIA. Ngara District: Bugarama, Bushubi, 16 Feb. 1961, *Tanner* 5830!

DISTR. **U** 1–4; **K** 2, 5; **T** 1; West and Central Africa, Sudan, Ethiopia, Zambia

HAB. Wet grasslands surrounding rocky outcrops, and water-filled depressions on rocks; 1200–1500 m

USES. None recorded on specimens from our area

CONSERVATION NOTES. Lesat Concern (LC); widespread

SYN. *Bonnaya trichotoma* Oliv. in Trans. Linn. Soc. 29: 121, t. 122E (1875). Type: Uganda, Madi, Dec. 1862, *Grant* s.n. (K!)

Ilysanthes trichotoma (Oliv.) Urb. in Bot. Deutsch. Bot. Ges. 4: 435 (1884)

I. schweinfurthii Engl.in E.J. 23: 504 (1897); Skan in F.T.A. 4(2): 350 (1906)

I. albertina S.Moore in Journ. Bot. 45: 331 (1907). Type: Uganda, near Nkusi [Ngusi] R., Lake Albert, *Bagshawe* 1383 (BM!, holo.)

Lindernia madiensis Dandy, F.P.S. 3: 139 (1956). Type: as for *Bonnaya trichotoma* Oliv. (*nom. nov.* for *B. trichotoma*, non *L. trichotoma* Schlecht.)

NOTE. Dandy made the combination *Lindernia schweinfurthii* as the name *L. trichotoma* had previously been used by Schlechter in 1924 (E.J. 59: 107) for a different species. Engler cited three syntypes with his description of *Ilysanthes schweinfurthii* and I (D.P.) have chosen *Schweinfurth* 3991 housed at Kew as the lectotype. Dandy clearly thought *L. schweinfurthii* and *L. madiensis* were distinct.

28. **Lindernia wilmsii** (*Engl. & Diels*) *Philcox* in Bol. Soc. Brot. ser. 2, 60: 268 (1987) & in F.Z.: 8(2): 72 (1990); Fischer in Trop. and subtrop. Pflanzenw. 81: 279, t. 132, 133 (1992) & F.A.C. Scrophulariaceae: 162, pl. 68 (1999). Type: South Africa, Transvaal, Lydenburg District, Hells Gate on Spitzkop, *Wilms* 900 (B†, holo.; K!, P iso.)

Erect annual, 2.5–7 cm tall, simple or branched towards base; stems quadrangular, glabrous to slightly minutely pubescent below, rigid. Basal leaves clustered, linear-spatulate, 4–7(–12) mm long, 1 mm wide, narrowing into a petiole-like base, apex obtuse, margins entire, subglabrous or minutely pubescent; cauline leaves opposite, narrowly elliptic to almost linear, 1.5–5 mm long, 0.8–1 mm wide, acute to subobtuse, margins entire, subglabrous, venation markedly prominent beneath. Flowers solitary, axillary; pedicels up to 10 mm long, slender, glabrous to minutely pubescent. Calyx (1.6–)3–4 mm long, glabrous to minutely glandular-pubescent, somewhat prominently ribbed; lobes deltate, 0.5–1.2 mm long, acute or obtuse, glabrous on margins or minutely ciliate. Corolla various shades of blue and white, (3.5–)5–7(–8) mm long; upper lip bifid; lower lip 3-lobed, lobes rounded; tube ± 3–4 mm long, externally glabrous. Stamens 2, fertile; filaments up to ± 1 mm long; staminodes 2, clavate, 0.3–0.5 mm long, erect or more frequently slightly curved outwards, sparsely to densely clothed with minute clavate hairs which extend to palate between staminodes, short antero-lateral ridges barely evident. Capsule elliptic-ovate in outline, 3.5–6 mm long, 1.5–2 mm broad, acute.

UGANDA. Acholi District: Paimol, Chua, Mar. 1935, *Eggeling* 1763!; Busoga District: Lake Victoria, Lolui Is., 11 May 1965, *G. Jackson* 111565!; Masaka District: Bukoto County, 2–3 km S of Kasokero, 12 May 1969, *Lye et al.* 2849!
KENYA. Trans Nzoia District: NE Mt Elgon, Sept. 1954, *Tweedie* 1192!
TANZANIA. Bukoba District: W of Bukoba, *Rose* 10015!; Musoma District: near Campi ya Pofu [Mpofu], 30 Mar. 1962, *Greenway et al.* 10565!
DISTR. **U** 1, 3, 4; **K** 3; **T** 1; Congo-Kinshasa, Burundi, Angola, Zambia, Mozambique, Zimbabwe, South Africa
HAB. In shallow wet soil, on or around granite rock outcrops; 900–2300 m
USES. None recorded on specimens from our area
CONSERVATION NOTES. Least Concern (LC); widespread

SYN. *Ilysanthes wilmsii* Engl. & Diels in E.J. 26: 123 (1898); Hiern in Fl. Cap. 4, 2: 366 (1904)
I. muddii Hiern in Fl.Cap. 4, 2: 366 (1904). Type: South Africa, Transvaal, *Mudd* s.n. (K!)

29. **Lindernia serpens** *Philcox*, **nom. nov**. Type: Tanzania, Bukoba District: Karagwe, *Grant* 211 (K!, holo.)

Annual herb 3–10(–12) cm tall; stems slender, branched, creeping or ascending, rooting at nodes, subquadrangular at times obscurely so, glabrous or sparsely minutely pubescent; Leaves sessile, elliptic or linear oblong, 2–9(–12) mm long, 0.5–2.5 mm wide, apex subobtuse, margins entire, minutely pubescent to subglabrous, at times appearing semi-succulent, venation scarcely evident. Flowers solitary in axils of upper leaves; pedicels 5–8(–13) mm long, slender to filiform, subglabrous, spreading, not reflexed in fruit. Calyx 2.5–4 mm long, glabrous; lobes ovate, ± 1 mm long, obtuse, not prominently veined. Corolla blue, pale mauve, through lilac and pink to cream or white, 3.5–5.5(–8) mm long; tube 2–3(–4) mm long; upper lip erect, emarginate; lower lip broadly 3-lobed, convex. Staminodes very minute. Capsule 7–12 mm long, acute.

UGANDA. Acholi District: near Paimol Rest Camp, 11 May 1961, *Lind* 2995!; Ankole District: Mwirasandu, Nov. 1948, *Eggeling* 5839!; Mengo District: Entebbe, 5 Apr. 1960, *Napper* 1516!

KENYA. Northern Frontier District: Gurika Hill, near Lesirikan, foothills W of N end of Ndotos, 18 Nov. 1977, *Carter & Stannard* 525!; Uasin Gishu District: Ol Dane Sapuk, 27 July 1951, *Greenway* 8536!; Fort Hall District: 6 km SW of junction of Thika and Fort Hall roads, 15 May 1978, *Gilbert & Thulin* 1757!

TANZANIA. Bukoba District: Kabirizi, Oct. 1931, *Haarer* 2237!; Songea District: Nangurukuru, about 26 km E of Songea, 8 Apr. 1956, *Milne-Redhead & Taylor* 9494!; Masasi District: Chironga Hill, near Masasi, 8 Mar. 1991, *Bidgood et al.* 1839!

DISTR. **U** 1, 2, 4; **K** 1–6; **T** 1, 8; Rwanda, Burundi, Ethiopia

HAB. Damp places and pools in shallow soil among and over rocky outcrops, lake shores and riversides mostly in open country; (400 m– fide U.K.W.F.) 1000–2500 m

USES. None recorded on specimens from our area

CONSERVATION NOTES. Least Concern (LC)

SYN. *Bonnaya pusilla* Oliv., Trans. Linn. Soc. 29: 121, t. 122 A (1875), *nom. illegit.*, *non* Benth. (1835)

Ilysanthes pusilla Urban, Ber. Deutsch. Bot. Ges. 2, 435 (1884), *non L. pusilla* Boldingh, Zakfl. Landbouwstr. Java: 165 (1916)

Lindernia philcoxii Eb.Fisch. in Trop. and subtrop. Pflanzenw. 81: 295, t. I41 (1992) & F.A.C. Scrophulariaceae: 166, pl. 70 (1999), *nom. illegit.*; U.K.W.F.: 257 (1994). Type as for *Lindernia ugandensis* (Skan) Philcox

NOTE. The name *Lindernia serpens* has been made for this species. The name originally used, *Lindernia philcoxii*, is illegitimate as when Fischer made the name he cited an earier valid name, *Ilysanthes ugandensis* Skan in synonymy. This he should have used as the basionym for a new combination. However, having seen the type of *I. ugandensis*, I (D.P.) consider it to be distinct from the above and have included it in this treatment under the new combination *Lindernia ugandensis* (Skan) Philcox. The epithet 'pusilla' cannot be used here as the name has already been used by Boldingh (1916) without citing any basionym in synonymy, and from the scant description it cannot compare with any certainty with our species.

30. **Lindernia hartlii** *Eb.Fischer & Hepper* in B.J.B.B. 59: 447, fig. 2 (1989); Fischer in Trop. and subtrop. Pflanzenw. 81: 298 (1992) & F.A.C. Scrophulariaceae: 168, pl. 71 (1999). Type: Tanzania, Mwanza District: SW of Lake Victoria, 64 km S of Nungwe-Emin Pasha gulf, *Morgan* 82 (BM!, holo.)

Erect annual, ± 3 cm tall, simple or branched; stems quadrangular, glabrous. Basal leaves absent; leaves, opposite, sessile, lanceolate, 4–5 mm long, 1–2 mm wide, apex shortly acuminate, pilose, margins entire, glabrous. Flowers solitary, axillary; pedicels 1.8–2.5 cm long. Calyx ± 3 mm long, pilose; lobes lanceolate, 0.5–1.2 mm long, acute, ciliate. Corolla blue-violet with yellow mark on lower lip; upper lip 4 mm long; lower lip 3-lobed, lobes 4 mm long, rounded; tube 4–5 mm long, throat glandular-pubescent. Stamens 2, fertile; filaments 2 mm long; staminodes 2, 1 clavate, 1 mm long, slightly curved outwards, with yellow papillae. Capsule not seen.

TANZANIA. Mwanza District: SW of Lake Victoria, 64 km S of Nungwe–Emin Pasha gulf, 5 Jan. 1937, *Morgan* 82!

DISTR. **T** 1; Congo-Kinshasa

HAB. In open *Brachystegia-Berlinia* (miombo) woodland; ± 1350 m

CONSERVATION NOTES. Data Deficient (DD)

NOTE. The above description is solely from Fischer & Hepper's original description. So far, the taxon is known only from 2 collections, one from Tanzania and the other from Congo-Kinshasa. This species was not initially included in Philcox's treatment of *Lindernia* for FTEA; it was possibly over-looked (S.A.G.).

31. **Lindernia nana** (*Engl.*) *Roessler* in Mitt. Bot. Staatss. Munchen 5: 691 (1965); Philcox in K.B. 8 (2): 69 (1990); Fischer in Trop. and subtrop. Pflanzenw. 81: 263, t. 126, 127 (1992) & F.A.C. Scrophulariaceae: 152, pl. 64 (1999) & in Fl. Ethiop. & Eritr. 5: 274 (2006). Type: Angola, Huilla, *Welwitsch* 5788 (B†, holo; K!, iso.)

Erect annual or perennial, 2.2–5.5 cm tall; stems leafy, quadrangular, minutely pubescent. Leaves opposite, the upper sessile, ovate, 4.5–11(–16) mm long, 1–5.5(–9) mm wide, apex obtuse, margins sparsely crenate-dentate, glabrous; lower leaves petiolate, spatulate, 7–15 mm long including petiole, 2–4 mm wide, apex obtuse; petioles where present, 2–6 mm long. Flowers solitary in axils of upper leaves; pedicels 0.5–1.5 mm long. Calyx 3–4 mm long, glabrous, strongly keeled or not; lobes ovate-lanceolate, 1.3–1.8 mm long, acute. Corolla white with pink or pale blue markings, ± 6–9 mm long; upper lip bifid; lower lip trilobed, lobes rounded. Stamens 2; filaments ± 0.8 mm long; staminodes 2, gibbous with minute filaments ± 0.3 mm long. Capsule narrowly ellipsoid, 5–8 mm long, 2–2.3 mm broad, acuminate.

UGANDA. Toro District: Queen Elizabeth National Park, Mweya Peninsula, 17 June 1969, *Lock* 69/185!

KENYA. Northern Frontier District: Moyale, 24 Aug. 1952, *Gillett* 13750!; Laikipia District: near Rumuruti, *Powys* 74!

TANZANIA. Kondoa District: Kolo, 24 km N of Kondoa, 15 Jan. 1962, *Polhill & Paulo* 1178!

DISTR. **U** 2; **K** 1, 3; **T** 5; Ethiopia, Congo-Kinshasa, Rwanda, Angola, Zambia, Zimbabwe, Namibia, South Africa

HAB. In shallow soil covering rocks and in seepage areas surrounding rocky outcrops; 900–1550 m

USES. None recorded on specimens from our area

CONSERVATION NOTES. Least Concern (LC)

SYN. *Ilysanthes nana* Engl., E.J. 23: 505 (1897); Hiern, Cat. Afr. Pl. Welw. 1: 764 (1898) & in Fl. Cap. 4, 2: 365 (1904); Skan in F.T.A. 4(2): 347 (1906)

24. GLOSSOSTIGMA

Arnott in Nova Acta Acad. Leop.-Carol. 18: 355 (1836), *nom. conserv.*

Peltinela Raf., Atl. Journ.: 199 (1833)

Small creeping perennial herbs. Leaves opposite or whorled, entire. Flowers axillary, solitary, ebracteolate. Calyx campanulate, 3-4-toothed. Corolla tube equalling the calyx. Stamens 2 (in Flora area) or 4; filaments erect; anthers bithecate, thecae parallel, diverging at base; staminodes absent. Stigma dilated above into a broad stigmatic lamella. Fruit capsular, 2-locular; dehiscence loculicidal, many-seeded, covered by the persistent calyx. Seeds compressed.

3 Old World species, one in Africa and tropical Asia, the other two in Australasia.

All species are moss-like herbs of wet places and can be easily overlooked.

Glossostigma diandra (*L.*) *Kuntze*, Rev. Gen. P1. 2: 461 (1891); Chevalier in Bull. Mus. Hist. Nat. Paris, ser. 2, 4: 587 (1932); Hepper in F.W.T.A. ed. 2, 2: 366 (1963). Lectotype: Herb. Linn. No. 794.2 (LINN), designated by Yamazaki in Leroy (ed.), Fl. Cambodge Laos Viêt-Nam 21 : 158 (1985)

Mat-forming minute creeping herb; stems filiform, rooting at the nodes. Leaves paired or fascicled, spatulate, ± 10 mm long, 1–1.5 mm wide. Flowers axillary, solitary; pedicels 3–9(–18) mm long. Calyx ± 1 mm long, shortly 3-toothed, glabrous, pale brown and papery. Corolla white; tube slender, ± 1.5 mm long, ± 2-lipped, 5-lobed. Stamens 2, inserted on the corolla tube. Stigma slightly exserted, reddish, dilated and flattened above. Capsule as long as calyx, dehiscent. Fig. 26, p. 92.

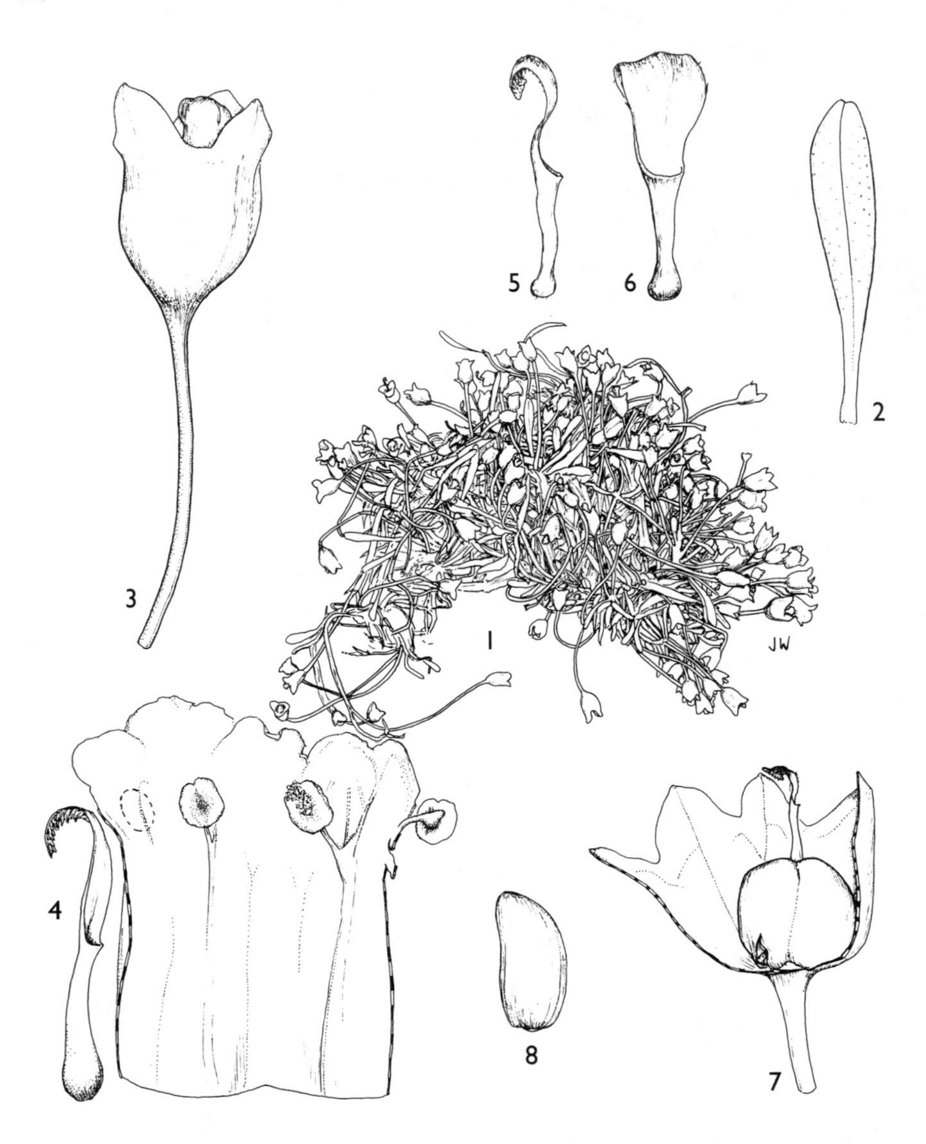

FIG. 26. *GLOSSOSTIGMA DIANDRA* — **1**, habit × 2; **2**, leaf × 8; **3**, flower × 16; **4**, corolla tube opened and style × 40; **5**, **6**, style side and front view × 40; **7**, calyx opened with capsule × 16; **8**, seed × 90. 1, 2, 3, 4, 7, 8, from *Gillett* 13688; 5, 6, from *Podor* 1426. Drawn by Juliet Williamson.

KENYA. Northern Frontier District: Sololo, 3 Aug. 1952, *Gillett* 13688!

DISTR. **K** 1; Mauritania, Senegal, Mali, Niger, Chad, Cameroon; India, Sri Lanka

HAB. On clayey mud at edge of ditches, dams, pools and streams; ± 740 m

USES. None recorded on specimens from our area.

CONSERVATION NOTES. Assessed as of Least Concern (LC) on its large distribution range, but for Flora area known from a single collection in a habitat that is prone to degradation

SYN. *Limosella diandra* L., Mant.: 252 (1767)

Microcarpaea spathulata Hook., Bot. Misc. 2: 101, suppl. t. 4 (1831); Benth., Scroph. Ind.: 31 (1835)

Glossostigma spathulatum (Hook.) Arnott in Nova Acta Acad. Leop.-Carol. 18: 355 (1836); Hook.f., Fl. Brit. Ind. 4: 288 (1884), *nom. illegit.*

25. LIMOSELLA

L., Sp. Pl.: 631 (1753); Glück in N.B.G.B. 12: 77 (1934) & in E.J. 66: 488–566, t. 6–8 (1934)

Small annual or perennial aquatic herbs, with slender stolons rooting to form new plants, glabrous. Leaves fascicled or in rosettes, erect or floating, long-petiolate, cylindrical or with distinct lamina. Flowers axillary, pedicels often deflexed in fruit. Calyx campanulate, (4–)5-toothed, teeth equal or calyx ± 2-lipped. Corolla tube cylindrical, (4–)5-lobed, almost actinomorphic; lobes ovate or ovate-triangular to lanceolate, ± pilose on the upper side. Stamens 4 (2 short, 2 long). Ovary 2-locular; style variable in length; stigma capitate, papillose, ± 2 lobed. Fruit capsular, dehiscing loculicidally by 2-valves. Seeds few to numerous, minute, longitudinally multi-angular with fine transverse striations.

About 15 species throughout the world, mainly in temperate or cooler areas such as tropical mountains.

All the species appear to be variable, and their taxonomy is imperfectly understood. Part of the problem lies is the presence of ecotypes found in deep water, shallow water and mud. This account basically follows the treatment by Glück (op. cit.), but a full revision of the genus is still required.

1. Leaves terete or subulate 2
   Leaves with distinct lamina 3
2. Calyx 2.5–4.5 mm long .... 1. *L. macrantha*
   Calyx 1–1.5 mm long .... 5. *L.* sp. A
3. Lamina base distinct, ± rounded, subcordate or peltate, leaves usually floating on surface of water; capsules exserted, subacute .... 2. *L. capensis*
   Lamina cuneate at base to attenuate into petiole, leaves rarely (*L. africana*) floating on surface, 3-veined 4
4. Flowers short-pedicellate with pedicels to 1 cm long .... 3. *L. africana*
   Flowers long-pedicellate with pedicels (1–)2–5 cm long .... 4. *L. maior*

1. **Limosella macrantha** *R.E.Fries* in Acta Hort. Berg. 8: 48, fig. 2 (1924); Glück in N.B.G.B. 12: 72 (1934) & in E.J. 66: 546, 555 (1934); Verdcourt in J. E. Afr. Nat. Hist. Soc. 22(94): 65, fig. 1c (1953); A.V.P.: 167 (1957); U.K.W.F.: 255 (1994); Fischer, F.A.C. Scrophulariaceae: 64, pl. 23 (1999) & in Fl. Ethiop. & Eritr. 5: 265 (2006). Type: Kenya, Aberdare Mts, *R.E. & T.C.E. Fries* 2691 (UPS, holo.; K!, iso.)

Herb with fibrous roots and stolons, sometimes mat-forming. Leaves basal, linear-subulate, terete, 2–5 cm long, ± 1 mm wide, sheathing at base, obtuse at apex. Flowers axillary, numerous among leaves; pedicels 5–15 mm long, ± reflexed in fruit. Calyx 2.5–4.5 mm long; lobes ± 1.5 mm, triangular. Corolla white or pale blue, 5–8 mm long, lobes equal, oblong, rounded at apex, sparsely pilose inside. Stamens with slender filaments; style filiform, equalling corolla tube; stigma papillose-capitate. Capsule ovoid, obtuse, shorter than calyx.

UGANDA. Mbale District: Mt Elgon, 21 Mar. 1951, *G.H.S. Wood* 912! & 14 Mar. 1997, *Wesche* 1094!
KENYA. Aberdare Mts, 1 Apr. 1922, *R.E. & T.C.E. Fries* 2691! Trans-Nzoia District: Kipsare Hill, 25 Dec. 1967, *Gillett* 18480!; Mt Kenya, below Meru Lodge, 12 Dec. 1985, *Beentje* 2427!
TANZANIA. Arusha District: Mt Meru crater, 28 Dec. 1966, *Richards* 21832! & 25 Dec. 1970, *Richards & Arasululu* 26578!; Moshi District: Kilimanjaro, saddle between Kibo and Mawenzi, 23 June 1948, *Hedberg* 1343!
DISTR. **U** 3; **K** 3, 4; **T** 2; Ethiopia; Yemen
HAB. Edges of lakes, pools, moist depressions, in the upper forest zone and afro-alpine belt; 2500–4500 m
USES. None recorded on specimens from our area.

CONSERVATION NOTES. Least Concern (LC); only a few flowering specimens from Flora area; possibly overlooked and undercollected

2. **Limosella capensis** *Thunb.*, Prodr.: 104 (1800) & Fl. Cap., ed. Schultes: 480 (1823); Hiern in Fl. Cap. 4(2): 358 (1904); Skan in F.T.A. 4(2): 353 (1906); Glück in N.B.G.B. 12: 77 (1934) & in E.J. 66: 548, 559 (1934); Verdcourt in J. E. Afr. Nat. Hist. Soc. 22(94): 65, fig. 1d,e (1953); Philcox in F.Z.: 8(2): 75 (1990); U.K.W.F.: 255 (1994); Fischer in Fl. Ethiop. & Eritr. 5: 266 (2006). Type: South Africa, *Thunberg* s.n. (UPS!, holo., microfiche)

Small, tufted aquatic herb rooting in mud or floating, often with stolons. Leaves yellow-green, usually floating or erect on mud, oblong-ovate, 1–3 cm long, 0.3–1.5 cm wide, base rounded or subcordate to peltate, apex rounded, margins entire, ± 5-veined; petioles (1–)5–10 cm long, slender to narrow. Flowers axillary; pedicels up to 0.5–2 cm long. Calyx ± 2 mm long, 5-lobed, (2 + 3); lobes broadly ovate, 0.5–1 mm. Corolla white to pale blue, ± 3 mm long; lobes oblong, obtuse. Stamens with short filaments. Capsule ovoid, subacute, exserted beyond calyx. Seeds oblong, acute, multi-angular, finely ribbed.

UGANDA. Mbale District: Sebei, Kaburonon, 26 Aug. 1975, *Norman* 274!
KENYA. Northern Frontier District: Maralal to Baragoi, 11 Nov. 1978, *Hepper & Jaeger* 6725!; Laikipia District: Rumuruti, Aug. 1975, *Powys* 83! 85!; Kiambu District: Muguga, 1 May 1952, *Verdcourt* 641!
TANZANIA. Musoma District: Moru, 27 Apr. 1964, *Greenway* 10113!; Masai District: Ol Doinyo Loldadwenye, 27 Jan. 1962, *Newbould* 5913!
DISTR. **U** 3; **K** 1, 3, 4, 6; **T** 1, 2; Ethiopia, Zambia, Malawi, Mozambique, Zimbabwe, Botswana, South Africa
HAB. Muddy pools, dams, slow streams, temporary rock pools; 1800–2800 m
USES. None recorded on specimens from our area
CONSERVATION NOTES. Least Concern (LC); common and widespread

NOTE. An extremely variable, ill-defined species. In South Africa plants with minute leaves and long pedicels are frequent. In FTEA area this form has not been recorded, and the distinction from *L. africana* Glück is often obscure.

3. **Limosella africana** *Glück* in E.J. 66: 540, 558, t.8 (1934) & in N.B.G.B. 12: 76 (1934); Verdcourt in J. E. Afr. Nat. Hist. Soc. 22(94): 65, fig. 1f (1953); A.V.P.: 166, 318 (1957); Hepper in F.W.T.A. ed. 2, 2: 366 (1963); U.K.W.F.: 255 (1994); Fischer in Fl. Ethiop. & Eritr. 5: 264 (2006). Types: South Africa, Cape, Greenpoint, Ex herb. Brehm s.n. & SW Africa, *Dinter* 4704 (?HBG, syn.)

Small herb with stolons, often forming mats. Leaves fasciculate or rosulate, (0.5–)1–4(–10) cm long, lamina ovate to linear lanceolate, (2–)6–11(–10) mm long, up to 10 mm wide, base cuneate, apex subacute, margins entire, ± 3-veined. Flowers axillary, amongst leaf bases, erect in flower, reflexed in fruit; pedicels up to 1 cm long. Calyx ± 2.5 mm long, lobes broadly ovate. Corolla white, bluish or pale rose, ± 3 mm long, 5-lobed; tube twice as long as lobes, lobes obtuse. Stamens 4. Capsule ovoid, ± 3.5 mm long and ± 2 mm in diameter. Seeds ovoid, 0.3 mm long, 0.2 mm broad, acute at each end.

UGANDA. Toro District: Lake Bujuka, 23 Mar. 1948, *Hedberg* 416!; Mt Elgon, Kapchorwa, 9 Sept. 1954, *Lind* 291! & Bugishu, Sasa Camp, 16 Apr. 1950, *Forbes* 280!
KENYA. Elgeyo-Marakwet District: Cherangani Hills, 2 Jan. 1971, *Mabberley* 570!; Laikipia District: Aberdare Mts, 15 Mar. 1922, *R.E. & T.C.E. Fries* 2416!; Mt Kenya, 4 Apr. 1975, *Hepper & Field* 4858!
TANZANIA. Kilimanjaro, 17 June 1948, *Hedberg* 1258!
DISTR. **U** 2, 3; **K** 3, 4; **T** 2; Cameroon, Ethiopia, Southern African mountains; Yemen
HAB. Upland pools, stream banks, edges of waterfalls and boggy places, 2200–2500 m

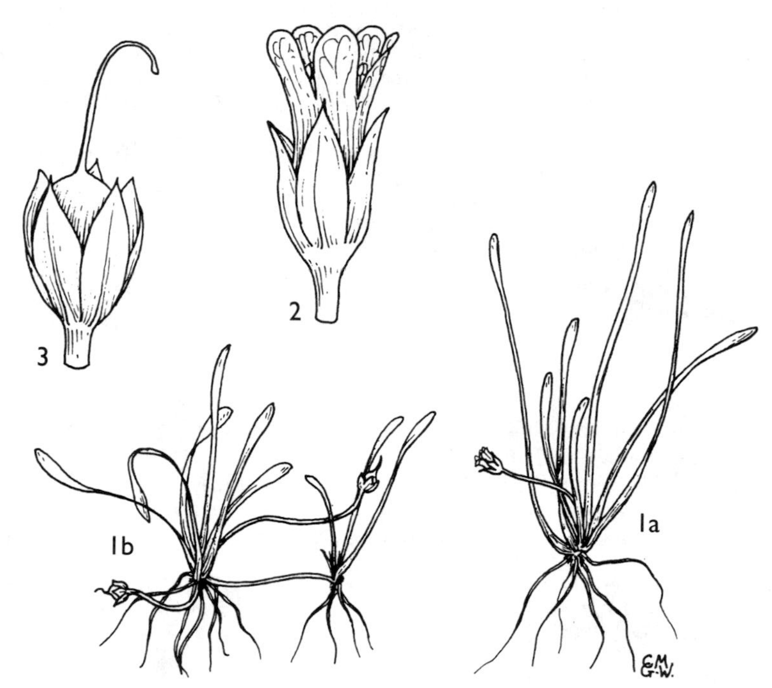

FIG. 27. *LIMOSELLA MAIOR* — **1a & b**, habit × ⅔; **2**, flower × 6; **3**, fruit × 6. All from *Fries* 2639. Drawn by Christine Grey-Wilson. Reproduced with permission from F.Z.

USES. None recorded on specimens from our area

CONSERVATION NOTES. Least Concern (LC); common

SYN. *L. aquatica* sensu auctt. *non* L.; Skan in F.T.A. 4(2): 352 (1906); R.E. Fries in Acta Hort. Berg. 8: 48 (1924); Peter in Abh. Ges. Wiss. Göttingen Math. Phys. Kl. N.F. 13(2): 127 (1928); Hauman in Bull. Acad. Belg. C1. Sci., ser. 5, 19: 705 (1933); U.K.W.F.: 553 (1974)

NOTE. Hedberg (A.V.P.: 318 (1957)) commented on the fine distinctions Glück recognised between *L. aquatica* L. and his *L. africana*. Although the taxonomy is still unsatisfactory I have maintained the use of the name *L. africana* pending further research. However, I do not recognise forma *terrestris* Glück and forma *natans* Glück (op. cit. 1934) as taxonomically distinct. Clear distinctions between small plants of *L. capensis* and L. *africana* have not been found.

4. **Limosella maior** *Diels* in E.J. 26: 122 (1898); Hiern in F1. Cap. 4(2): 357 (1904); Skan in F.T.A. 4(2): 353 (1906); Glück in N.B.G.B. 12: 77 (1934) & in E.J. 66: 547, 559 (1934) (as *major*); Verdcourt in J. E. Afr. Nat. Hist. Soc. 22(94): 65, fig. 1a,b (1953); Philcox in F.Z.: 8(2): 73 (1990); Fischer, F.A.C. Scrophulariaceae: 62, pl. 22 (1999) & in Fl. Ethiop. & Eritr. 5: 265 (2006) (as *major*). Type: South Africa, near Pretoria, Dec. 1883, *Wilms* 977 (B†, holo.; K!, iso.)

Aquatic herb with or without stolons. Leaves radical, arcuate or erect, 3–15 cm long, lamina spatulate to oblong-spatulate, 1–4.5 cm long, 3–16 mm wide, attenuate at base into a stout petiole, apex obtuse to subacute, margins entire, ± 3-veined. Flowers axillary; pedicels (1–)2–5 cm long, erect in flower, reflexed in fruit. Calyx 3–4 mm long, 5-lobed; lobes ovate, 1–1.5 mm, acute. Corolla white or mauve (especially on the outside), 4–5(–6) mm long; lobes oblong and ± pubescent on inner surface. Filaments white, slender, exserted; anthers dark blue; style nearly as long as corolla, inserted to one side of ovary. Capsule ovoid, 5 mm long. Fig. 27.

UGANDA. Mt Ruwenzori, Bujuku Hut–Irene Hut, Jan. 1969, *Lye* 1304!
KENYA. SE Mt Elgon, June 1964, *Tweedie* 2848!; Naivasha District: Kedong valley, 21 Jan. 1953, *Greenway & Hemming* 8768!; Masai District: Enesambulai valley, *Greenway & Kanuri* 14558!
TANZANIA. Dodoma District: Bereke Kurasoni, 14 Jan. 1974, *Richards & Arasululu* 28675!; Mbeya District: Mt Mbeya, 13 May 1956, *Milne-Redhead & Taylor* 10217!; Njombe District: upper R. Ruhudje, May 1931, *Schlieben* 956!
DISTR. **U** 2; **K** 3, 5, 6; **T** 5, 6, 7; Congo-Kinshasa, Eritrea, Ethiopia, Malawi, Mozambique, Zimbabwe, Botswana, South Africa
HAB. Upland seepage areas and muddy slow-moving streams: 1800–2700 m
USES. None recorded on specimens from our area
CONSERVATION NOTES. Least Concern (LC)

NOTE. forma *terrestris* Glück in N.B.G.B 12: 77 (1934) is of no taxonomic value since it refers to small plants occurring on mud.

5. **Limosella sp. A**

Minute aquatic herb, forming mats with interlocking creeping stolons and fibrous roots. Leaves subulate, 2–7 mm long. Flowers basal; pedicels up to 5 mm, reflexing or not. Calyx 1–1.5 mm, 5-lobed; lobes ovate, ± 1 mm, acute. Corolla white, with violet markings on outside of lobes or at base of lobes, ± 2 mm long, oblong, obtuse. Capsule subglobose, ± 1.5 mm. Seeds oblong, acute, multi-angular, finely ribbed.

KENYA. Kiambu District: Lari swamp, 19 Aug. 1967, *Kabuye & Wood* 105!
TANZANIA. Rungwe District: R. Kiwere, 25 Oct. 1947, *Brenan & Greenway* 8216!; Njombe District: Ndumbi R., 19 Oct. 1956, *Richards* 6620!
DISTR. **K** 4; **T** 7; see note
HAB. Mud or wet gravel by streams; 2300–2700 m
USES. None recorded on specimens from our area
CONSERVATION NOTES. Known only from a few collections in Flora area and that too more than 40 years ago. In Kenya, road and other development near the Lari swamp has greatly altered the area where this plant may be endangered. It is here assessed as Vulnerable, Vu B1a.

SYN. *L. aquatica* L. var. *tenuifolia* Hook.f., F1. Antarct. 2: 334 (1846); Skan in F.T.A. 4(2): 352 (1906). Type: Bioko, Clarence Peak, *Mann* 597 (?B, holo.)

NOTE. The identity of these plants is uncertain. The Kenya and Tanzania specimens may represent two separate taxa. They appear to fall into the *L. subulata* complex to which *L. aquatica* var. *tenuifolia* and *L. longiflora* Kuntze (Gen. P1. (3)2: 235 (1898)) appear to belong. But the latter South African species has its leaves somewhat expanded toward the apex and the tiny flowers on slender pedicels are nearly as long as the leaves.

26. **SIBTHORPIA**

L., Sp. P1.: 631 (1753) & Gen. Pl. ed. 5: 279 (1754)

Perennial small prostrate herbs; stems slender, rooting at the nodes. Leaves alternate, circular or reniform, crenate to incised. Flowers axillary, l-several; pedicellate. Calyx 5-lobed. Corolla small, regular or slightly irregular. Stamens 3–5; filaments filiform. Nectaries present or absent. Seeds small, white-tuberculate or smooth black.

5 species in Europe, tropical Africa, C and S America.

Sibthorpia europaea *L.*, Sp. P1.: 631 (1753); Hedberg in Bot. Not. 108: 168 (1955) & A.V.P.: 168 (1957) & in Caryologia 28: 253, fig. 6 (1975); Hepper in F.W.T.A. ed. 2, 2: 356, fig. 285 (1986); Philcox in F.Z.: 8(2): 75, t. 27 (1990); U.K.W.F.: 257 (1994); Fischer, F.A.C. Scrophulariaceae: 210, pl. 87 (1999) & in Fl. Ethiop. & Eritr. 5: 284 (2006). Lectotype: "Alsine spuria pusilla repens foliis Saxifragae aureae" in Plukenet, Phytographia, t. 7, f. 6, 1691, designated by Hampshire in Jarvis & al. (ed.), Regnum Veg. 127 : 88 (1993)

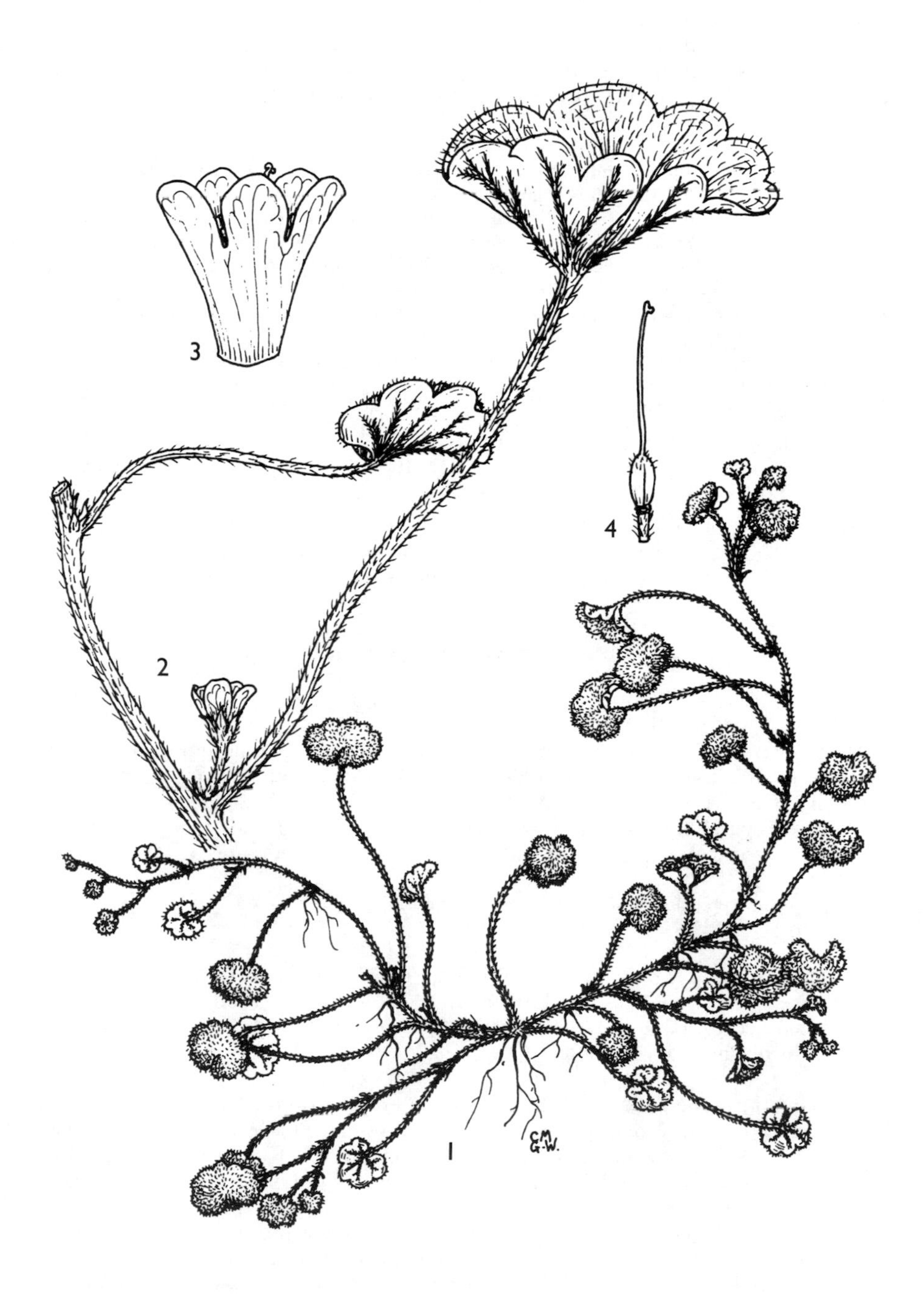

FIG. 28. *SIBTHORPIA EUROPAEA* — **1**, habit × ²⁄₃; **2**, portion of stem with leaves × 4; **3**, corolla × 12; **4**, gynoecium × 12. All from *Wild* 1390. Drawn by Christine Grey-Wilson. Reproduced with permission from F.Z.

Prostrate perennial; stems creeping, often matted, rooting at the nodes, ± pilose. Young distal leaves reniform, ± 5 mm in diameter, very shortly petiolate; older leaves reniform, up to 15 mm long and 20 mm wide, margins crenate-dentate, pilose on both surfaces; petiole ± 2 cm long (up to 7 cm in shaded locations). Flowers solitary; pedicels 2–5 mm, much shorter than the petiole. Calyx ± 2 mm long, usually 5-lobed, erect in fruit, pubescent. Corolla pale pink or mauve, slightly longer than calyx. Stamens 4 or 5, included. Capsule obovoid, shorter than calyx, pilose at apex. Fig. 28, p. 97.

UGANDA. Ruwenzori, Bujuku valley, 1 Apr. 1948, *Hedberg* s.n!; Kigezi District: Mt Muhavura, 3 Oct. 1948, *Hedberg* 2063!; Mt Elgon, Jan. 1918, *Dummer* 3521!

KENYA. Mt Elgon, 3 Mar. 1948, *Hedberg* 270!; Nyeri District: Aberdare National Park, Chania Falls, 9 Apr. 1975, *Hepper & Field* 4950!; Mt Kenya, 28 Feb. 1970, *Faden & Evans* 70/101!

TANZANIA. Kilimanjaro, 21 June 1948, *Hedberg* 1334!; Mt Meru crater, 30 July 1970, *Vesey-FitzGerald* 6787!; Rungwe District: Mwakaleli, 8 May 1975, *Hepper, Field & Mhoro* 5419!

DISTR. **U** 2, 3; **K** 2–4; **T** 2, 7; Cameroon south to Zimbabwe; Ethiopia; Mediterranean and western Europe

HAB. Damp places in mountains such as shady banks along roads; 2000–3500 m

USES. None recorded on specimens from our area

CONSERVATION NOTES. Least Concern (LC); widespread and common

SYN. *S. prostrata* Salisb., Ic. Stirp. P. 11 (1791), *nom. illegit.*
S. europaea L. var. *glabra* Skan in F.T.A. 4(2): 354 (1906). Type: Uganda, Butemba, *Scott-Elliot* 7848 (K!, holo.)

NOTE. Hedberg (in Bot. Not. 108: 168 (1955)), followed by a number of later authors, indicated 793.1 (LINN) as type but it is not annoted by Linnaeus and is not original material for the name (http://www.nhm.ac.uk/jdsml/research-curation/projects/linnaean-typification/)

27. SCOPARIA

L., Sp. P1.: 116 (1753)

Herbs or small shrubs. Leaves opposite or verticillate, glandular-punctate. Flowers axillary, several together, ebracteolate. Calyx 4- or 5-lobed, imbricate, glandular-punctate. Corolla rotate, 4-lobed, throat densely bearded. Stamens 4, ± equal; anthers with parallel or divergent cells; style longer than the calyx; stigma truncate or emarginate. Fruit capsular, septicidal, valves entire. Seeds numerous.

20 species in tropical America; one a pantropical weed.

Scoparia dulcis *L.*, Sp. P1.: 116 (1753); Skan in F.T.A. 4(2): 354 (1906); Hepper in F.W.T.A. ed. 2, 2: 356 (1963); Philcox in F.Z.: 8(2): 77 (1990); Fischer, F.A.C. Scrophulariaceae: 36, pl. 11 (1999) & in Fl. Ethiop. & Eritr. 5: 257 (2006). Type: West Indies, Herb. Clifford (BM, holo.)

Herb or a branched sub-shrub, 30–60(–90) cm high, woody at base; stems obtusely angled, glabrous. Leaves opposite, often 3 per node, elliptic-lanceolate to linear-lanceolate, 1–4 cm long, 0.3–11 cm broad, long-cuneate at base into a short petiole, acute or obtuse at apex, margins coarsely serrate in the upper half, glabrous, densely glandular-punctate beneath. Flowers 2–4 in all upper axils of leaf-like bracts; pedicels filiform, 4–6 mm, glabrous. Calyx 2–3 mm long, lobed nearly to the base; lobes ovate, ± 1.5 mm broad, shortly ciliate towards the apex. Corolla white, 4-lobed, lobes ovate, 2.5–3.5 mm, with conspicuous white hairs in throat. Capsule ovoid, yellowish brown, 2–4 mm, beaked by ± persistent style. Fig. 29, p. 99.

UGANDA. West Nile District: Omugo, 10 Aug. 1953, *Chancellor* 158! & Ocodri [Ochodri], May 1940, *Eggeling* 3923! & Arua, Sept. 1940, *Purseglove* 1020!

FIG. 29. *SCOPARIA DULCIS* — **1**, habit × ²⁄₃; **2**, part of flowering branch × 4; **3**, flower dissected × 6; **4**, corolla × 6; **5**, dehiscing capsule × 8. 1 from *Robinson* 5412; 2–4 from *Fanshawe* 6827; 5 from *Rensburg* 2325. Drawn by Christine Grey-Wilson. Reproduced with permission from F.Z.

KENYA. Kwale District: Vanga, Nov. 1929, *Graham* 2207! & Mkongani forest, Shimba Hills, 10 May 1968, *Magogo & Glover* 1034!; Kilifi District: Arabuko-Sokoke Forest, *Harvey, Mwachala & Vollesen* 57!

TANZANIA. Mwanza District: Mabale, 9 May 1952, *Tanner* 781!; Tanga District: Machui, 10 May 1965, *Faulkner* 3518!; Uzaramo District: Dar es Salaam, 1 July 1967, *Harris* 673!; Zanzibar: Marahubi, 19 June 1962, *Faulkner* 3055!

DISTR. **U** 1; **K** 7; **T** 1, 3–6, 8; **Z**; **P**; widespread in tropical Africa

HAB. In cultivated and waste ground, gardens and coconut groves especially on sandy soil; 0–1300 m

USES. None recorded on specimens from our area

CONSERVATION NOTES. Least Concern (LC); widespread

28. **VERONICA**

L., Sp. P1.: 9 (1753)

Annual or perennial herbs, sometimes woody at the base. Leaves opposite. Bracts sometimes foliose, usually alternate. Flowers in terminal or lateral (then main axis continuing vegetative growth) racemes; terminal racemes sometimes elongated with foliose bracts; bracteoles absent. Calyx deeply divided into 4(–5) often unequal segments, the fifth posterior segment when present usually smaller. Corolla rotate to campanulate, with 4 (rarely 5) lobes; tube short, rarely exceeding the calyx; upper and lower lobes often narrower. Stamens 2, inserted on the corolla tube at the sides of the upper lobe, exserted; anther thecae divergent or parallel, obtuse, confluent at the apex. Style with subcapitate stigma. Fruit a loculicidal and sometimes also pseudosepticidal capsule, more or less compressed at right angles to the septum, usually with hairs and venose surface. Seeds few or many, ellipsoidal or suborbicular, affixed by the inner flat or concave surface, reticulate, sometimes almost smooth, or rugulose on the back, often with a thickened or keeled margin.

A large genus of some 450 species mostly in northern temperate countries and often occurring as weeds of cultivation. The shrubby species of New Zealand and Australia have often been placed in the genus *Hebe* Juss.

It is surprising that only one of the temperate European annual "speedwells" of cultivated ground and lawns has been reported from the Flora area. Others likely to occur are *V. polita* L., *V. peregrina* L., and *V. arvensis* L. Several species are cultivated as ornamental shrubs in upland gardens and parks: *V. longifolia* L., *V. salicifolia* Forst.f., *V. speciosa* A.Cunn., *V. spicata* L., and the shrubby species often referred to *Hebe*.

1. Inflorescence terminal with leaf-like bracts, flowers appearing solitary in their axils 2
   Inflorescence with lateral bracteate racemes with the main axis continuing vegetative growth 5
2. Herbs of ± aquatic and marshy habitats; stems hollow, ± quadrangular; flowers in long racemes; capsule elliptic .... 2. *V. anagallis-aquatica*
   Herbs not of marshy habitats; stems solid, terete; racemes 2–15-flowered, not long; capsule obcordate 3
3. Stems ascending, ultimately erect in the flowering part; leaves with entire margins; calyx glandular-pilose .... 6. *V. glandulosa*
   Stems decumbent to prostrate to trailing; leaves with crenate margins; calyx eglandular 4

4. Leaves ovate-orbicular, ± 5 × 4 mm; pedicel ± 2 mm; corolla rotate; capsule with lobes not diverging .... 5. *V. gunae*
   Leaves ovate, 5–25 × 3–18 mm; pedicel 15–20 mm; corolla not rotate; capsule with lobes diverging at an angle of about 120° .... 7. *V. persica*
5. Pedicel ± 1 mm long; corolla 1–2 mm long .... 3. *V. javanica*
   Pedicel more than 1 mm long; corolla more than 2 mm long 6
6. Leaves sessile or almost so, glabrous or puberulous on margins near base only; pedicel not filiform ...... 1. *V. serpyllifolia*
7. Leaves peiolate with petiole 4–18 mm long, pilose to subglabrous on both sides; pedicel filiform .... 4. *V. abyssinica*

1. **Veronica serpyllifolia** *L.*, Sp. Pl.: 12 (1753); Walters & Webb. in Fl. Europ. 3: 243 (1972); Mart. Ort. & Rico in J.L.S. 135: 179 (2001). Lectotype: Europe, Herb. Linn. 26. 30 (LINN) designated by Cramer in Dassanayake, Rev. Handb. Fl. Ceylon: 3, 437 (1981)

Perennial herb (5–)8–30(–40) cm tall; stems ascending, rooting, sometimes branched in the prostrate part, with short eglandular, rarely some glandular hairs, internodes usually longer than leaves. Leaves sessile to shortly petiolate, orbicular to elliptic or oblanceolate, 0.8–2.5 cm long, 0.5–1.3 cm wide, margins entire to crenate, glabrous or puberulous on margins near base. Inflorescence of terminal bracteate racemes; flowers (8–)15–25 appearing solitary in the axils of ovate to lanceolate bracts; pedicels 3–8 mm long. Calyx-segments more or less elliptic to ovate, 2–3(–5.5) mm long, sometime some glandular and eglandular hairs at the margins. Corolla blue to white with bluish veins; tube 5–6(–8) mm long; lobes rounded, ± 3 mm long; style 2–3 mm long. Capsule compressed, two-lobed, 2.5–3.5 mm long, 4–6 mm wide, with a persistent long style.

UGANDA. Toro District: Mt Ruwenzori National Park, 18 Feb. 1997, *Lye* 22525!
DISTR. **U** 2; not recorded elsewhere in the Flora area, but widespread in montane to alpine regions worldwide
HAB. Damp places at forest edges and meadows; ± 4050 m
USES. None recorded on specimens from our area.
CONSERVATION NOTES. Even though known only from the above collection in Flora area, the species is weedy and widespread; here assessed as of Least Concern (LC)

2. **Veronica anagallis-aquatica** *L.*, Sp. Pl. :12 (1753); Skan in F.T.A. 4(2): 357 (1906) as *V. anagallis*; Walters & Webb. in Fl. Europ. 3: 248 (1972); Philcox in F.Z.: 8(2): 80 (1990); U.K.W.F.: 258, pl. 111 (1994); Fischer, F.A.C. Scrophulariaceae: 206 (1999) & in Fl. Ethiop. & Eritr. 5: 283 (2006). Lectotype: Europe, Herb. Linn. No.5.7 (S) designated by Fischer in F.R. 108: 115 (1977)

Perennial herb usually 30–120 cm high, ± aquatic; stems erect, ± quadangular, simple or well-branched, sometimes rooting at the base, hollow, glabrous to sparsely glandular-pubescent. Leaves linear-lanceolate to ovate-lanceolate, the lower broader than upper ones, ± sessile and amplexicaul, 2–8(–10) cm long, up to 3 cm wide, margins irregularly serrate near the apex, glabrous. Inflorescence lateral with flowers numerous in long racemes at most axils, up to 3 times as long as subtending leaves; pedicels in fruit 3–4 mm long, filiform, spreading or curved upwards, sparsely glandular-pubescent. Calyx-segments oblong, 3 mm long in fruit, subacute. Corolla blue with violet veins to white with reddish veins, ± 5 mm in diameter. Capsule about as long as calyx, elliptical, glabrous, sinus obscure; persistent style much longer than sinus.

UGANDA. Mt Elgon, Oct. 1961, *Tweedie* 2237!

KENYA. Northern Frontier District: Mt Kulal, 18 Nov. 1978, *Hepper & Jaeger* 6912!; Elgeyo District: Cherangani Mts, 23 May 1949, *Maas Geesteranus* 4796!; Masai District: Aitong Spring, 4 Apr. 1961, *Glover, Gwynne & Samuel* 445!
TANZANIA. Musoma District: Loliondo, 9 Nov. 1953, *Tanner* 1739!; Arusha District: Mt Meru, 9 May 1965, *Richards* 20383!; Mpwapwa District: Bokeko, 19 Dec. 1860, *Grant* s.n.!
DISTR. **U** 3; **K** 1, 3, 4, 6, 7; **T** 1, 2, 5; widespread in upland tropical Africa, SW Arabia, and almost throughout the northern temperate zones
HAB. In permanent wet places with shallow running water; 480–2400 m
USES. None recorded on specimens from our area
CONSERVATION NOTES. Least Concern (LC); widespread and common

NOTE. The species is considerably variable in its vegetative characters probably depending on the wetness of its habitat. This is shown by the gradation from tall, lush plants in rich muddy places to much smaller ones growing in drying mud nearby.
This species belongs to the Section *Beccabunga* which consists of a group of similar species, but in East Africa apparently only represented by *V. anagallis-aquatica.*

3. **Veronica javanica** *Blume,* Bijdr. Fl. Ned. Ind.: 742 (1826); Skan in F.T.A. 4(2): 358 (1906); Philcox in F.Z.: 8(2): 82, t. 30/A (1990); U.K.W.F.: 258, pl. 111 (1994); Fischer, F.A.C. Scrophulariaceae: 205, pl. 86 (1999) & in Fl. Ethiop. & Eritr. 5: 281 (2006). Type: Java, Gede Mts, type specimen unspecified, but probably *Blume* s.n. (L, holo.)

Annual herb, simple or usually well-branched, (4–)10–15(–27) cm high, up to 20 cm across, decumbent, pubescent to pilose. Leaves broadly ovate, 1–2.5 cm long, up to 2 cm wide, truncate-cuneate at the base, subacute at the apex, coarsely serrate, pilose mainly on the veins beneath; petioles 0–3 mm long. Inflorescence of lateral racemes, lax, 1.5–3 cm long, extending somewhat in fruit, 2–5-flowered; bracts oblong-lanceolate, 4 mm long, obtuse; pedicels ± 1 mm long. Calyx segments oblong, 3 mm long in flower, enlarging to 4–5 mm in fruit, obtuse, ciliate. Corolla pale blue or pink purple, 1–2 mm long, soon falling. Fruits broadly obcordate, 2–3 mm long, ciliate. Fig. 30, 1–4, p. 103.

UGANDA. Kigezi District: Kachwekano farm, Kigezi, Mar. 1950, *Purseglove* 3343!; Ankole District: near Buhira and Nkoga swamps, 26 Sept. 1957, *Tallantire* 87!; Toro District: Fort Portal, 25 Feb. 1932, *Hazel* 179!
KENYA.; Mt Elgon, *Tweedie* 1448!; Kericho District: SW Mau forest, 14 Aug. 1949, *Maas Geesteranus* 5782!; Teita District: Taita Hills, Ngerenyi, 18 Sept. 1953, *Drummond & Hemsley* 4400!
TANZANIA. Arusha District: Arusha National Park, 18 Aug. 1969, *Vesey-FitzGerald* 6396!; Morogoro District: Mlali to Mgota road, 31 Dec. 1970, *Wingfield* 2317!; Mbeya District: Mbeya, 4 May 1975, *Hepper & Field* 5281!
DISTR. **U** 2; **K** 3–5, 7; **T** 2, 3, 6–8; Ethiopia, extending through much of tropical Asia
HAB. Damp grassy places, ditches and cultivation, usually near human habitations; 900–2100 m
USES. None recorded on specimens from our area.
CONSERVATION NOTES. Least Concern (LC); widespread

SYN. *V. wogerensis* A.Rich., Tent. Fl. Abyss. 2: 126 (1851). Type: Ethiopia, Wogera, *Schimper* 730 (K!, iso.)
V. chamaedryoides Engl., P.O.A. C: 358 (1895). Types: Tanzania, Kilimanjaro, Marangu [Marongo], *Volkens* 585, (B†, syn.; K!, isosyn.) & 760 (B†, syn.)

4. **Veronica abyssinica** *Fresen.* in Bot. Zeit. 2:356 (1844); Skan in F.T.A. 4(2): 358 (1906); Römpp in F.R., Beih. 50: 141 (1928); A.V.P.: 169 (1951); Hepper in F.W.T.A. ed. 2, 2: 355 (1963); Cribb & Leedal, Mount. Fl. S Tanz.: 119, pl. 29C (1982); Philcox in F.Z.: 8(2): 82, t. 30/B (1990); U.K.W.F.: 258, pl. 111 (1994); Fischer, F.A.C. Scrophulariaceae: 204, pl. 85 (1999) & in Fl. Ethiop. & Eritr. 5: 281 (2006). Type: Ethiopia, Simien, *Rüppell* s.n. (FR, holo.)

Trailing perennial herb with prostrate stems up to 80(–150) cm, rooting at the nodes, sometimes profusely branched, sparsely to rather densely pubescent in two

Fig. 30. *VERONICA JAVANICA* — **1**, habit × ²⁄₃; **2**, flower × 8; **3**, flower dissected × 8; **4**, capsules × 4. 1 from *Drummond* 4962; 2–4 from *Fries et al.* 3783. *VERONICA ABYSSINICA* — **5**, habit × ²⁄₃; **6**, flowers × 4; **7**, corolla opened × 4; **8**, capsules × 4. 1 from *Drummond* 4962; 2–4 from *Fries et al.* 3788; 5–8 from *Phillips* 1114A. Drawn by Christine Grey-Wilson. Reproduced with permission from F.Z.

lines. Leaves broadly ovate, 1–5 cm long, 0.5–3.5 cm wide, rounded to subcordate at base, subacute to obtuse, margins rather coarsely crenate, pilose to subglabrous on both sides; petioles 4–9(–18) mm long. Inflorescence of lateral racemes, 2–5 cm long, 2–4(–8)-flowered; bracts linear, 4 mm long; pedicels filiform, ± 5 mm long. Calyx 4–5 mm long in flower, 6–7 mm long in fruit; segments elliptic-oblong, ciliate. Corolla blue; tube white; lobes ± 6 mm long. Filaments white; anthers yellow. Capsule broadly obovate, emarginate, compressed, pilose to nearly glabrous. Fig. 30, 5–8, p. 103.

UGANDA. Karamoja District: Mt Moroto, 11 June 1970, *Katende & Lye* 400!; Kigezi District: Behungi, 1 Dec. 1930, *Burtt* 2926!; Mbale District: Mt Elgon, Bulambuli, 12 Nov. 1933, *Tothill* 2335!

KENYA. Trans-Nzoia District: Cherangani, 6 Dec. 1958, *Symes* 516!; Kiambu District: Muguga forest, 4 Sept. 1965, *Kokwaro & Kabuye* 347!; Teita District: Taita Hills, Mt Vuria, 9 Feb. 1966, *Gillett & B.L. Burtt* 17071!

TANZANIA. Lushoto District: W Usambara Mts, 11 June 1953, *Drummond & Hemsley* 2889!; Njombe District: Kitulo Plateau, 6 May 1975, *Hepper & Field* 5313!; Songea District: Matengo Hills, 1 Mar. 1956, *Milne-Redhead & Taylor* 8937!

DISTR. **U** 1–4; **K** 2–7; **T** 1–3, 6–8; extending to the mountains of Cameroon, Congo-Kinshasa, Rwanda, Burundi, Sudan, Ethiopia, Malawi

HAB. Montane forest, stream banks, bushy grassland; 1200–3100 m (–3900 m fide U.K.W.F.)

USES. None recorded on specimens from our area

CONSERVATION NOTES. Least Concern (LC)

SYN. *V. petitiana* A.Rich., Tent. Fl. Abyss. 2: 127 (1851). Type: Ethiopia, Wodgerat, *Petit* s.n. (P!, holo.)

NOTE. Robust forms with larger, deep blue flowers occur in Tanzania (**T** 7) at the edge of the Kitulo Plateau, Njombe District (*Hepper & Field* 5345) and at some other places.

5. **Veronica gunae** *Engl.* in Abh. Preuss. Akad. Wiss. 1891(2): 380 (1892); Skan in F.T.A. 4(2): 361 (1906); A.V.P.: 170 (1957); U.K.W.F.: 258 (1994); Fischer in Fl. Ethiop. & Eritr. 5: 281 (2006). Type: Ethiopia, Mt Guna, *Steudner* 811 (B†, holo.)

Prostrate herb, up to 40 cm, with numerous stems rooting from nodes; stems unbranched for 10–15 cm, nearly glabrous to pubescent in two lines; nodes 3–6 mm apart. Leaves ovate-orbicular, ± 5 mm long, 4 mm wide, rounded or cuneate at base, apex obtuse, margins minutely crenate, rather thick, usually ciliate or pubescent on the midrib beneath and on the short petiole. Inflorescence of terminal racemes, lax (flowers appearing solitary in leaf axils); pedicels ± 2 mm long. Calyx lobes 4, spatulate-lanceolate, ± 3 mm long, obtuse, ciliate. Corolla blue, rotate, 9 mm long. Capsule obcordate, ± 3 mm long, ± 2 mm wide.

UGANDA. Mt Elgon, Sasa Trail, *Wesche* 88!

KENYA. Mt Elgon, 14 May 1948, *Hedberg* 941!; Mt Kenya, Teleki Valley, 4 Aug. 1948, *Hedberg* 1813; North Nyeri District: Mt Kenya, above Meru Mt Kenya Lodge, 17 Jan. 1985, *Townsend* 2233!

DISTR. **U** 3; **K** 3, 4; Ethiopia

HAB. Rocky places and paths in alpine zone of high mountains; 2700–4300 m

USES. None recorded on specimens from our area

CONSERVATION NOTES. Least Concern (LC)

NOTE. This species is not clearly distinct from *V. glandulosa* and may simply represent an alpine variant of the latter.

6. **Veronica glandulosa** *Benth.* in DC., Prodr. 10: 482 (1846); Skan in F.T.A. 4(2): 359 (1906); A.V.P.: 170 (1957); U.K.W.F.: 258, pl. 111 (1994); Fischer, F.A.C. Scrophulariaceae: 202, pl. 84 (1999) & in Fl. Ethiop. & Eritr. 5: 281 (2006) Type: Ethiopia, Demerki, Mt Bachit, *Schimper* II: 1149 (K!, holo.; BM, P!, S, iso.)

Perennial herb with numerous wiry branches 10–45 cm, ascending from base and ultimately erect, rooting at lower nodes, ± glandular-pilose on the younger parts, shortly pubescent in two lines or nearly glabrous on the old stems; nodes 5–16 mm apart. Leaves oblong to ovate, 5–11(–17) mm long, 3–4(–9) mm wide, rounded or cuneate at base into a short petiole, apex obtuse, margins entire. Inflorescence of terminal racemes, lax. Pedicels 2–3 mm long. Calyx 4-lobed (or 5 with a minute posterior segment); lobes oblanceolate, 4 mm long, glandular-pilose. Corolla pale blue, white or pink, 4–5-lobed, 7–9 mm in diameter; lobes ± 5 mm long. Capsule obcordate, ± 4 mm long, ± 3 mm wide, compressed, sparingly glandular-pubescent on upper part, shorter than calyx. Seeds yellowish, angular-ovoid, nearly smooth.

UGANDA. Toro District: Mt Ruwenzori, June 1948, *J. Adamson* 33!; Mt Elgon, Sasa Hut, 16 June 1970, *Lye* 5724! & Mt Elgon, 9 Oct. 1961, *Rose* 10200!

KENYA. Mt Elgon, 26 Sept. 1957, *Irwin* 363!; Aberdare Mts, 15 Oct. 1970, *Mabberley* 350!; Mt Kenya, Teleki valley, 6 Aug. 1948, *Hedberg* 1839!

TANZANIA. Mt Kilimanjaro, near Peters Hut, 16 June 1948, *Hedberg* 1213! & near Peters Hut, 23 Feb. 1934, *Greenway* 3766!; Mt Meru, June 1968, *King* 2819

DISTR. **U** 2, 3; **K** 3, 4; **T** 2; Rwanda, Ethiopia

HAB. Borders of montane forest and in alpine zone; 2000–4100 m

USES. None recorded on specimens from our area

CONSERVATION NOTES. Least Concern (LC)

SYN. *V. myrsinoides* Oliv. in Hook. Ic. Pl.: 16, t. 1509 (1886) & Trans. Linn. Soc., Bot. 2:343 (1887); Skan in F.T.A. 4(2): 360 (1906). Type: Tanzania, Kilimanjaro, *Johnston* 144 (K!, holo.; BM!, iso.)

V. keniensis R.E.Fries in Acta Hort. Berg. 8: 56 (1924). Type: Kenya, Mt Kenya, *Fries & Fries* 1278a (UPS!, holo.)

V. aberdarica R.E.Fries in Acta Hort. Berg. 8: 57 (1924). Type: Kenya, Aberdares, *Fries & Fries* 2636 (UPS, holo.; BR, K!, iso.)

V. linnaeoides R.E.Fries in Acta Hort. Berg. 8: 58 (1924). Type: Kenya, Mt Kenya, *Fries & Fries* 1278 (UPS, holo.; BR, K!, S, iso.)

V. battiscombei R.E.Fries in Acta Hort. Berg. 8: 58 (1924). Type: Kenya, Aberdares, *Fries & Fries* 2389 (UPS, holo.; BR, K!, S, iso.)

NOTES. O. Hedberg, who knows this plant very well in the field, considers that the variants described and named as distinct taxa cannot be maintained, and we agree with this conclusion. This extreme variation may be accounted for at least in part by the wide range of high altitude habitats, from moist dark bamboo forests to open frosty alpine slopes.

7. **Veronica persica** *Poir.*, Nov. Encycl. Méth. Bot. 8: 542 (1808); Walters & Webb in Fl. Europ. 3: 250 (1972); Fischer in Fl. Ethiop. & Eritr. 5: 282 (2006). Type: Persia, cult. in Paris (P-Lam., holo.)

Annual herb; stems 10–60 cm long, trailing, rooting at some nodes and internodes, pubescent. Leaves ovate, the lower often wider than long, 5–25 mm long, 3–18 mm wide, upper leaves smaller and narrower, base rounded to cuneate into a short petiole, apex obtuse, margins shallowly crenate. Inflorescence of terminal racemes, lax, flowers few to numerous, appearing solitary in upper axils of foliose bracts; pedicels slender, 15–20 mm long. Calyx segments ovate-lanceolate, 6–7 mm long, not or only slightly overlapping near the base. Corolla deep blue, 9–15 mm in diameter, soon falling; style 2–3 mm long. Capsule obcordate, 4–5 mm long, 7–10 mm wide, the lobes wide and diverging at an angle of about 120°.

KENYA. Naivasha District: Lake Naivasha, 22 Feb. 1967, *S.F. Polhill* 216!

DISTR. **K** 3; a European weed now introduced worldwide

HAB. A weed of cultivation; ± 2000 m

USES. None recorded on specimens from our area.

CONSERVATION NOTES. Even though known only from the above collection in Flora area, the species is weedy and widespread; here assessed as of Least Concern (LC)

29. **THUNBERGIANTHUS**

Engl. in E.J. 23: 509 (1897); F.T.A. 4(2): 439 (1906)

Climbing plants with slender woody stems. Leaves opposite, petiolate. Flowers solitary or paired, axillary. Calyx campanulate, 5-lobed. Corolla slightly bilabiate, 5-lobed; lobes suborbicular. Stamens 4, 2 long, 2 short, inserted on corolla above base; anther bithecal, thecae unequal. Ovary 2-locular; style slender, curved, thickened towards apex. Capsule ovoid, as long as calyx.

Two species, one endemic to São Tomé and the other to central tropical Africa.

Thunbergianthus ruwenzoriensis *R.Good* in J.B. 66: 38 (1928); Troupin, Fl. pl. lign. Rwanda: 648, fig. 223 (1982). Type: Uganda, Ruwenzori Mts, *Humphreys* 505 (BM!, holo.)

Climbing shrub up to 12 m high; stems quadangular, grooved, scabrid with minute reflexed prickles. Leaves opposite, ovate to narrowly or broadly triangular, 4–9 cm long, 2.5–6 cm broad, base truncate to cordate, apex acuminate, margins usually coarsely toothed near the base, glabrous above, aculeate on veins beneath; petioles 1–3 cm long, aculeate. Flowers 1–2 in each distal axil forming a pendent inflorescence; pedicels 3–5 cm long; bracteoles 2, linear, 2–4 cm long. Calyx tube 8 mm long; lobes broadly deltoid, 8 mm long, acute, sparsely aculeate. Corolla pink to mauve, tube ± 3.5 cm long, cylindrical, basal part 5 mm in diameter, widening to throat to about 2.5 cm in diameter; lobes 1.5–2.5 cm long and broad. Stamens 2 + 2; anthers 8 mm long. Ovary glabrous; style up to 6 cm long; stigma globose, 4–5 mm in diameter. Capsule ovoid, ± 1 cm in diameter, dehiscing loculicidally with 4 valves. Fig. 31, p. 107.

UGANDA. Ruwenzori, Bujuku valley, 1950, *Osmaston* 3643! & Mubuku valley, July 1938, *Eggeling* 3805!; Kigezi District: Rutenga, Aug. 1949, *Purseglove* 3075!
DISTR. **U** 2; Congo-Kinshasa, Rwanda, Burundi
HAB. Submontane and montane forest; 1300–2100 m
USES. None recorded on specimens from our area
CONSERVATION NOTES. Least Concern (LC)

SYN. *T. ruwenzoriensis* R.Good forma *macrocalyx* R.Good in J.B. 66: 39 (1928); Type: Uganda, Ruwenzori, *Scott Elliot* 7919 (BM!, holo.; K!, iso.)

30. **MELASMA**

Berg., Desc. Pl. Cap.: 162 (1767)

Erect perennial herbs, scabrid or hispid. Leaves opposite, occasionally alternate within inflorescence, sessile or subsessile. Flowers solitary in axils of leaves or bracts, in lax or congested terminal racemes, pedicellate, bi-bracteolate. Calyx 5-lobed, subequal, campanulate, 10-ribbed, inflated in fruit; lobes erect, valvate. Corolla tubular-campanulate, 5-lobed, larger than calyx; lobes imbricate, widely spreading. Stamens 4, subequal, didynamous, included; anther thecae parallel, distinct, apiculate or acuminate; style clavate, flattened, persistent, exserted from the persistent calyx. Capsule loculicidal, included in calyx. Seeds numerous, small, straight or curved, outer testa transparent, extended at each end beyond seed nucleus, truncate.

Genus of about 20 species, from tropical Africa and America.

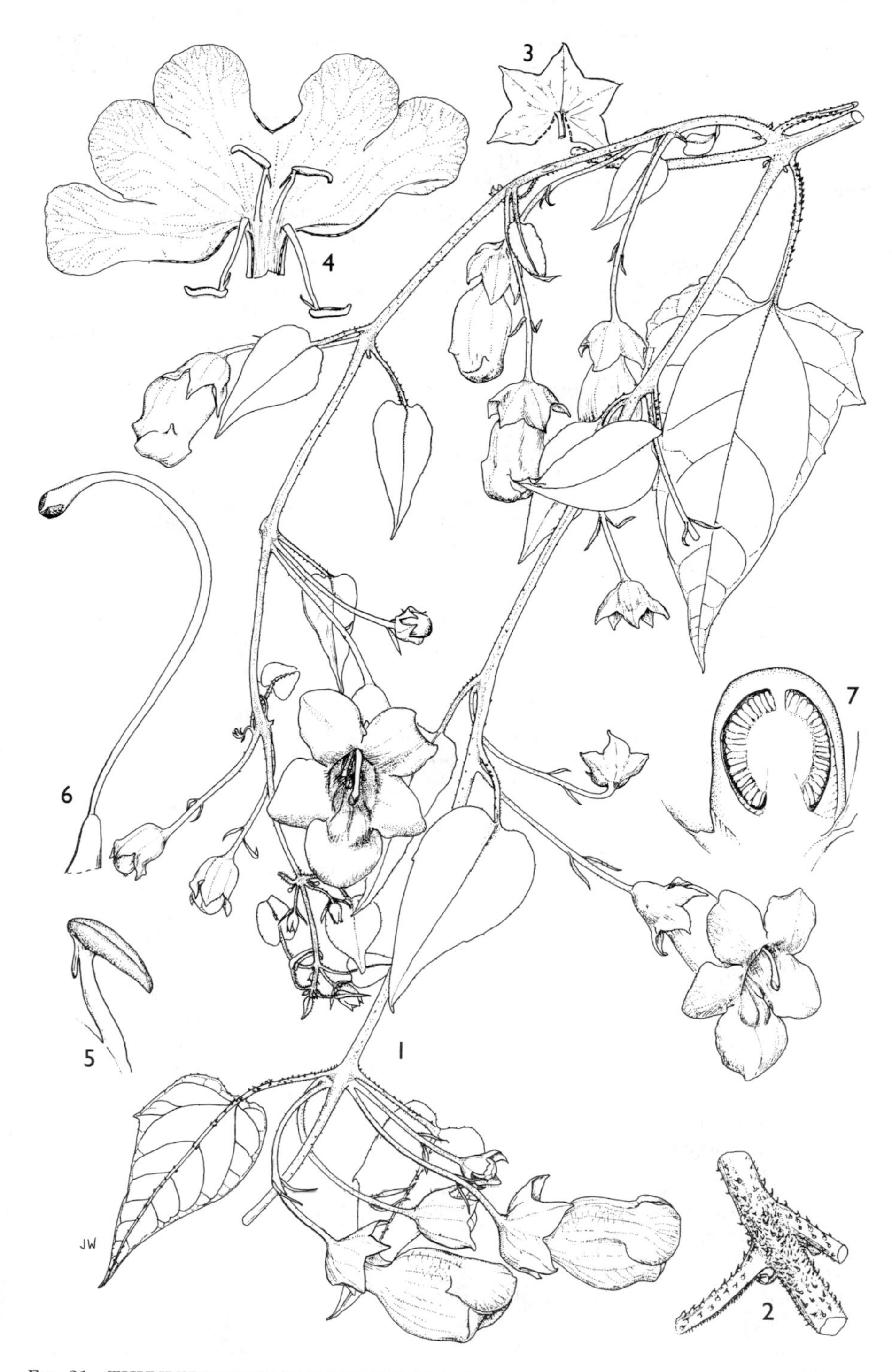

FIG. 31. *THUNBERGIANTHUS RUWENZORIENSIS* — **1**, habit × $^2/_3$; **2**, details of part of stem × 3; **3**, calyx opened × $^2/_3$; **4**, corolla opened × $^2/_3$; **5**, stamen × $^1/_2$; **6,** ovary and style × $1^1/_2$; **7,** LS ovary × 6. 1, 3, 4, 6 from *Purseglove* P2657 sheet 1& 2; 2 from *Eggeling* 3805; 5 from *Scott Elliot* 7919; 7 from *Bridson* 385. Drawn by Juliet Williamson.

Melasma calycinum (*Hiern*) *Hemsl.* in F.T.A. 4(2): 362 (1906); Philcox in F.Z.: 8(2): 84, t. 31/A (1990). Type: Angola, Huilla, *Welwitsch* 1263 (LISC, holo.; BM, K!, iso.)

Erect perennial herb, up to 60 cm, blackening on drying; stems several arising from a woody base, ± branched above, scabrid to pubescent thoughout; roots orange. Leaves oblong-ovate to ovate-lanceolate, 20–45(–84) mm long, and 6–20(–35) mm wide, base narrowing into a short petiole or leaves sessile, apex obtuse to emarginate, margins entire to irregularly and shallowly dentate to crenate, strongly 3-veined from the base, scabrid on both surfaces. Flowers in lax terminal racemes; bracts lanceolate, 3–4 mm; bracteoles linear-lanceolate, 3–4 mm, arising 2–3 mm below the calyx; pedicels elongating in fruit to 50 mm. Calyx campanulate, 9–15(–25) mm long, 10-veined; lobes triangular, acute, ± 3 mm, hispid-scabrid outside and inside. Corolla yellow to orange or white, red- or purple-veined, 20–30 mm, ± glabrous to shortly glandular-pubescent on the outer surface; corolla lobes rounded, glabrous to minutely pubescent. Capsule ovoid, 9–15 mm long. Fig. 32, p. 109.

TANZANIA. Iringa District: near Ifunda, Nov. 1928, *Haarer* 1601!; Ufipa District: Chapota, Dec. 1949, *Bullock* 2059!; Songea District: 12 km E of Songea by Nonganonga stream, Jan. 1956, *Milne-Redhead & Taylor* 8296!

DISTR. **T** 4, 7, 8; Angola, Zambia, Malawi, Zimbabwe

HAB. Swamps and marshy places in grassland and secondary *Brachystegia* woodland; 950–2000 m

USES. None recorded on specimens from our area

CONSERVATION NOTES. Least Concern (LC); common

SYN. *Velvitsia calycina* Hiern, Cat. Afr. Pl. Welw. 1: 771 (1898)
Melasma nyassense Melch. in N.G.B.G. 15: 124 (1940). Type: Tanzania, Songea District: Matenga, *Zerny* s.n. (W, holo.; EA!, photo)

NOTE. A collection from a woodland in Tanzania: Kakombe Valley, Gombe Stream Reserve, *Pirozynski* 229, shows lax racemes and large calyces (± 25 mm); similar specimens but with terminal infloresceness are known from Zambia and Angola and have been identified as *M. calycinum*. These are possibly variants of the typical *M. calycinum*.

31. **ALECTRA**

Thunb., Nov. Gen. Pl.: 81 (1784); Gen. Pl. 2: 966 (1876)

Annual or perennial herbs, with a depleted root system, parasitic on roots of grasses and other herbs; plants turning black on drying; stems simple or branched. Leaves opposite, subopposite or alternate, reduced to scales in some species, sessile or shortly petiolate. Inflorescence spicate; flowers solitary in axils of upper leaves or leaf-like bracts. Calyx campanulate, 5-lobed, 5–10-veined. Corolla campanulate, slightly longer than the calyx, longitudinally veined, soon withering, persistent in calyx adhering to ovary and remaining over the capsule. Stamens 4, didynamous, inserted below the middle of the corolla, included; filaments all bearded or only the two longer bearded, or all glabrous; anthers coherent or connivent in pairs. Ovary glabrous; style curved, clavate in the upper half. Capsule loculicidal. Seeds numerous, linear or clavate.

About 30 species distributed from tropical Africa to South Africa and Madagascar, tropical Arabia, N India to S China and Thailand, Philippines, Mesoamerica and Caribbean to Colombia and Guyana, and Brazil.

1. Leaves absent; plant entirely white .... 10. *A. alba*
   Leaves present; plant green 2
2. All filaments glabrous or nearly so 3
   All or the lower two filaments bearded 6
3. Anthers apiculate with or without a few hairs present just below the anther 4
   Anthers not apiculate 5

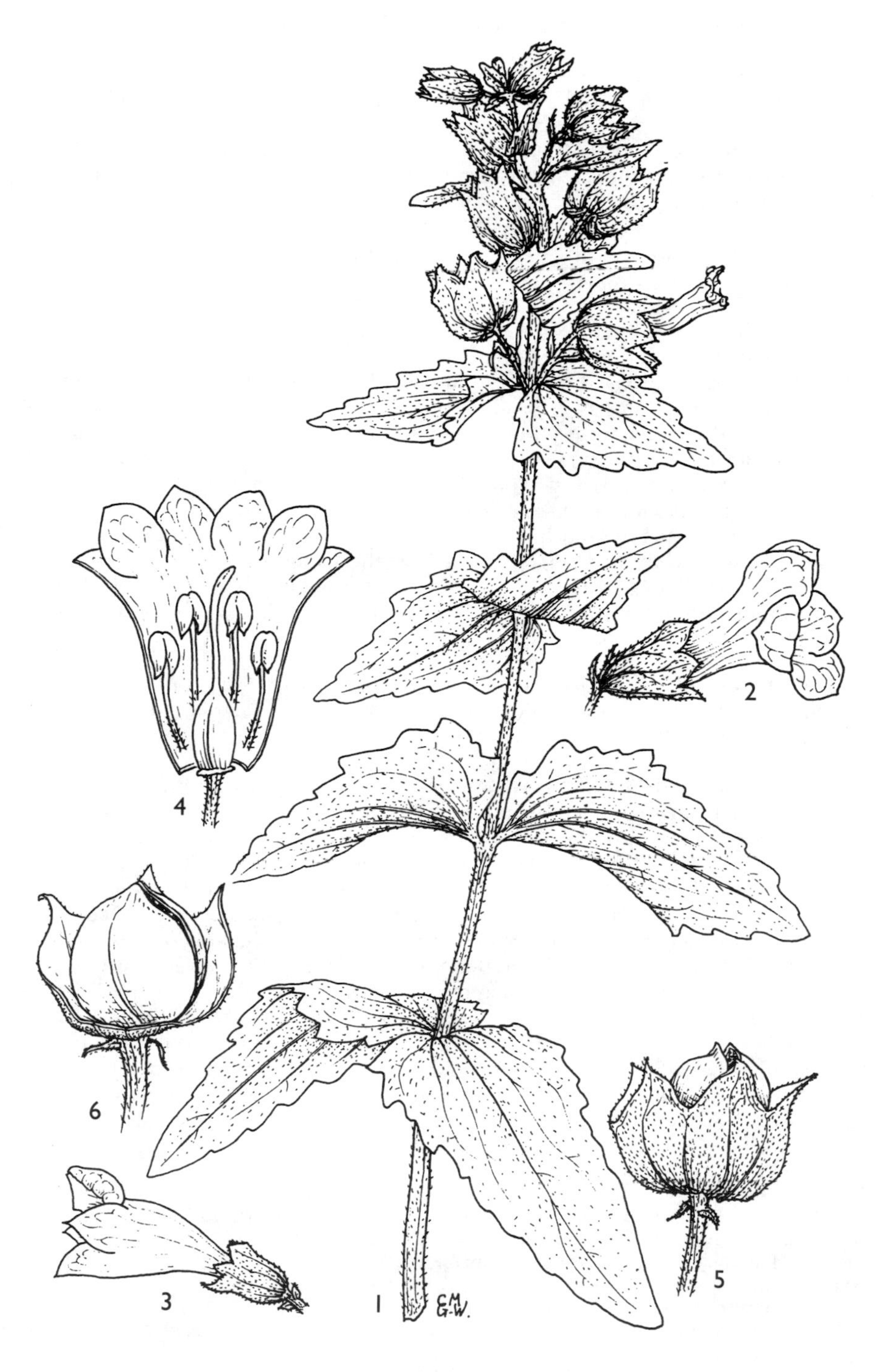

Fig. 32. *MELASMA CALYCINUM* — **1**, habit × $^{2}/_{3}$; **2, 3**, flower × $^{1}/_{3}$; **4**, flower opened × $^{1}/_{3}$; **5**, capsule × 1; **6**, capsule with part of calyx removed × 4. 1 from *Bramley* 2568; 2 from *Richards* 7002; 3, 4 from *Lawton* 1226; 5, 6 from *Richards* 27445A. Drawn by Christine Grey-Wilson. Reproduced with permission from F.Z.

4. Annual herbs; leaves ovate to ovate-lanceolate, (2–)3–11 (–15) mm wide, 3–5-veined; petiole 0.5–2 mm .... 5. *A. asperrima*
   Perennial herbs; leaves sessile, oblong-lanceolate to linear-lanceolate, 2–3.5 mm wide, 1-veined .... 1. *A. rigida*
5. Leaves well developed, not fleshy or thick, 3–5-veined; hairs without bulbous bases; calyx 5-veined; filaments with a fews hairs present underneath the anther ...... 6. *A. vogelii*
   Leaves thick, fleshy and/or scale-like, usually adpressed to the stem, not well developed; calyx veins obscure; filaments completely glabrous .... 8. *A. parasitica*
6. Anthers apiculate 7
   Anthers not apiculate 9
7. Calyx stipitate-glandular; corolla 2–3× the length of calyx, stipitate-glandular outside .... 2. *A. glandulosa*
   Calyx not glandular; corolla less than 2× the length of calyx, not glandular 8
8. Leaves 3(–5)-veined; upper two filaments glabrous ...... 4. *A. sessiliflora*
   Leaves 1-veined; all filaments bearded .... 3. *A. dolichocalyx*
9. Leaves spreading, lower leaves petiolate; flowers usually in the upper part of stem; calyx lobes with margins not thickened, often densely hirsute .... 7. *A. picta*
   Leaves somewhat reduced, embracing the stem or spreading; flowers ± throughout the stem; calyx lobes with margins often thickened, scabrid .... 9. *A. orobanchoides*

1. **Alectra rigida** (*Hiern*) *Hemsl.* in F.T.A. 4(2): 368 (1906); Philcox in F.Z.: 8(2): 85 (1990); Fischer in Fl. Ethiop. & Eritr. 5: 287 (2006). Type: Angola, *Welwitsch* 5797 (LISC, holo.; BM, K!, iso.)

Perennial herb 20–70(–85) cm high; stems erect, simple or branched about midway, terete to angular, purplish, pilose to hispid with retrorse hairs; roots orange. Leaves sessile, erect, opposite, oblong-lanceolate to linear-lanceolate, 10–30 mm long, 2–3.5 mm wide, 1-veined, hispid on both surfaces with usually bulbous-based coarse hairs on thickened margins, fine retrorsely hairy otherwise, margins entire with 2–3 obscure teeth on each side, apex rounded. Inflorescence of terminal spike-like racemes, sometimes lax; flowers solitary in the axils of leaf-like bracts; pedicel 0–0.5 mm. Bracts oblong-ovate to ovate, ± 10 mm long, with 1–3-veins prominent, entire, margins with tubercle-based retrorse hairs, retrorse hairs on the veins, ± glabrous otherwise; bracteoles ovate, 5–18 mm long, acute or obtuse. Calyx 7–8 mm long, 10-veined, ± glabrous to finely and sparsely pubescent with ± retrorse hairs on the margins and veins; lobes 2.5–3 mm, ovate, acute. Corolla pale yellow with purplish veins, about twice the length of the calyx, lobed above with rounded, ± equal lobes. Stamens unequal; filaments glabrous; anthers apiculate. Capsule oblong-ovoid, ± 5 mm long, ± 3 mm in diameter, glabrous.

TANZANIA. Kigoma District: Lubalisi village to Ntakatta, 8 June 2000, *Bidgood, Leliyo & Vollesen* 4620!; Ufipa District: Tatanda, 26 April 1997, *Bidgood et al.* 3497!; Songea District: W of Songea, by Kimarampaka stream, *Milne-Redhead & Taylor* 9844A!

DISTR. **T** 4, 8; Angola, Zambia, Malawi, Zimbabwe

HAB. Wet and boggy grassland; 950–1650 m

USES. None recorded on specimens from our area

CONSERVATION NOTES. Least Concern (LC) due to its wide distribution range, however known in the Flora area from three collections only

SYN. *Melasma rigida* Hiern, Cat. Afr. Pl. Welw. 1: 767 (1898)

2. **Alectra glandulosa** *Philcox* in Bol. Soc. Brot., sér. 2, 60: 269 (1987) & in F.Z.: 8(2): 86, t. 32 (1990). Type: Zambia, 42 km from Mwinilunga on road to Solwezi, Mundwizi Dambo, *Philcox, Pope, Chisumpa & Ngoma* 10350 (K!, holo.; BR, LISC, MO, NDO, SRGH, iso.)

Annual herb 9–23 cm high; stems erect to decumbent, branched and spreading, leafy, terete to angular, ± with furrows, retrorse pilose to hispid. Leaves all opposite, sessile, lanceolate-oblong to lanceolate to linear-lanceolate, 10–45 mm long, 1.5–5 mm wide, apex acute or obtuse, base cuneate, margins with 1–3(–5) teeth on each side, usually 1-veined, but sometimes with 2 obscure lateral veins, retrorse hispid-pilose, short hispid to scabrid on margins and midrib beneath, glabrous above. Flowers solitary in the axils of leaf-like bracts; bracts similar to leaves, reducing in size upwards; bracteoles linear, 5–6 mm, acute, hispid; pedicels 1–3(–4) mm long, slender. Calyx 6–8 mm long, campanulate, prominently 10-veined, stipitate-glandular and hispid, especially on veins; lobes subequal, broadly triangular, 1–2 mm, obtuse, margins hispid. Corolla sulphur yellow with purple to reddish veins, 12–24 mm long, lobed above with rounded lobes, stipitate-glandular outside. Stamens ± unequal; filaments bearded; anthers apiculate. Capsule ovoid, 8–10 mm long, ± 7 mm in diameter, glabrous.

TANZANIA. Mpanda District: Mpanda–Uvinza road, Uzondo Plateau, 29 May 2000, *Bidgood, Leliyo & Vollesen* 4508!; Ufipa District: M'Wimbe dambo, 21 Apr. 1962, *Richards* 16345!; Singida District: 38 km from Sekenke road, 1 May 1962, *Polhill & Paulo* 2266!

DISTR. **T** 4, 5; Zambia

HAB. Seasonally wet or flooded grassland, seepages; 1450–1700 m

USES. None recorded on specimens from our area

CONSERVATION NOTES. Known in the Flora area from a few collections from Tanzania; here assessed as Least Concern (LC) based on its distribution in Zambia

3. **Alectra dolichocalyx** *Philcox* in Bol. Soc. Brot., ser 2, 60: 269 (1987) & in F.Z.: 8(2): 86, t. 33, (1990). Type: Zambia, Kabulamwanda, 120 km N of Choma, *Robinson* 1238 (K!, holo.; SRGH, iso.)

Annual herb (5–)8–30 cm high; stems purplish, erect to ascending to ± prostrate, simple or usually branched, pilose. Leaves opposite, sessile to very shortly petiolate, linear to linear-lanceolate, 15–60 mm long, 2–5 mm wide, apex acute to rounded, base subcordate, margins dentate with 3–7 teeth on each side of leaf, 1-veined, glabrous above, pilose scabrid on veins beneath and thickened margins. Flowers solitary in the axils of leaves or leaf-like bracts; bracts similar to leaves, as long as or longer than calyx; bracteoles linear, 10–12 mm, 1–1.5 mm wide, pilose on margins; pedicels 1.5–2.5 mm, elongating to 4 mm in fruit. Calyx 10–20 mm long, 10-veined, long pilose on veins and margins of lobes; lobes narrow triangular, 5–10 mm, acute. Corolla pale yellow, 15–22 mm, lobed above with rounded, equal lobes. Stamens unequal; filaments bearded; anthers apiculate. Capsule ovoid, 8–11 mm long, 7–8 mm in diameter, somewhat compressed, glabrous.

UGANDA. Mengo District: Wabusana–Kalungi, July 1956, *Langdale-Brown* 2229!

KENYA. N Kavirondo District: Webuye Falls [Broderick Falls], Aug. 1965, *Tweedie* 3074! & near Bungoma, Sept. 1968, *Tweedie* 3584!

TANZANIA. Mpanda District: 11 km on Karema road from Mpanda–Uvinza road, May 1997, *Bidgood et al.* 3903!; Ulanga District: about 3 km from Mlahi, May 1977, *Vollesen* in MRC 4598!

DISTR. **U** 4; **K** 5; **T** 4, 6; Zambia

HAB. Wet grassland, seepages, seasonally flooded grassland, edges of water holes; 250–1650 m

USES. None recorded on specimens from our area

CONSERVATION NOTES. Least Concern (LC)

4. **Alectra sessiliflora** (*Vahl*) *Kuntze*, Rev. Gen. Pl. 2: 458 (1891); Hepper in K.B. 14: 405 (1960); Philcox in F.Z.: 8(2): 86 (1990); U.K.W.F.: 258, pl. 111 (1994); Fischer in Fl. Ethiop. & Eritr. 5: 286 (2006). Type: South Africa, Cape of Good Hope, *Bulow* s.n., type not designated

Annual herb 13–38 cm high; stems erect, simple or sparingly branched above, occasionally with numerous lateral branches, terete or quadrangular with furrows, variously pilose to hispid with patent or retrorse hairs. Leaves sessile to shortly petiolate, oblong-lanceolate to lanceolate or ovate to broadly ovate to suborbicular, 12–32 mm long, 6–12(–18) mm wide, lower leaves rounded at the base and apex, upper cuneate at base and acute at apex, margins coarsely acutely toothed, 3(–5)-veined from the base, hispid above and on veins beneath. Flowers solitary in leaf axils forming congested inflorescense; bracts leaf-like, coarsely toothed, hispid to subglabrous; pedicels 0–0.5 mm. Calyx subequal, 6–8 mm long, lobes 3–5 mm, acutely acuminate, ciliate on the veins and margins, 10-veined. Corolla pale yellow with purplish veins, 8–10 mm, soon withering and persisting. Stamens unequal; upper 2 filaments glabrous, lower 2 bearded; anthers apiculate. Capsule globose, ± 5 mm in diameter, glabrous. Fig. 33, p. 113.

UGANDA. West Nile District: near Omugo rest camp, Aug. 1953, *Chancellor* 139!; Teso District: Foot of Kyeri Rock, Oct. 1932, *Chandler* 970!; Mengo District: Mawokota, Mar. 1905, *E. Brown* 191!

KENYA. Mt Elgon, Dec. 1967, *Mwangangi* 411!; Kericho District: Koiwa, Oct. 1940, *Bally* 1197!; Teita District: Mbololo forest, May 1985, *Taita Hills Expedition* 356!

TANZANIA. Lushoto District: Chambogo Forest Reserve, 6 July 1987, *Kisena* 555!; Mbeya District: Mporoto ridge, 2 June 1992, *Gereau, Mwasumbi & Kayombo* 4516!; Njombe District: Poroto Mts, Kitulo Plateau, Ndumbi Valley, 24 Mar. 1991, *Bidgood et al.* 2139!

DISTR. **U** 1–4; **K** 3–7; **T** 1–4, 6–8; West Africa from Senegal to Sudan and Ethiopia and south to Angola and Botswana; S Asia

HAB. Montane and submontane grassland, rocky grassy slopes, riverine and secondary forest, forest margins, swampy and moist places; 0–2750 m

USES. None recorded on specimens from our area

CONSERVATION NOTES. Least Concern (LC); widespread

SYN. *Gerardia sessiliflora* Vahl, Symb. Bot. 3: 79 (1794)

A. senegalensis Benth. in DC., Prodr. 10: 339 (1846). Type: Senegal, *Leprieur* '9 & 10' (K!, holo.)

Melasma indicum Wettst. var. *monticolum* Engl. in E.J. 30: 402 (1901). Type: Tanzania, Njombe District: Luhigi R., *Goetze* 807, 1137 (BM!, syn.)

Alectra communis Hemsl. in F.T.A. 4(2): 372 (1906). Type: Malawi, locality not known, *Buchanan* 520 (K! lecto.; BM, iso.)

A. sesiliflora Vahl var. *monticola* (Engl.) Melch. in N.B.G.B. 15: 126 (1940), & 438 (1941); Hepper in K.B. 14: 406 (1960); Philcox in F.Z. 8 (2): 86 (1990)

A. sessiliflora Vahl var. *senegalensis* (Benth.) Hepper in K.B. 14: 406 (1960); Philcox in F.Z. 8(2): 86 (1990)

NOTE. An extremely variable species with an Africa-wide distribution. The species has been variously separated into varieties based on leaf-shape and pubescence. I (S.A.G.) have seen a lot of material both from W and E Africa and have found it very difficult to separate the varieties as recognized by Hepper (op. cit.) and Philcox (op. cit.). The type variety, recognized on the pubescence of the calyx and on the filaments being either all glabrous or only the longer two bearded is restricted to Zambia, Malawi, Zimbabwe, S Africa and Madagascar.

5. **Alectra asperrima** *Benth.* in DC., Prodr. 10: 340 (1846); Hemsl. in F.T.A. 4(2): 369 (1906); Philcox in F.Z.: 8(2): 90 (1990); Cuccuini & Nepi in Thulin (ed.), Fl. Somal.: 3: 284 (2006); Fischer in Fl. Ethiop. & Eritr. 5: 285 (2006). Type: Ethiopia, Mt Ambazion near Geraz, *Schimper* 1094 (K!, holo.; K!, P!, iso.)

Annual herb (9–)15–45 cm high; stems purplish, erect, simple or with several branches arising about midway, generally terete with furrows, pilose to hispid with fine and bulbous-based retrorse hairs; roots orange. Leaves opposite, ovate to ovate-lanceolate, (5–)10–15(–25) mm long, (2–)3–11(–15) mm wide, base rounded or

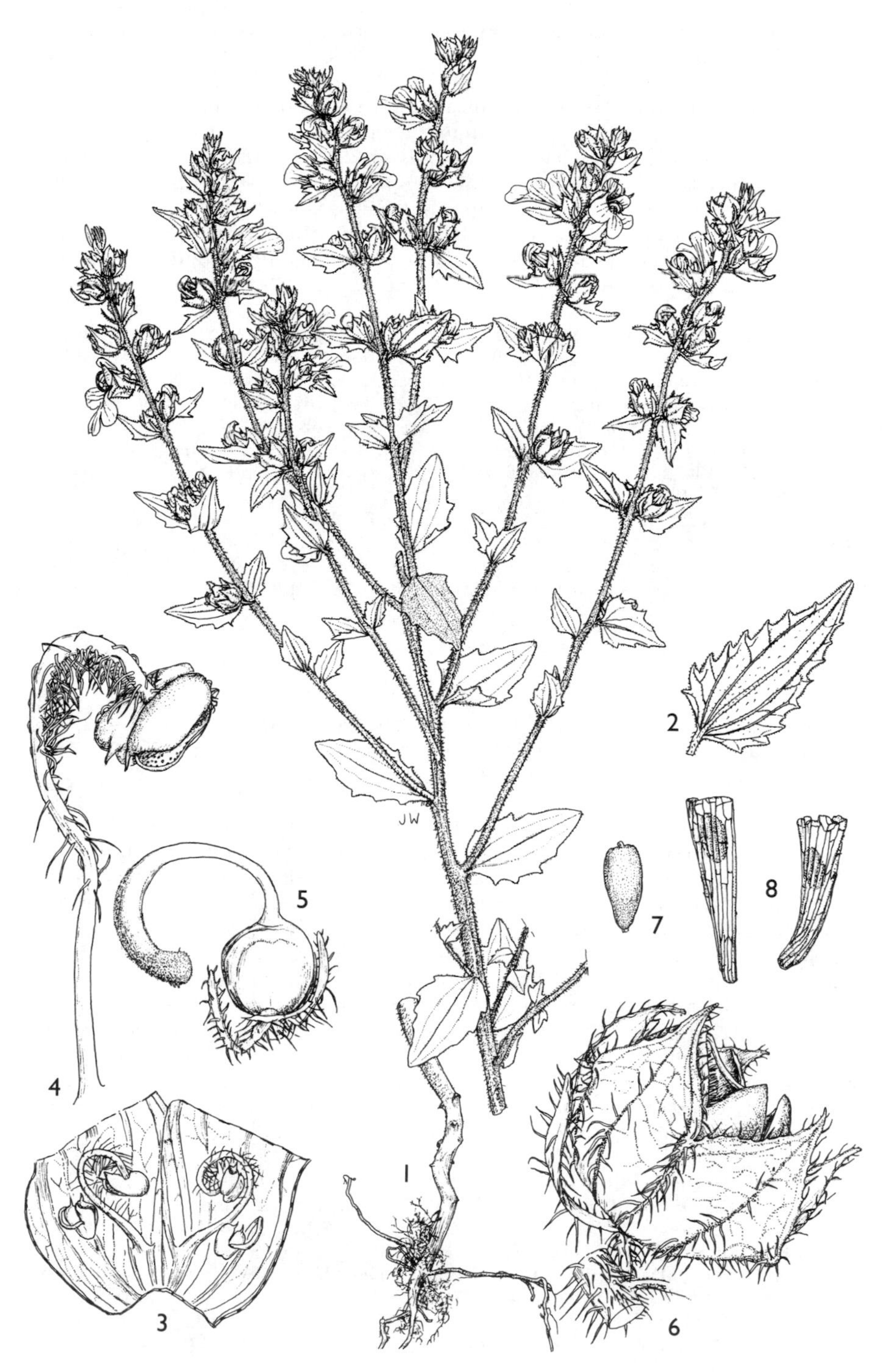

FIG. 33. *ALECTRA SESSILIFLORA* — **1**, habit × ²⁄₃; **2**, detail of leaf × 1; **3**, corolla opened × 4; **4**, stamen × 12; **5**, carpel × 4; **6**, capsule × 4; **7**, seed × 24; **8**, seed × 40. 1, 3, 4, 5 from *Verdcourt, Polhill & Lucas* 3029; 2 from *Whyte* s.n. dated Nov. 1889; 7, 8 from *Greenway* 6520. Drawn by Juliet Williamson.

caudate, apex rounded to acute, margins entire with 2–3 coarse teeth on each side, thick, with bulbous-based coarse hairs, 3–5-veined, hispid to ± glabrous on surfaces, hairs coarse or fine, ± retrorse; petiole 0.5–2 mm. Inflorescence of terminal racemes, generally compact; flowers solitary in the axils of leaf-like bracts; bracts oblong-ovate to ovate, 10–12 mm long, with 1–3 prominent veins, entire or with 1–3 teeth, teeth sometimes acuminate, margins with bulbous-based hairs; bracteoles ovate, acute or obtuse; pedicels 0.5–2 mm. Calyx 6–8 mm long, 10-veined, ± glabrous to pubescent with ± bulbous-based hairs on the margins and veins; lobes ovate, 2.5–3 mm, acute. Corolla yellow with purplish veins, 12–16 mm long, lobed above with rounded, ± equal lobes. Stamens unequal; filaments glabrous; anthers shortly apiculate. Capsule ovoid, 5–6 mm long, ± 3 mm in diameter, glabrous.

UGANDA. Karamoja District: Mt Debasien, Jan. 1936, *Eggeling* 2692!; Busoga District: Busembatia–Iganga road, Busenbatia Busoga Plantation, 17 Dec. 1952, *G.H.S. Wood* 560!; Mengo District: Kyadondo, Bukoto Valley, 24 March 1991, *Rwaburindore* 3196!

KENYA. W Suk District: Sondang Hills, Jan. 1935, *Thorold* 2761!; Mt Elgon, Oct./Nov. 1930, *Major Lugard* s.n.!; Nyeri District: 10 km E of Nanyuki, 22 Feb. 1969, *Magor* 22!

TANZANIA. E Kilimanjaro, May 1927, *Haarer* 650!; Iringa District: Mt Image, N of Morogoro road, 2 March 1962, *Polhill & Paulo* 1641!; Songea District: ± 2.5 km E of Johannesbruck, 20 April 1956, *Milne-Redhead & Taylor* 9784!

DISTR. **U** 1, 3, 4; **K** 2–5; **T** 2, 3, 7, 8; Ethiopia, Somalia, Zambia, Malawi, Zimbabwe

HAB. Wet and upland grassland; 1200–2900 m

USES. None recorded on specimens from our area

CONSERVATION NOTES. Least Concern (LC); common and widespread

SYN. *Glossostylis asperrima* Hochst. in sched. (Ethiopia, Mt Ambazion near Geraz, Schimper Iter Abyss. Sectio secunda No. 1094 (1842–43) *nom. nud.*

6. **Alectra vogelii** *Benth.* in DC., Prodr. 10: 339 (1846); Hemsl. in F.T.A. 4(2): 368 (1906); Philcox in F.Z.: 8(2): 91 (1990); Fischer in Fl. Ethiop. & Eritr. 5: 287 (2006). Type: Nigeria, Mount Patteh [Quorro], *Vogel* 186 (K!, holo. & iso.)

Annual herb (8–)20–50 cm high; stems erect, simple or with several branches arising from the base or above, terete to angular, with furrows, pilose to hispid; base of stem and roots orange-yellow. Leaves opposite, ovate-lanceolate to lanceolate to linear-lanceolate, 7–25 mm long, 3–8 mm wide, base cuneate, apex acute, entire in the apical part, and with 3–4 coarse or obscure teeth on each side, 3–5-veined, hispid; petiole ± 2 mm. Inflorescence of terminal racemes, generally compact; flowers solitary in the axils of leaf-like bracts; bracts linear to linear-lanceolate, 9–13 mm long, entire or with 1–3 blunt teeth, with 1–3 prominent veins; bracteoles linear, 5–6 mm, acute, hispid; pedicels 1–2 mm. Calyx campanulate, 5–6 mm long, 5-veined, veins not prominent, hispid; lobes ovate, 2–3 mm, acute, ciliate. Corolla yellow with purple veins, 10–12 mm long, lobed above with rounded lobes. Stamens ± unequal; filaments glabrous with a few hairs present right below the anther; anthers not apiculate. Capsule ovoid, 5–6 mm long, ± 5 mm in diameter, glabrous.

UGANDA. Teso District: Serere Farm, 5 Sept. 1948, *Jameson* 58! & Dec. 1931, *Chandler* 202!

KENYA. Machakos District: Kapiti Plains, slopes of Mwami Hill, 22 June 1957, *Bally* 11518!; Embu District: Embu, 10 Oct. 1928, *J. McDonald* 1272!; S Kavirondo District: ICIPE Mbita Point Field St., 30 Nov. 1981, *Gachathi & Opon* 98/81!

TANZANIA. Shinyanga District: Shinyanga, Nov. 1938, *Koritschoner* 2162!; Tanga District: 6 km E of Korogwe, 20 July 1953, *Drummond & Hemsley* 3404!; Dodoma District: Manyoni, 24 Apr. 1962, *Polhill & Paulo* 2156!

DISTR. **U** 3; **K** 3–5, 6; **T** 1–8; Guinea, Ghana, Nigeria, Congo-Brazzaville, Zambia, Malawi, Mozambique, Zimbabwe, Botswana

HAB. Mostly in cultivated land, parasitic on legumes, also grassy river banks and grassland; 300–2500 m

USES. None recorded on specimens from our area

CONSERVATION NOTES. Least Concern (LC); common and widespread

7. **Alectra picta** (*Hiern*) *Hemsl.* in F.T.A. 4(2): 368 (1906); Philcox in F.Z.: 8(2): 91 (1990). Type: Angola, Huilla, *Welwitsch* 5799 (LISC, holo.; BM, K!, iso.)

Annual herb 10–35 cm high; stems erect, simple or with several branches arising from the base or above, angular, with furrows, densely pilose to hispid with patent hairs. Leaves or leaf-like bracts opposite to alternate, spreading, ovate to lanceolate to linear-lanceolate, 5–10 mm long, 2–5 mm wide, sessile or leaf base narrowing into a short petiole, apex acute to subacute, margins entire in the apical half, and with 1–2 coarse or obscure teeth on each side, pilose to hispid on both surfaces, 3-veined. Inflorescence of terminal racemes, flowers solitary in the axils of leaf-like bracts in upper part of the stems; bracts linear to linear-lanceolate, 5–8 mm long, with 1–3 prominent veins, usually entire; bracteoles linear, ± 3 mm, acute, hispid. Pedicel 2–2.5 mm, hispid-pilose. Calyx campanulate, 6–7 mm long, 10-veined, hispid; lobes triangular, 3–3.5 mm, acute, pilose-ciliate. Corolla yellow to deep yellow, 12–14 mm, lobed above with subequal rounded lobes. Stamens ± unequal; filaments bearded; anthers not apiculate. Capsule subglobose, 5–7 mm long and wide, glabrous.

KENYA. Kwale District: Mwachi Forest Reserve, 17 May 1990, *Robertson & Luke* 6213!; Kilifi District: Kacharoroni, 17 Aug. 1988, *Luke* 1332!

TANZANIA. Kilosa District: Ruaha valley, 15 March 1986, *Bidgood & Lovett* 264!; Iringa District: Mlangali, 1 May 1989, *Kayombo* 450!; Mikindani District: Ruvuma R., near waterfall, 30 Mar. 1986, *De Leyser* 118!

DISTR. **K** 7; **T** 6, 7, 8; Angola, Zambia, Malawi, Mozambique, Zimbabwe, Botswana

HAB. Riverine *Brachystegia* woodland, and amongst shrubs and sedges at edges of river; 30–1000 m

USES. None recorded on specimens from our area

CONSERVATION NOTES. Least Concern, (LC); recorded only from a few specimens from the Flora region but not at threat as far as is known; elsewhere common

SYN. *Melasma picta* Hiern, Cat. Afr. Pl. Welw. 3: 770 (1898)

8. **Alectra parasitica** *A.Rich.*, Tent. Fl. Abyss. 2: 117 (1850); Hemsl. & Skan in F.T.A. 4(2): 366 (1906); Philcox in F.Z.: 8(2): 92 (1990); U.K.W.F.: 258, pl. 111 (1994); Cuccuini & Nepi in Thulin (ed.), Fl. Somal.: 3: 284 (2006); Fischer in Fl. Ethiop. & Eritr. 5: 288 (2006). Type: Ethiopia, Tacazze Valley, *Schimper* 1464 (P!, holo.; K!, iso.)

Annual herb 6–30 cm high; stems erect, slender, simple or with several branches arising from near the base, puberulous; roots orange-yellow. Leaves opposite to alternate, few and scale-like, thick, embracing the stem, sessile, ovate-lanceolate to linear-lanceolate, 4–10 mm long, 2–5 mm wide, apex acute to obtuse, base shortly cuneate, margins entire, veins obscure in dried specimens, puberulous to hispid on both surfaces. Inflorescence of terminal racemes; flowers solitary in the axils of leaf-like bracts in upper half of stems; bracts linear-lanceolate, 3–4 mm long, entire; bracteoles linear, ± 2 mm, acute, hispid; pedicels 1.5–2 mm, puberulous. Calyx campanulate, 5–6 mm long, puberulous to hispid, veins obscure in dried specimens; lobes triangular, 3.5–4 mm, acute, margins often thickened, scabrid. Corolla orange with crimson red to purplish veins, 10–15 mm, lobed above with subequal rounded lobes. Stamens ± unequal; filaments glabrous; anthers not apiculate. Capsule subglobose, 4.5–6 mm long and wide, glabrous.

UGANDA. Karamoja District: Mt Moroto, 4 Sept. 1956, *Bally* 10714!

KENYA. Northern Frontier District: Mt Kulal below Gatab lower airstrip, 30 Nov. 1978, *Hepper & Jaeger* 7156!; Elgeyo District: Cherangani Hills, Embotut R., near Tot, 29 Aug. 1969, *Mabberley & McCall* 276!; Masai District: Ngong Hills, Magadi road, 18 June 1939, *Bally* B2!

TANZANIA. Arusha District: Arusha, Engare Olmotoni Plain, 2 Feb. 1970, *Richards* 25326!; Singida District: 9 km from Makiungu along Makiungu–Misughaa road, 29 April 2002, *Kuchar* 24493!; Mbeya District: Unyamwanga, 5 Apr. 1932, *Davies* 167!

DISTR. **U** 1; **K** 1–3, 6; **T** 2, 3, 5, 7; Ethiopia, Somalia, Zambia, Malawi; SW Arabia

HAB. Grassland, *Brachystegia* woodland, *Acacia* bushland, swamps; 1200–1500 m

USES. None recorded on specimens from our area
CONSERVATION NOTES. Least Concern (LC); common and widespread

9. **Alectra orobanchoides** *Benth.* in DC., Prodr. 10: 340 (1846); Philcox in F.Z.: 8(2): 92 (1990). Type: South Africa, near Durban, *Drège* s.n. (K!, holo.)

Annual herb 8–45 cm high; stems often dark crimson to purple-red, erect, simple, but with several branches arising from near the base, terete with ridges, puberulous to ± glabrous; roots and basal part of stem orange. Leaves opposite or alternate, sessile, somewhat reduced, embracing the stem or spreading, ovate-lanceolate to lanceolate to oblong-lanceolate, 4–10(–15) mm long, 2–5(–7) mm wide, base shortly cuneate, apex acute to obtuse, margins entire to shortly toothed, puberulous to hispid to hispid-scabrid on both surfaces, the larger leaves with 5 veins, obscure in smaller leaves. Inflorescence of terminal racemes, flowers solitary in the axils of leaf-like bracts, usually in the upper half of stems, but also whole stem floriferous; bracts lanceolate to linear-oblong, 3–6 mm long, entire; bracteoles linear, 2–5 mm, acute, shortly ciliate, ± hispid; pedicels 2.5–3 mm, puberulous. Calyx campanulate, 4–6 mm long, veins clear, puberulous to hispid; lobes purple tinged, triangular, 2–3 mm, obtuse, ciliate. Corolla yellow to orange with red-brown to purple veins, 8–12 mm long, lobed above with subequal rounded lobes, lobes often thickened on the margins, scabrid. Stamens ± unequal; filaments hairy; anthers not apiculate. Capsule ovoid to subglobose, 4.5–6(–9) mm long, 3–6 mm in diameter, glabrous.

UGANDA. Karamoja District: Mt Moroto, 6 Sept. 1956, *Hardy* & *Bally* 10779!; Teso District: Kyeri, Oct. 1932, *Chandler* 974!
KENYA. Northern Frontier District: Moyale, 8 Aug. 1952, *Gillett* 13536!; Mt Elgon, Oct. 1961, *Tweedie* 2230!; Kwale District: Shimba Hills, Risley's Ridge, 21 March 1991, *Luke & Robertson* 2765!
TANZANIA. Kilimanjaro, W end of Lake Chala, July 1968, *Bigger* 2001!; Dodoma District: E of Itigi Station, Apr. 1964, *Greenway & Polhill* 11485!; Lindi District: 12 km on Mnazimoja–Mtwara road, 2 Mar. 1991, *Bidgood, Abdallah & Vollesen* 1753!
DISTR. **U** 1, 3; **K** 1, 3, 7; **T** 2–5, 8; Angola, Zambia, Malawi, Mozambique, Zimbabwe, Botswana, Namibia, South Africa
HAB. Grassland, woodland, montane scrub, rocky outcrops, lowland and fringe forest, river banks, plantations; 50–2000 m
USES. None recorded on specimens from our area
CONSERVATION NOTES. Least Concern (LC); widespread and common

SYN. *Melasma orobanchoides* (Benth.) Engl., P.O.A. C.: 359 (1895)
Alectra kilimandjarica Hemsl. in F.T.A. 4(2): 365 (1906). Syntypes: Tanzania, Lushoto District: Usambara Mts, Amboni, *Holst* 2803! & Lake Chala, *Volkens* 325 (K!, syn.)
Alectra kirkii Hemsl. in F.T.A. 4(2): 366 (1906). Type: Mozambique, Kongone, mouth of Zambesi R., 1 July 1859, *Kirk* s.n. (K!, holo.)

NOTE. A widespread and variable species, with several species recognized by various authors based on Hemsley's treatment in F.T.A. I (S.A.G.) have followed Philcox (in F.Z.) using a broad species concept.

10. **Alectra alba** (*Hepper*) *B.L.Burtt* in Edinb. J. Bot. 56(3): 459–460 (1999). Type: Rwanda, Cyamudongo Forest, *Fischer* 575 (K!, holo.; BR, MJG, iso.)

Parasitic herb, entirely white, drying black; stems simple, aerial portion erect, (5–)10–40 cm high, lower part decumbent with numerous rootlets, glabrous. Leaves absent. Bracts lanceolate, 2–6 mm long, acute. Flowers opposite, in a lax terminal raceme; pedicels 1.5–2 mm. Calyx campanulate, 6–9 mm long, 4–6 mm wide, glabrous or sometimes with a few hairs on the main veins; lobes 2–3 mm long, acuminate. Corolla white; tube 12–13 mm long, lobes unequal, 3–8 mm long. Stamens inserted $^1/_3$ up corolla tube; filaments ± 6 mm long; anthers 2-thecate, all similar, rounded. Stigma linear. Capsule not seen.

UGANDA. Toro District: Itwara Forest, Fort Portal, 29 Jan. 1945, *Greenway & Eggeling* 7054!
KENYA. N Kavirondo District: Kakamega Forest, 23 Feb. 1978, *Cunningham-van Someren* in EA16258!
DISTR. **U** 2; **K** 5; Rwanda, Burundi
HAB. Montane forest floor leaf litter, parasitic on tree roots; 1500–2200 m
USES. None recorded on specimens from our area
CONSERVATION NOTES. So far known only from a few collections, and from the above cited collections in the Flora area, here assessed as Data Deficient (DD)

SYN. *Harveya alba* Hepper in K.B. 47: 729 (1992); U.K.W.F.: 261 (1994); Fischer in Natur und Umwelt Ruandas: 77, 78, fig. 59 (1992)

32. **HARVEYA**

Hook., Ic. Pl.: t. 118 (1837); Hemsley & Skan in F.T.A. 4(2): 435 (1906)

Herbs, saprophytic or parasitic on roots, often drying black; stems erect or ascending, simple, reduced or little-branched. Leaves opposite or alternate, usually at least the lower reduced to scales or absent. Flowers usually large and conspicuous, solitary, axillary, often in terminal spikes or racemes, sessile or pedicellate, bi-bracteolate or ebracteolate. Calyx tubular or tubular-campanulate, 5-dentate or 5-lobed to above the middle. Corolla tube usually narrow at base and more enlarged above and towards throat, curved; lobes 5, erect-spreading, entire or with undulate-crenate margins. Stamens 4, didynamous, usually included or barely exserted; anthers bithecal, thecae parallel or transverse, normally one perfect, acuminate or sharply mucronate or rounded at base, the other empty, longer, subulate-acuminate. Style incurved; stigma clavate or oblong, tongue-shaped. Capsule loculicidal. Seeds numerous, oblong-truncate.

Some 40 species mainly in Southern Africa, with some in eastern Africa with most endemic to Tanzania; one in Madagascar and another in Comoro Islands.

1. Plants pilose with multicellular glandular hairs; stems green, 10–60 cm high; in grassland and woodland 2
   Plants glabrous or sparsely pubescent; stems white (drying black), 2–10(–18) cm long; in forest; [corolla lobes red, white or purple] 4
2. Open flowers 3–4 cm across; corolla tube white, 5–6 cm long; corolla lobes purple; in woodland .... 1. *H. versicolor*
   Open flowers 0.5–2 cm across 3
3. Stems stout, 10–60 cm high; upper bracts leafy, often longer than the flower (calyx + corolla); corolla greenish yellow, lobes pinkish; in grassland .... 2. *H. obtusifolia*
   Stems slender, 10–12 cm high; upper bracts shorter than the calyx; corolla white, lobes rose to dark pink .... 3. *H. huillensis*
4. Corolla lobes white or purple/mauve 5
   Corolla lobes red, 8–15 mm long, corolla tube 4.5–5 cm long 8. *H. tanzanica*
5. Bracts narrowly lanceolate, acuminate; calyx lobes acute to acuminate; corolla white, centre sometimes dark purple 4. *H. buchwaldii*
   Bracts ovate to ovate-oblong, ± acute 6
6. Calyx ± 20 mm long, 5-angled, ciliate on veins; corolla lobes white, yellow inside 7
   Calyx 11–12 mm long, not angled, glabrous; corolla lobes purple .... 7. *H. kenyensis*
7. Calyx lobes acuminate .... 5. *H. liebuschiana*
   Calyx lobes acute .... 6. *H.* sp. *A*

1. **Harveya versicolor** *Engl.*, P.O.A. C.: 362 (1895); Hemsley & Skan in F.T.A. 4(2): 436 (1906). Types: Tanzania, Waly, *Böhm* 97 (B†, syn.); Bukome, *Stuhlmann* 3437, 3425 (B, syn.)

Green semi-parasitic herb, drying purple-black; stem simple, erect, 12–40 cm high, sparsely to densely glandular-pilose. Leaves opposite or sub-opposite, sessile, lanceolate to oblong-lanceolate, lower leaves small, 5–10 mm long, increasing in length upwards to 40 mm long, 8 mm wide, apex obtuse, margins entire, glandular-pilose. Flowers large, solitary, usually in the upper axils, open flowers 3–4 cm across; bracts leafy, nearly as long as calyx; pedicels 5–19(–36) mm long, glandular-pilose. Calyx tubular-campanulate, 15–30 mm long, glandular-pilose; lobes narrowly triangular, up to 10 mm long, margins glandular-pilose, 10-veined. Corolla tube white outside, yellow inside, 50–60 mm long, 4–10 mm in diameter; lobes purple, rounded, 10–20 mm long and wide. Filaments glandular-pilose. Stigma oblong. Capsule and seed not seen.

TANZANIA. Ufipa District: 5 km on Namanyene–Karonga road, 4 Mar. 1994, *Bidgood et al.* 2613!; Mpanda District: Kapapa, Msaginya R., 18 Sept. 1970, *Richards & Arasululu* 25975!; Mbeya District: Usangu, Ilongo, 11 Feb. 1979, *Cribb, Grey-Wilson & Mwasumbi* 11400!

DISTR. **T** 1, 4, 5, 7; not known elsewhere

HAB. Marshy grassland and *Brachystegia* woodland; 950–1650 m

USES. None recorded on specimens from our area

CONSERVATION NOTES. Least Concern (LC); apparently common

2. **Harveya obtusifolia** (*Benth.*) *Vatke* in Abhandl. Natrw. Ver. Bremen 9: 130 (1885); Engl., P.O.A. C.: 362 (1895); Hemsley & Skan in F.T.A. 4(2): 437 (1906); Philcox in F.Z.: 8(2): 95 (1990); U.K.W.F.: 262 (1994); Fischer in Thulin (ed.), Fl. Somal.: 3: 581 (2006). Type: Madagascar, *Lyall* 228 (K!, holo.; K!, P! iso.)

Green-stemmed parasitic herb; stem simple, erect, 10–60 cm high, ± 3.5 mm thick, sparsely glandular-pilose. Leaves few, opposite or subopposite, sessile, lanceolate, lower leaves scale-like or small and increasing in size upwards to 50 mm long and 15 mm wide, apex obtuse, margins entire, glandular-pilose. Flowers solitary in upper axils of leafy bracts; bracts elliptic to elliptic-lanceolate, often longer than calyx and corolla; pedicels 4–11 mm long, glandular-pilose. Calyx tubular-campanulate, 10–20 mm long, ± glandular-pilose; lobes triangular, unequal, 6–8 mm long. Corolla greenish yellow, tube curved, ± 20 mm long, lobes pinkish, suborbicular, 5–10 mm in diameter. Filaments pilose at the apex. Stigma ± exserted. Capsule ovoid, 10 mm long, shortly beaked. Fig. 34, p. 119.

UGANDA. Kigezi District: Kabale, 6 July 1945, *A.S. Thomas* 4309!; Ankole District: Igara, Feb. 1939, *Purseglove* 578!; Mbale District: Budama at Molo, July 1926, *Maitland* 1206!

KENYA. Naivasha District: Longonot Estate, 29 Dec. 1961, *Kerfoot* 3374!; Machachos District: Lukenya Hill, 7 Apr. 1978, *Verdcourt* 5225A!; Kwale District: Lungalunga to Msambweni, 18 Aug. 1953, *Drummond & Hemsley* 3865!

TANZANIA. Musoma District: 5 km W of Eastern Boundary, 5 June 1962, *Greenway & Turner* 10701!; Tanga District: Amboni Caves, 1 Aug. 1953, *Drummond & Hemsley* 3598!; Morogoro District: Mtibwa Forest Reserve, Nov. 1953, *Semsei* 1427!; Zanzibar, 1908, *Last* s.n.!

DISTR. **U** 2–4; **K** 1, 3, 4, 7; **T** 1, 3, 6, 7; **Z**; Ethiopia, Eritrea, Somalia, Zambia, Zimbabwe, Malawi; Madagascar, Yemen

HAB. Among grass in thickets, bushland, fields and orchards; 350–1800 m

USES. None recorded on specimens from our area

CONSERVATION NOTES. Least Concern (LC); widespread and common

SYN. *Gerardia obtusifolia* Benth. in Hook., Comp. Bot. Mag. 1: 211 (1836). Type: Madagascar, *Bojer* s.n. (P, holo.)
Sopubia obtusifolia (Benth.) G.Don, Gen. Syst. 4: 560 (1838)
Aulaga obtusifolia (Benth.) Benth. in DC., Prodr. 10: 523 (1846)

FIG. 34. *HARVEYA OBTUSIFOLIA* — **1**, habit × $^{2}/_{3}$; **2**, flower side view × 1; **3**, flower with opened calyx × $1^{1}/_{2}$; **4**, corolla opened × 2; **5**, **6**, stamens (upper and lower) showing anther detail × 8. All from *Drummond & Hemsley* 3598. Drawn by Juliet Williamson.

Harveya helenae Buscal. & Muschl. in Engl., E.J. 49: 493 (1913); Fischer in Fl. Ethiop. & Eritr. 5: 285 (2006). Type: see note below

NOTE. In the protologue for *H. helenae* the type locality is given as Broken Hill which is in Zambia, but as for all other species described by Buscalioni & Muschler in 1913, this is incorrect. The type is from Galeb, Eritrea (see E.J. 53: 372 (1915)); this also notes that *H. helenae* was removed from the same plant as the type of *H. foliosa, Schweinfurth* 1730 & 1316, which (the later name) was never validly published.

3. **Harveya huillensis** *Hiern*, Cat. Afr. Pl. Welw. 1: 780 (1898): Hemsley & Skan in F.T.A. 4(2): 437 (1906); Philcox in F.Z.: 8(2): 95 t. 34b (1990). Type: Angola, *Welwitsch* 5840 (LISC, holo.; BM!, iso.)

Slender green stemmed semi-parasitic herb; stems simple, 10–12 cm high, densely pilose. Leaves opposite, sessile, ovate-oblong or narrowly obovate, 8–19 mm long, 2–6 mm wide, the lower ones smaller, apex obtuse, margins entire, pilose. Flowers 1–4, solitary in the upper axils, large for the size of the plant; bracts shorter than calyx; pedicels 4–9 mm long. Calyx tubular, 16–25 mm long, glandular-pilose, lobes subequal, ± 10 mm long. Corolla white, tube curved, (15–)25–30 mm; lobes rose or dark pink, rounded, 5–10(–30) mm long, ± 10 mm in diameter. Capsule subspherical, ± 7 mm long.

TANZANIA. Iringa District: Iheme, 31 km S of Iringa, 24 Feb. 1962, *Polhill & Paulo* 1600A!; Mafinga to Madibira, 9 Mar. 1986; *Congdon in Bidgood* 185!; Songea District: 17 km W of Songea, 15 Feb. 1956, *Milne-Redhead & Taylor* 8747A

DISTR. **T** 3 (fide Iversen 1991: The Usambara mountains, NE Tanzania: phytogeography of the vascular plant flora, Uppsala), 6–8; Angola, Zambia, Malawi, Mozambique

HAB. *Brachystegia* woodland; 1400–1800 m

USES. None recorded on specimens from our area

CONSERVATION NOTES. Least Concern (LC); at its northern most limit in the Flora area, apparently common in F.Z. area

SYN. *H. schliebenii* Melchior in N.B.G.B. 11: 683 (1932). Type: Tanzania, Ulanga District: Mbangala, *Schlieben* 1808 (B†, holo.; BM!, P!, iso.)

NOTE. Two collections from Tanzania, Udzungwa Mt Nature Park, Mt Luhomero pt. 132, swamp, 2040 m, *Luke et al.* 6825 & 1440 m, montane forest, *Luke et al.* 6762, come closest to *H. huillensis* but differ in their slender stems reaching up to 60 cm. With more material this may prove to be a distinct taxon.

4. **Harveya buchwaldii** *Engl.*, E.J. 23: 517 (1897); Hemsl. & Skan in F.T.A. 4(2): 438 (1906). Type: Tanzania, Usambara, *Buchwald* 474 (B†, holo.). Neotype: Tanzania, Lushoto District: E Usambaras, Kwamkoro–Kihuhwi, *Greenway* 7919 (K!, chosen here)

White parasitic perennial, drying black; stems clustered, 5–40 cm high, simple, pilose to glabrous. Leaves ovate, reduced to small scale leaves, ± 5 mm long, pilose to glabrous. Flowers in racemes; bracts narrow-lanceolate, 4–12 mm long, 1–2 mm wide, acute to long-acuminate, glabrous. Calyx tubular, 8–35 mm long, 6–14 mm wide, sparsely to densely pilose; lobes triangular, unequal, 3–8 mm long, acute to acuminate. Corolla white, centre sometimes dark purple, 30–45 mm long; lobes rounded, ± 10 mm. Stamen pairs unequal, included, shorter than the style. Stigma clavate. Mature capsule not seen.

TANZANIA. Lushoto District: Kwamsambia, 16 Sept. 1986, *Ruffo & Mmari* 2361!; Tanga District: Sigi Forest Reserve, 11 Feb. 1971, *Mabberley* 710!; Morogoro District: Nguru Mts, Dikurura and Chazi valleys, 14 Sept. 1988, *Pócs* 88182/Z!

DISTR. **T** 3, 6; not known elsewhere

HAB. Montane and submontane forest; 900–1150 m

USES. None recorded on specimens from our area

CONSERVATION NOTES. Although known from several specimens, the area where found is under pressure with logging and habitat degradation, assessed here as Near Threatened (NT)

NOTE. Some pilose specimens (*Greenway* 3290, 3380, *Williams* 23) from the same area have smaller corollas.

5. **Harveya liebuschiana** *Skan* in F.T.A. 4(2): 437 (1906). Type: Tanzania, Usambaras, *Liebusch* s.n. (B†, holo.). Neotype: Lushoto District: Tanzania, W Usambaras, between Mkuzi & Kifunglio, *Drummond & Hemsley* 2096 (K!, chosen here)

White parasitic perennial, drying black; stems clustered, ascending, 6–15 cm high, simple, almost glabrous. Leaves reduced to ovate scales 3–5 mm long. Flowers large for the size of plant, in racemes; bracts lanceolate to broadly ovate, 15–22 mm long, 5–12 mm broad, ± acute, 1–3-veined; pedicels 3–12 mm long, sometimes with a bracteole above midway, glabrous. Calyx tubular, 5-angled, tube ± 20 mm long, 12 mm wide; lobes unequal, triangular, ± 5 mm long, acuminate, ciliate on veins. Corolla white, yellow inside, 38–68 mm long, tube narrowed at base; lobes spreading to reflexed, 10–12 mm, subacute. Filaments inserted 20 mm from base; anthers acute. Capsule obovoid, 10 mm in diameter; style ± persistent. Seeds oblong, 1.1 mm long, 0.3 mm wide, reticulate.

UGANDA. Bunyoro District: Budongo Forest, 1905, *Dawe* 795! & 5 Sept. 1995, *Poulsen, Nkuutu & Dumba* 930!

TANZANIA. Lushoto District: Mazumbai, 13 July 1980, *Tanner* 232!; Tanga District: E Usambaras, Lutindi Forest Reserve, 11 May 1987, *Iversen, Persson & Pettersson* 87/3436!; Morogoro District: Ulugurus, Tanana, Feb. 1935, *Bruce* 816!

DISTR. **U** 2; **T** 3, 6; not known elsewhere

HAB. Dense shade in montane forest, and tea estate: 1000–1600 m

USES. None recorded on specimens from our area

CONSERVATION NOTES. Least Concern (LC); not uncommon

6. **Harveya** sp. **A**

Annual, white parasitic ?perennial, drying black; stems simple, erect, 8–20 cm high, sparsely glandular-pubescent. Leaves (bracts) fleshy, brittle when fresh, opposite, broadly ovate, 2–23 long, 15–16 mm wide, base cuneate, apex subacute, minutely glandular-pubescent. Flowers white, in racemes, solitary in the axils of leafy bracts; bracteoles 1 or 2 or absent, attached halfway on the pedicel, oblanceolate to linear-lanceolate; pedicels thick and fleshy, 7–10 mm long, glandular-pubescent. Calyx tubular to campanulate, 18–23(–25) mm, obscurely 5-angled, unequally 5-lobed; lobes triangular, ± 5 mm, acute. Corolla white with yellow throat, 40–45 mm long (incl. lobes); tube narrow below, expanding above, 30–33 mm, densely glandular-pubescent ouside and inside; lobes rounded, 7–11 mm, reflexed. Filaments white, 10–12.5 mm, attached where corolla tube becomes narrow, glandular. Ovary globose, ± 5 long, ± 4.5 mm wide; style longer than the corolla tube, bent downwards at the apex; stigma clavate. Capsule subrotund, ± 10 mm long, ± 9 mm in diameter, glabrous. Seeds oblong, ± 5 mm, reticulate.

KENYA. Embu District: Kiang'ombe Hill, 8 July 2005, *Ngugi, Mwachala & Kirika* 279! & *Mwachala & Ngugi* 616! & 6 June 2003, *Ngugi & Kirika* 40!

DISTR. **K** 4; not known elsewhere

HAB. At edge of moist forest (in shade), riverine; 1350–1600 m

CONSERVATION NOTES. So far only known from two populations both present in an ungazetted forested area, close to saw pits and encroaching development; here assessed as Critically Endangered (CR B1a) with extent of occurrence estimated to be less than 100 km² and so far known to exist at only a single location.

NOTE. The two populations known so far from the above locality are about 100 m apart – one underneath *Phoenix reclinata* (*Mwachala & Ngugi* 616) and the other (*Ngugi, Mwachala & Kirika* 279) under *Croton sylvaticus*.

7. **Harveya kenyensis** *Hepper* in K.B. 44: 163, fig. 2 (1989). Type: Kenya, Kitui District: Ukamba, *Bally* 1611 (K!, holo.; EA!, iso.)

Perennial white parasite on tree roots; stem stout, 3–5 cm long, ± 1 cm thick, glabrous. Leaves absent or reduced to scales; scales broadly ovate, 5–8 mm long and wide, acute, glabrous; bracts violet, ovate to lanceolate, up to 10 mm long, 4 mm wide. Flowers large, up to 6 cm long, in racemes; pedicels up to 20 mm, with a bracteole just below the calyx, glabrous. Calyx tubular, 11–12 mm long, shortly toothed; teeth 2–3 mm, ± acute, glabrous. Corolla tube white, throat yellow, expanding beyond calyx, 4–5 cm long; lobes purple, rounded, ± 10 mm long. Filaments glabrous, sterile anther locule long-acuminate; style club-shaped, scarcely exserted. Mature capsule not seen.

KENYA. Teita District: Mwatate, 15 July 1960, *Leach & Baylis* 10250! & Mt Kasigau, 28 June 2000, *Muasya* 2097!; Kitui District: Ukamba, Kanziko, 23 Jan. 1942, *Bally* 1611!
DISTR. **K** 4, 7; not known elsewhere
HAB. Thickets and forest, shaded ground; 30–450 m
USES. None recorded on specimens from our area
CONSERVATION NOTES. Known only from the collections cited above where the area is under some pressure and habitat degradation; assessed here as Near Threatened (NT)

NOTE. A collection from Tanzania, Pangani District: Mkwaja Ranch, *Frontier Tanzania* 3035!, comes closest to *H. kenyensis*, but has a longer calyx (17 mm) and corolla lobes (15 mm). The corolla tube is densely pubescent unlike the sparsely pubescent corolla of *H. kenyensis*. With more material this may qualify as a new taxon.

8. **Harveya tanzanica** *Hepper* in K.B. 45: 374 (1990). Type: Tanzania, Morogoro District: Mkobwe, *Drummond & Hemsley* 1868 (K!, holo.; EA!, iso.)

Perennial root parasite, drying black; stems 1–5 cm long with numerous nodes. Leaves absent or reduced to scales ± 2 mm long, the upper ovate, ± 7 mm long, glabrous or ciliate at margins; bracts scarlet, ovate, 7–10 mm long, 3–5 mm broad, ± acute, ciliate. Calyx scarlet, tubular; tube 12–20 mm long, glabrous to pubescent on the veins, teeth shortly triangular, ± 3 mm long, ciliate. Corolla red or deep scarlet outside, shading to pale yellow at base, deep scarlet inside changing abruptly to yellow at mouth of tube; tube 45–50 mm long, lobes rounded, 8–15 mm long. Stamens nearly equal; filaments ± 30 mm long, inserted at same level, pubescent at base; anthers with perfect locule acute, sterile locule long-apiculate; style club-shaped. Capsule and seed not seen.

TANZANIA. Morogoro District: Nguru Mts, Mkobwe, 29 Mar. 1953, *Drummond & Hemsley* 1868! & Turiani, 20 Aug. 1971, *Schlieben* 12239! & Manyanga Forest Reserve, 18 Sept. 1960, *Paulo* 791!
DISTR. **T** 6; not known elsewhere
HAB. Wet upland forest floor; 1250–1600 m
USES. None recorded on specimens from our area
CONSERVATION NOTES. Although the species is not uncommon it is very localised and the habitat is under pressure being converted into agricultural land; assessed here as Vulnerable (VU B2a)

SYN. *H. coccinea* Hepper in K.B. 44: 163, fig. 1 (1989), *non* (Harvey) Schlechter (1899). Type same as above; see Burtt in Notes R.B.G. Edinb. 43: 378 (1968)

33. **BUCHNERA**

L., Sp. Pl. 2: 630 (1753) & Gen. Pl., ed. 5: 278 (1754)

Annual or perennial, often hemi-parasitic, herbs, usually scabrid and rigid, usually becoming black upon drying; stems simple or branched from the base or below the inflorescence, glabrous to hispid-scabrid often with callus-based hairs. Leaves alternate, subopposite to opposite on stem, rosulate below, sessile to shortly petiolate; lower leaves usually broader with upper narrower and passing into bracts. Inflorescence terminal, spicate or with flowers in terminal or axillary globose or subglobose clusters, or flowers solitary-axillary; flowers subtended by one bract and usually two bracteoles. Calyx tubular, 10- (rarely 7- or 8-) veined or ribbed, sometimes obscurely so, 4–5-dentate. Corolla tubular, cylindric, slightly curved to arcuate; limb subequally 5-lobed, spreading. Stamens 4, included, slightly didynamous; anthers monothecal, dorsifixed. Stigma narrowly cylindric; style apex clavate, entire. Capsule loculicidal. Seeds numerous.

A genus of some 100 species from the tropics and subtropics, mostly from the Old World.

1. Calyx 4-lobed or occasionally with a rudimentary fifth lobe represented by bristles 2
   Calyx 5-lobed 5
2. Stems minutely retrorse-pubescent with large multi-cellular hairs present; calyx densely hirsute to pilose; corolla tube pilose externally — 7. *B. rungwensis*
   Stems glabrous to shortly pubescent or hispid, multi-cellular hairs lacking; calyx glabrous to subglabrous or hispid only on veins 3
3. Inflorescence of compact heads of spikes, arranged corymbosely; lower bracts lanceolate, upper broadly ovate to orbicular .... 1. *B. foliosa*
   Inflorescence spicate, terminal or lateral; bracts ovate to elliptic-rhombic or broadly reniform, at times lower somewhat trilobed with lobes or margins dark-coloured 4
4. Bracts broadly reniform, mucronate, pubescent, hispid-ciliate .... 5. *B. bangweolensis*
   Bracts ovate to elliptic-rhombic or 3-lobed, mucronate, glabrous or minutely glandular .. 6. *B. quadrifaria*
5. Calyx tube glabrous, except with occasional hairs on veins; lobes glabrous to ciliate 6
   Calyx tube hispid to pilose throughout 9
6. Corolla deep pink to magenta, densely minutely pubescent externally; tube arcuate above midway .... 18. *B. usuiensis*
   Corolla yellow to bluish, lilac, purple or white, glabrous externally; tube straight 7
7. Leaves often with clusters of small leaves or sterile branches in axils; inflorescence of many long spreading branches each of laxly-flowered spikes or racemes .... 17. *B. speciosa*
   Clusters of small leaves or sterile branches absent; inflorescence of slender spikes 8

8. Bracts broadly ovate to ovate-elliptic, obtuse, concave, closely appressed to calyx, minutely toothed or hispid on margins; bracteoles absent . . . . 9. *B. humpatensis*
   Bracts lanceolate, acuminate, ciliate or finely pubescent; bracteoles linear . . . . 14. *B. lastii*
9. Corolla tube externally glabrous or minutely glandular 10
   Corolla tube externally pilose 14
10. Inflorescence terminal, globose, hemispheric to cylindric, densely compactly flowered, (0.8–)1.5–3.8(–14) cm long; corolla tube glabrous or minutely glandular; plant not drying black . . . 3. *B. cryptocephala*
    Inflorescence terminal or lateral, spiciform or of widely spreading branches, ± laxly flowered, 4–25(–35) cm long; corolla glabrous; plants often drying black 11
11. All leaves narrowly linear, 0.5–2.3 mm wide; corolla tube markedly arcuate . . . . 12. *B. tenuifolia*
    Upper leaves linear to linear-lanceolate, up to 2.5 mm wide, lower elliptic to laneolate to lanceolate-ovate, ≥ 5 mm wide; corolla tube never arcuate 12
12. Woody herb; stems distichously minutely pubescent; inflorescence densely spicate, 4–10(–14) cm long, rarely lax at fruiting, axis with few small bristly hairs present in addition to minute pubescence . . . . 13. *B. scabridula*
    Annual or perennial herbs; stems pubescent throughout; inflorescence laxly spicate, (2–)6–25(–40) cm long, axis subglabrous to hispid-pubescent 13
13. Stems with branches ascending, not widely spreading, subglabrous to appressed hispid-pubescent; inflorescence of long slender laxly-flowered spikes; corolla tube 5.5–6.5 mm long; capsule narrowly oblong, ± 1.8 mm broad . . . . 15. *B. leptostachya*
    Stems with branches widely spreading, long pilose to hispid-pilose; inflorescence of terminal spikes, flowers initially crowded at apex, soon becoming lax; corolla tube 6–8 mm long; capsule ovoid, ± 3 mm broad . . . . 16. *B. hispida*
14. Inflorescence of axillary sessile or short pedunculate compound heads . . . . 2. *B. nuttii*
    Inflorescence terminal at end of branches only 15
15. Cauline leaves tightly inrolled when dried; inflorescence quadrangular; upper bracts with apices bi- or trifid; bracteoles absent . . . . 10. *B. ruwenzoriensis*
    Cauline leaves not inrolled when dried; inflorescence cylindric to hemispheric to globose, not always clearly quadrangular; upper bracts with undivided apex; bracteoles present 16

16. Leaves with coarse very long white hispid, patent or antrorse, callus-based hairs; calyx irregularly or unevenly 5-lobed, sometimes apparently 4-lobed with fifth lobe much reduced or absent . 8. *B. randii*
    Leaves and calyx appressed hispid-scabrid, lacking long white hairs; calyx regularly and equally 5-lobed 17
17. Inflorescence of cylindric spikes > 5 cm long, initially crowded but soon becoming lax .... 16. *B. hispida*
    Inflorescence of compact cylindric to hemispheric to globose spikes, 5 cm long, not becoming lax 18
18. Spikes cylindric, ± quadangular, ± 8 mm broad; calyx lobes unequal, pilose .... 11. *B. welwitschii*
    Spikes globose, hemispheric to cylindric, 10–18 mm broad; calyx equally 5-lobed, densely hispid-pilose .... 4. *B. capitata*

1. **Buchnera foliosa** *Skan* in F.T.A. 4(2): 389 (1906); Philcox in F.Z. 8(2): 99 (1990). Type: Malawi, Tanganyika Plateau, Chitipa [Fort Hill], *Whyte* s.n. (K!, holo.)

Annual, 40–60 cm or more tall, stout; stems erect to spreading, branches opposite, mostly above, obscurely quadrangular, markedly leafy, white, long, subappressed-hispid, to subglabrous in parts. Leaves opposite, lower leaves oblong, the upper lanceolate, 20–65(–90) mm long, (3–)6–25(–42) mm wide, narrowing at base, the lower obtuse at apex, the upper acute at apex, margins entire or remotely toothed, 3–5-veined, sparsely appressed-pubescent above, hispid with short hairs beneath (especially on major veins). Inflorescence of compact heads of spikes, corymbosely arranged, quadrangular, markedly so in fruit, 10–30 mm long, 10–22 mm wide; bracts varied, lowest lanceolate, to ± 25 mm long, upper broadly ovate to orbicular 5–8 mm long, 3–4 mm wide, apex acute to acuminate, occasionally keeled at apex, hispid; bracteoles 4–5 mm long, 0.6–1 mm wide, distinctly keeled, hispid. Calyx subquadrangular, 5.5–7 mm long, 4-lobed, 4-veined, hispid mainly on veins, split on upper and lower sides to ± 2.5 mm; lobes narrowly deltate-acute, ± 1.5 mm long. Corolla blue to occasionally purple or white; tube slender, straight, cylindric, 6(–8) mm long, externally very sparsely pilose or pubescent, sometimes appearing glabrous; limb 2.5–4 mm diameter; lobes obovate or obovate-cuneate, 1.5–2(–3) mm long, 1.5–2 mm wide. Capsule ovoid, ± 4.5 mm long, shortly beaked.

TANZANIA. Rungwe District: Mwaga–Tukuyu, 24 Sept. 1932, *Geilinger* 2606!
DISTR. **T** 7; Angola, Zambia, Malawi
HAB. In *Brachystegia* woodland, and grassland; 1000–1600 m
CONSERVATION NOTES. Least concern (LC); the species is known from a single collection in the Flora area, but as far as is known not rare or under threat in its other areas of distribution

2. **Buchnera nuttii** *Skan* in F.T.A. 4(2): 388 (1906); Philcox in F.Z. 8(2): 100 (1990); U.K.W.F.: 259 (1994). Type: Zambia, Urungu, Fwambo, 1600 m, *Nutt* s.n. (K!, holo.)

Perennial, up to 150 cm tall; stems erect, rigid, branched usually in upper part or occasionally simple, branches spreading, hispid-scabrid-pilose with white ascending hairs. Leaves opposite or subopposite, lanceolate, 18–50(–90) mm long, 2–10(–30) mm wide, narrowed at base, apex acute, margins entire with one or two obscure, shallow lobes or toothed, 3–5-veined, more or less appressed pilose-scabrid above, hispid-scabrid beneath, especially on veins. Inflorescence of densely flowered, irregular, sessile or shortly pedunculate, axillary compound heads up to 40 mm long, 25 mm broad, individual flowers not becoming lax when fruiting; bracts ovate-lanceolate, 3–4 mm long, 2–2.5 mm wide, acuminate, keeled, subglabrous, ciliate;

bracteoles narrowly lanceolate, 1.5–4 mm long, 0.5–1.5 mm wide, keeled, ciliate. Calyx 5-lobed, 3.5–4.5 mm long, sparsely pubescent; lobes ovate-lanceolate, 0.8–1.5 mm long, acuminate, ciliate. Corolla white or occasionally lavender, with a mauve or bluish tinge, frequently with lobes bordered with mauve or blue, fragrant; tube rather wide, funnel-shaped, 4–5(–9) mm long, straight, externally hirsute-pilose, varying in density from sparse immediately below sinuses to somewhat denser throughout, throat densely pilose with purplish hairs spreading well onto limb; limb 3–4 mm in diameter; lobes obovate, ± 1.5(–3) mm long, 1–1.8 mm wide. Capsule broadly ovoid-globose, 2.5–3 mm long, ± 2 mm wide.

KENYA. West Suk/Elgeyo District: Cherangani Hills, Oct. 1937, *Gardner* in F.D. 3723!; Trans-Nzoia District: Sandum's Bridge, Nov. 1965, *Tweedie* 3204!; South Kavirondo District: Suna, Sept. 1933, *Napier* 5353!

TANZANIA. Ngara District: Mbuba, Busubi [Bushubi], 6 Sept. 1960, *Tanner* 5134!; Mpanda District: Pasagulu, 10 km N of Kasogi, 8 Aug. 1959, *Harley* 9204!

DISTR. **K** 2, 3, 5; **T** 1, 4; Zambia, Malawi

HAB. Grassland and open *Brachystegia* woodland; 1300–2000 m

CONSERVATION NOTES. Least Concern (LC); common

3. **Buchnera cryptocephala** (*Baker*) *Philcox* in K.B. 40: 606 (1985) & in F.Z. 8(2): 99 (1990). Type: Zambia, Fwambo, Lake Tanganyika, *Carson* 59 (K!, holo.)

Annual, (7–)20–60(–180) cm tall, erect, drying green or brown, not black; stems simple or less frequently branched above, moderately to densely leafy throughout, scabrid. Leaves opposite or alternate, ovate-elliptic to oblong, 12.5–65(–90) mm long, 1.5–13(–16) mm wide, base cuneate, apex obtuse, shortly mucronate or not, margins entire, 3(–5)-veined, hispid-scabrid. Inflorescence terminal, globose, hemispheric to cylindric, densely compactly flowered, not quadrangular, (8–)15–38(–140) mm long, 8–35 mm wide, usually subtended by numerous upper leaves; bracts lanceolate, ovate or elliptic-lanceolate, (5.5–)7.5–9 mm long, (1.5–)2–3 mm wide, keeled, acuminate, pilose-hispid to subglabrous, ciliate; bracteoles linear to linear-lanceolate, (4.5–)6–8 mm long, 0.5 mm wide, hispid to pilose. Calyx 5-lobed, (5–)7–8.5 mm long, prominently 10-veined, hispid-scabrid especially on veins; lobes ovate or ovate-lanceolate to linear-triangular, 1.5–2.8 mm long, hispid-ciliate. Corolla blue to purple, or white; tube to ± 10 mm long, externally glabrous to minutely glandular, throat pilose; lobes subequal, broadly obovate, to 4.5 mm long, 4 mm wide. Capsule ovoid-ellipsoid, 4.3–5 mm long, 2–2.5 mm broad, glabrous.

UGANDA. Kigezi District: Rukiga [Ruchigga], *Bagshawe* 402! & Maziba, Dec. 1944, *Purseglove* 1606!; Masaka District: 1–2 km S of Mityebili, 21 Feb. 1971, *Lye* 5901!

TANZANIA. Ngara District: Mu Rukarazo, Bugufi, 22 Aug. 1960, *Tanner* 5100A!; Ufipa District: Nsanga Mt, Malonje Plateau, 13 Mar. 1959, *Richards* 12104!; Songea District: Lupembe Hill, 20 May 1956, *Milne-Redhead & Taylor* 10261!

DISTR. **U** 2, 4; **T** 1, 4, 7, 8; Zambia, Malawi, Zimbabwe

HAB. Wet grassland, woodland and swamps; 600–2600 m

CONSERVATION NOTES. Least Concern (LC); widespread

SYN. *Lobostemon cryptocephalum* Baker in K.B. 1894: 30 (1894)
Buchnera pulchra S.Moore in J.L.S. Bot. 37: 190 (1905) & in F.T.A. 4(2): 383 (1906), & F.P.U.: 136 (1962). Type: Uganda, Kigezi District: Rukiga [Ruchigga], *Bagshawe* 402 (K!, holo.)

4. **Buchnera capitata** *Benth.* in DC., Prodr. 10: 495 (1846); Engl., P.O.A. C: 359 (1859); Skan, F.T.A. 4(2): 381 (1906); F.P.U.: 136 (1962); Hepper in F.W.T.A. ed. 2, 2: 369 (1963); Philcox in F.Z. 8(2): 100 (1990); U.K.W.F.: 259, pl. 111 (1994); Fischer in Fl. Ethiop. & Eritr. 5: 292 (2006). Type: Madagascar, *Bojer* s.n. (K!, holo.)

Annual, 30–50(–95) cm or more tall, erect; stems simple or occasionally branched above, somewhat furrowed, glabrous to shortly pubescent above. Leaves opposite or upper alternate, oblong-lanceolate to linear obovate-oblong, 10–60(–80) mm long, 2–14(–26) mm wide, subamplexicaul, lower and basal leaves narrowed below into petiole-like base, apex obtuse, margins entire to shallowly crenate, occasionally shortly and remotely toothed, sparsely to more or less densely hispid-scabrid above, minutely hispid on margins and veins beneath otherwise glabrous particularly in upper leaves, (1–)3–5-veined, sometimes obscurely so. Inflorescence terminal, globose, hemispheric to cylindric, densely compactly flowered, not quadrangular, 6–45 mm long, 10–18 mm broad; bracts oblong-lanceolate or lanceolate, 3–5 mm long, ± 1 mm wide, acuminate, pilose; bracteoles linear to filiform, 3–4.5(–6) mm long, pilose. Calyx ± 4.5–6 mm long, 5-lobed, densely hispid-pilose, obscurely 10-veined; lobes subulate, 1.5–2(–3) mm long, frequently hispid-scabrid on margins, frequently widely spreading in fruit. Corolla usually white, occasionally blue to violet; tube 3.5–8(–10) mm long, externally pilose; lobes obovate to obovate-oblong, 2–3.5 mm long, 1.5–2 mm wide, rounded to shallowly emarginate. Capsule ovoid-cylindric, ± 2.75 mm long, up to 2 mm broad, obtuse. Fig. 35, p.128.

UGANDA. West Nile District: near Koboko [Kobboko], Mar. 1935, *Eggeling* 1849!; Ankole District: Igara, Mar. 1939, *Purseglove* 646!; Masaka District: 1.6 km from Katera on Katera–Kiebbe road, 1 Oct. 1953, *Drummond & Hemsley* 4518!

KENYA. Trans-Nzoia District: near Maboonde, 9.5 km W of Kitale, Dec. 1971, *Tweedie* 4205!

TANZANIA. Biharamulo District: Lusahanga, 15 Oct.1960, *Tanner* 5183!; Mpanda District: 11 km on Mpanda–Uruwira road, 9 Mar. 1994, *Bidgood et al.* 2736!; Songea District: by R. Luhira, N of Songea, 6 May 1956, *Milne-Redhead & Taylor* 10004!

DISTR. **U** 1–4; **K** 3; **T** 1, 4, 5, 7, 8; widespread in western, central and southern tropical Africa; Madagascar

HAB. Swamps, bogs and marshy grasslands; 900–2600 m

CONSERVATION NOTES. Least Concern (LC); common and widespread

5. **Buchnera bangweolensis** *R.E.Fries*, Wiss. Ergebn. Schwed. Rhod.-Kongo-Exped. 1: 291 (1916); Philcox in F.Z. 8(2): 104 (1990). Type: Zambia, Lake Bangweulu, *Fries* 892 (K!, lecto.)

Annual, 30–65(–90) cm or more tall, erect; stems slender, simple but occasionally branched within inflorescence, glabrous except for few patent hairs towards base. Leaves opposite or upper alternate, subappressed to stem, linear, 4–30 mm long, 0.2–1(–1.5) mm wide, apex acute, 1-veined, scabrid, hispid-ciliate. Spikes subquadrangular, compact, 8–25(–50) mm long, to 6.5 mm wide, not becoming lax in fruit; bracts broadly subreniform, 3.5–5 mm long, 3–4 mm wide, apex with a broad mucro, not keeled, shortly pubescent, hispid-ciliate; bracteoles linear-lanceolate, 2.5–3.5 mm long, 0.5–0.8 mm wide, acute, hispid-ciliate. Calyx 4-lobed, (2.5–)3.5–4 mm long, 4-veined; lobes lanceolate, 1.5–2 mm long, ciliate, erect, sinus ± 2–2.5 mm deep. Corolla pink to mauve or pale purple; tube 5.5–6 mm long, externally glabrous, slightly short-pilose at throat; limb 1.5–2 mm in diameter; lobes subobovate to oblong, ± 1 mm long, 0.7–0.8 mm wide. Capsule broadly ovoid, 2.8–3 mm long, 1.6–1.8 mm broad, slightly emarginate; beak to ± 0.4 mm long.

TANZANIA. Songea District: 11 km W of Songea, 11 Feb. 1956, *Milne-Redhead & Taylor* 8735! & 12 km E of Songea, by Nonganonga stream, 28 Dec. 1955, *Milne-Redhead & Taylor* 7788! & valley near R. Mtanda, 9.5 km SW of Songea, 21 June 1956, *Milne-Redhead & Taylor* 9841!

DISTR. **T** 8; Zambia, Zimbabwe

HAB. Peaty bogs and wet grasslands; 900–1600 m

CONSERVATION NOTES. Least Concern (LC); localised (but not uncommon) in the Flora area, and not known to be at threat

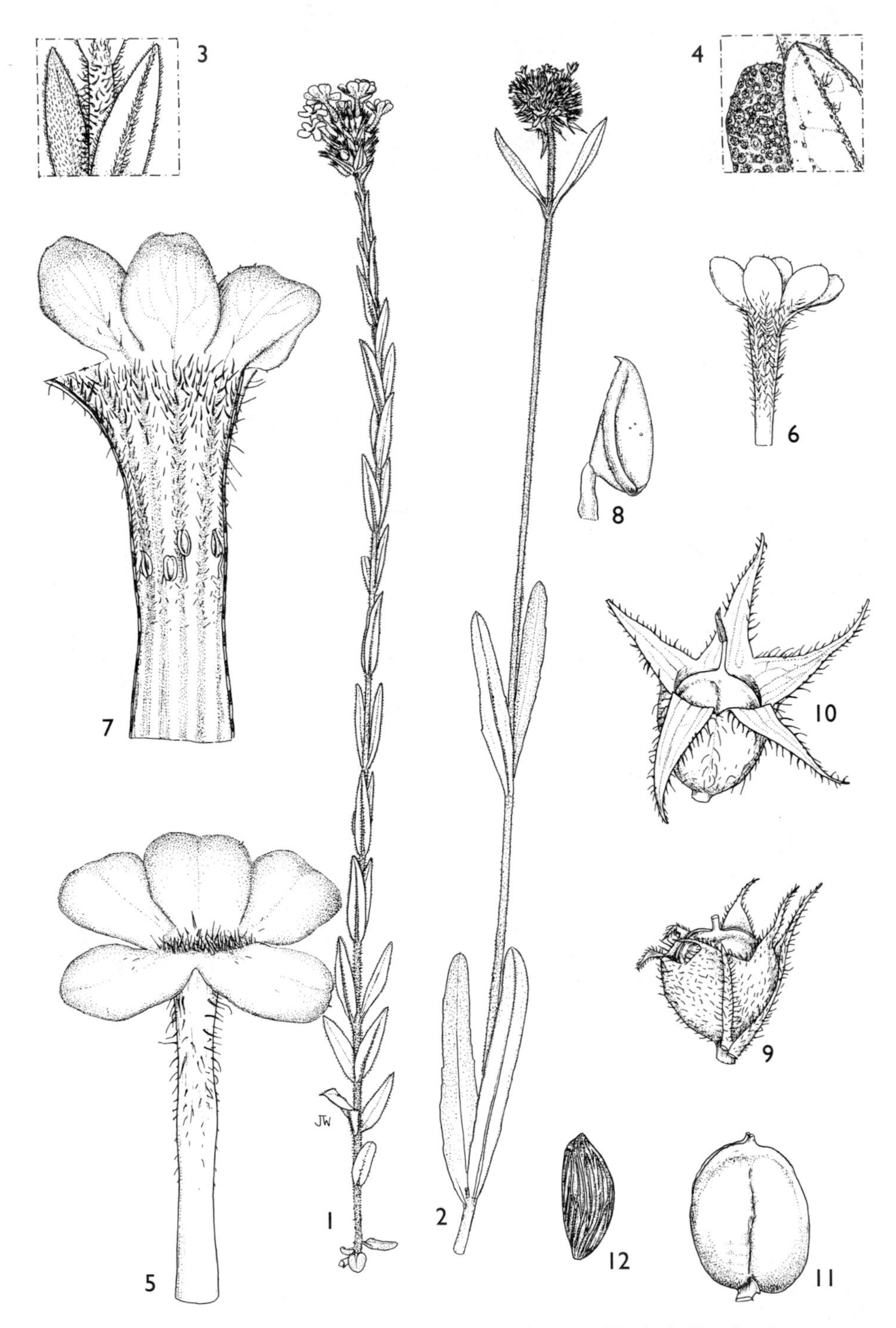

FIG. 35. *BUCHNERA CAPITATA* — **1**, habit with flowers × ²/₃; **2**, habit showing mature inflorescence × ²/₃; **3**, **4**, leaf and stem detail × 4; **5**, **6**, corolla showing variation × 5; **7**, corolla opened × 5; **8**, anther × 30; **9**, capsule with calyx and bracts × 6; **10**, capsule × 6; **11**, capsule with calyx removed × 6; **12**, seed × 30. 1, 3, from *Bidgood, Leliyo & Vollesen* 4451; 2, 4, 5–8, 10 from *Gilbert & Mesfin* 6435; 9, 11, 12 from *Festo & Bayona* 1742. Drawn by Juliet Williamson.

6. **Buchnera quadrifaria** *Baker* in K.B. 1895: 71 (1895); Skan in F.T.A. 4(2): 378 (1906); Cribb & Leedal, Mount. Fl. S Tanz.: 118, pl. 28E (1982); Philcox in F.Z. 8(2): 103 (1990). Type: Zambia, Fwambo, 1894, *Carson* 100 (K!, lecto.)

Annual, 20–68 cm tall; stems erect, slender, simple or more rarely branched, quadrangular, glabrous or occasionally shortly pubescent immediately below insertion of leaves, glabrous or scabrid on prominent angles. Leaves alternate or opposite, sparse, subulate or linear, (2.5–)5–8 mm long, 0.7–1 mm wide, apex acute to obtuse, margins entire, 1-veined, sparsely minutely scabrid to glabrous. Spikes quadrangular, densely compacted, (4–)10–20(–35) mm long, 4–5 mm broad; bracts broadly ovate to elliptic-rhombic, 3.5–5 mm long, 2.5–3 mm wide, acuminate, at times somewhat trilobed towards base with lobes or margins dark-coloured as with mid-vein, glabrous or minutely granular-glandular; bracteoles linear-lanceolate, 3–4 mm long, shallowly keeled, glabrous to finely glandular. Calyx 4-lobed, 3–3.5 mm long, deeply divided on lower side, glabrous to subglabrous; lobes subulate-lanceolate, 3–4 mm long, glabrous except for few small hairs on margins. Corolla blue to mauve, or purple or white; tube (4–)7–9 mm long, glabrous or glandular, throat pilose; limb 4–6 mm in diameter; upper lobes elliptic-ovate, lower oblong and up to 5.5 mm long 1.5–2.5 mm wide. Capsule broadly ellipsoid, 2.5–3 mm long, 1.5 mm broad, glabrous.

TANZANIA. Ufipa District: between Lake Tanganyika and Lake Rukwa, 1896, *Nutt* s.n.!; Iringa District: Dabaga Highlands, ± 16 km S of Dabaga on Kibengu road, 22 Feb. 1962, *Polhill & Paulo* 1566!; Njombe/Mbeya District: Elton Plateau, Lualalira, Jan. 1978, *Leedal* 4845!
DISTR. **T** 4, 7; Congo-Kinshasa, Burundi, Zambia, Malawi
HAB. Bogs, marshes, damp grasslands and open woodlands; 1200–2500 m
CONSERVATION NOTES. Least Concern (LC)

7. **Buchnera rungwensis** *Engl.* in E.J. 30: 403 (1901); Skan in F.T.A. 4(2): 384 (1906); Philcox in F.Z. 8(2): 105 (1990). Type: Tanzania, Mt Rungwe, on slopes of the highest peak, *Goetze* 1153 (B†, holo.). Neotype: Tanzania, Rungwe District, Mt Rungwe summit, *Stolz* 2439 (K!, chosen here)

Annual or perennial, 16–48 cm tall; stems erect, one to many arising from rootstock, simple or branched above, usually below inflorescence, retrorsely pubescent throughout, frequently with large, patent, multicellular hairs present. Leaves opposite to subopposite, oblong-elliptic to narrowly oblong- or elliptic-ovate, 8–35 mm long, 1–4(–6.5) mm wide, apex obtuse, obscurely 1–3-veined, glabrous to subglabrous above, densely hirsute-pilose beneath especially on major veins, ciliate; basal leaves when present, similar but broader or those at extreme base elliptic to lanceolate-elliptic, obtuse, narrowed at base into slender petiole up to 6 mm long. Spikes compact or with lowest flowers lax or somewhat remote at anthesis, 15–45(–80) mm long, 5–12 mm broad; bracts imbricate, lanceolate, (4.8–)8–12.5 mm long, 2.5–4 mm wide, densely pilose to subglabrous, sparsely ciliate or densely so; bracteoles linear-lanceolate, 3–6.5 mm long, 0.5(–1.3) mm wide, pilose, ciliate. Calyx 4-lobed, but occasionally with rudimentary fifth lobe represented by bristly tuft, 6.5–9.5(–10.5) mm long, obscurely 10-veined, densely hirsute-pilose with antrorse, spreading or subappressed hairs; lobes narrowly triangular or acutely lanceolate, (0.8–)1.5–2.5 mm long, sinus present on adaxial side. Corolla pale lilac or mauve to purple, or magenta; tube 7–10.5(–13) mm long, externally short pilose, throat pilose; limb 8–13 mm in diameter, lobes 3–6.5 mm long, ± 4.5 mm wide at most. Capsule oblong-cylindric, ± 6.5 mm long, 2 mm broad, tapering towards base, apex rounded to shortly beaked.

TANZANIA. Mbeya District: Elton Plateau, 25 Jan. 1961, *Richards* 14172!; Njombe District: Poroto Mts, Kitulo Plateau, Ndumbi Valley, 24 Mar. 1991, *Bidgood et al.* 2124!; Iringa District: Ludewa, Lusitu ridge, 6 Feb. 1988, *Lovett & Congdon* 3015!
DISTR. **T** 7; Zambia, Malawi

HAB. Montane and submontane grasslands; 2100–2800 m
CONSERVATION NOTES. Least Concern (LC)

SYN. *B. crassifolia* Engl. in E.J. 30: 403 (1901); Skan in F.T.A. 4(2): 385 (1906); Cribb & Leedal, Mount. Fl. S Tanz.: 117, pl. 28C (1982); Philcox in F.Z. 8(2): 104 (1990). Type: Malawi, without locality, *Whyte* s.n. (B†, holo; K!, iso.)

NOTE. In the absence of the holotype, presumably destroyed in Berlin during World War II and the apparent lack of any isotype material, I (D.P.) have chosen *Stolz* 2439 from the type area and housed at Kew as the neotype.

8. **Buchnera randii** *S.Moore* in Journ. Bot. 38: 467 (1900); Skan in F.T.A. 4(2): 387 (1906); Norlindh & Weimarck in Bot. Not. 98: 122 (1951); Philcox in F.Z. 8(2): 106 (1990). Type: Zimbabwe, Harare, July 1898, *Rand* 573 (BM!, holo.)

Annual, 10–75(–130) cm tall; stems erect, slender, simple or more usually branched especially above, pilose-hispid. Upper cauline leaves opposite, narrowly linear, 9–55 mm long, 3–7 mm wide, apex obtuse, entire, 1-veined, densely long white-hispid with callus-based, patent or antrorse hairs; middle and lower cauline leaves narrowly linear-obovate, up to 70 mm long, 12 mm wide, obtuse, entire or rarely coarsely dentate with one or two teeth on each margins, 3-veined, densely hispid; basal leaves where present, ovate to obovate-oblong, up to 35 mm long, 15 mm wide, obtuse, margins entire to coarsely dentate, hispid. Spikes terminal or lateral, markedly to somewhat obscurely quadrangular, densely compacted except occasionally lower flowers becoming more remote during fruiting, 5–38 mm long, 5–12 mm broad; bracts ovate-lanceolate, 4–6.5 mm long, 1.5–2.5(–3) mm wide, hispid-scabrid, long white-hispid usually on margins; bracteoles 2–4 mm long, ± 0.5–0.7 mm wide, subulate, white hispid-scabrid. Calyx 5-lobed, all lobes narrowly lanceolate, 4–6 mm long; frequently lobes 4, subequal, ± 1.6 mm long, fifth lobe much smaller, ± 0.8–1 mm long and narrower; white hispid-strigose, obscurely 10-veined. Corolla light blue, through dark-blue to mauve and purple, occasionally white; tube green to yellow-green, (5.5–)7–8 mm long, slender, externally pilose to pubescent at least in upper 2–3 mm, throat densely pilose, limb 2.25–3(–6) mm in diameter; lobes obovate-oblong, obtuse, subequal. Capsule ovoid, 3.5–4.5 mm long, 2–2.4 mm broad, glabrous.

TANZANIA. Mbeya District: Ruaha National Park, foot of Magangwe Hill, 10 May 1972, *Bjørnstad* 1756!; Songea District: valley near Mtanda, 9.5 km SW of Songea, 21 June 1956, *Milne-Redhead & Taylor* 10863! & 6.5 km E of Gumbiro, 8 May 1956, *Milne-Redhead & Taylor* 10019!
DISTR. **T** 7, 8; ?Congo-Kinshasa, Angola, Zambia, Malawi, Mozambique, Zimbabwe, Botswana
HAB. Grassland and open woodland; 600–1650 m
CONSERVATION NOTES. Least Concern (LC)

9. **Buchnera humpatensis** *Hiern*, Cat. Afr. Pl. Welw. 1: 777 (1898); Skan in F.T.A. 4(2): 393 (1906); Philcox in F.Z. 8(2): 108 (1990). Type: Angola, Huilla, Humpata and Nene, *Welwitsch* 5798 (LISC, holo.; K!, iso.)

Annual, 8–38 cm tall; stems erect, slender, branched with branches ascending, angular, glabrous. Leaves opposite, or upper at times alternate, lanceolate or linear-lanceolate, 10–40 mm long, 1.5–6 mm wide, narrowed at base, apex obtuse or occasionally subacute, margins entire or remotely minutely dentate, variously covered with callus scales especially on margins, 1-veined. Inflorescence of long, slender, terminal spikes, 5–12(–16) cm long, at maturity flowers very lax, opposite to almost opposite; bracts broadly ovate to ovate-elliptic, 2.5–4.5 mm long, 1.3–2 mm wide, obtuse, concave, closely appressed to calyx, minutely toothed or hispid on margins; bracteoles absent. Calyx 5-lobed, (2.5–)3–4 mm long, 5-veined, glabrous;

lobes ovate, 0.5–1.3 mm long, acute, at times somewhat unequal. Corolla pink or blue, lilac to purple; tube 2–2.6 mm long, whitish, glabrous; limb 1.3–1.5 mm in diameter; lobes narrowly obovate, ± 0.6–1 mm long. Capsule subglobular to ellipsoid, 1.9–2.3 mm long, 1.8–2.4 mm broad, somewhat emarginate.

TANZANIA. Iringa District: Great North Road, 6.4 km N of Iringa, 5 Feb. 1962, *Polhill & Paulo* 1363!
DISTR. **T** 7; Angola, Zambia, Malawi
HAB. Wet grasslands and pools; ± 1750 m
CONSERVATION NOTES. The species is known from a single collection in the Flora area, but as far as is known not rare nor at threat in its other areas of distribution and therefore assessed overall as of Least Concern (LC)

10. **Buchnera ruwenzoriensis** *Skan* in F.T.A. 4(2): 378 (1906); Philcox in F.Z. 8(2): 108 (1990); U.K.W.F.: 259 (1994). Type: Uganda, Ruwenzori, hillsides below 2133 m [7000 ft.], *Scott-Eliot* 7782 (K!, lectotype chosen here)

Annual, (10–)30–65 cm tall; stems erect, simple or occasionally branched above, patent to subappressed hispid-pilose. Leaves cauline, opposite or subopposite, narrowly linear-lanceolate, 18–40 mm long, 2.5–5 mm wide, apex acute or obtuse, margins entire, somewhat densely long pilose to hispid-pilose above, more sparsely so beneath, obscurely 1–3-veined being tightly inrolled when dried; lower and basal leaves elliptic to oblong, 25–35 mm long, 8–12 mm wide, obtuse, entire to sparsely crenate-dentate, scabrid above, hispid-pilose beneath mainly on veins, 3–5-veined. Spikes terminal or lateral, quadrangular, with flowers dense, compact, 10–26(–30) mm long; bracts lanceolate to ovate-lanceolate, 3–5.5 mm long, up to 1.8 mm wide, upper at least with the apices bi- or trifid, hispid to hispid-scabrid; bracteoles absent. Calyx 5-lobed, 4–4.5 mm long, 10-veined, densely long, soft pilose; lobes lanceolate, 0.6–1.3 mm long, uneven, frequently with one lobe very much shorter. Corolla blue to purple, or white; tube ± 5 mm long, externally pubescent; limb 1.6–2.3 mm in diameter; lobes obovate to circular, 1–1.3 mm long, ± 1 mm wide, margins slightly involute or crenate when dry. Capsule cylindric-ovoid, 4.5–5 mm long, 2 mm broad.

UGANDA. Ankole District: Igara County, Lubare Ridge, 31 May 1970, *Lye* 5475!; Mbale District: close to Kaburoron–Bukwa road, 9.6 km ESE of Kaburoron, N slopes of Mt Elgon, 12 Oct. 1952, *Wood* 478!; Mengo District: Jumba, Oct. 1916, *Dummer* 2984!
DISTR. **U** 2–4; Angola, Zambia
HAB. Rocky or grassy slopes, often on clayey soils; 1200–1900 m
CONSERVATION NOTES. Least Concern (LC)

11. **Buchnera welwitschii** *Engl.* in E.J. 18: 70, t. 3, fig. f in part (1894); Skan in F.T.A. 4(2): 386 (1906); Philcox in F.Z. (2): 112 (1990); Fischer in Fl. Ethiop. & Eritr. 5: 293 (2006). Type: Angola, Pungo Andongo, *Welwitsch* 5832 (K!, lectotype chosen here)

Annual, 12–44 cm tall; stems erect, simple, rarely branched, slender, white to fawn, sparingly patent to antrorse hirsute, denser at base. Leaves cauline, opposite to subopposite, oblong-linear to linear, 2–3(–4) pairs only, 12–40 mm long, 1–3(–8) mm wide, obtuse, white hirsute-scabrid, 1-veined; basal leaves elliptic-oblong to broadly ovate or subcircular, 17–40 mm long, 10–25 mm wide, apex obtuse, margins entire to crenate, hirsute-scabrid above, pale brown hispid-scabrid beneath especially on major veins, 3–5-veined. Spikes somewhat quadrangular, 8–30 mm long, to ± 8 mm broad, flowers crowded or compact, occasionally base of spike lax with flowers of lowest internode ± 45 mm distant from next above; bracts lanceolate, 3–5.5 mm long, 1–2 mm wide, concave, pilose; bracteoles linear to subulate, (2–)3.5–4.5 mm long, (0.3–)0.6 mm wide, pilose. Calyx 5-lobed, 4.5–5.5 mm long, pilose; lobes unequal, subulate, 0.8–1.5(–2.5) mm long. Corolla blue to bluish-lilac; tube 4.5–5.5(–8) mm long, externally pilose, densely so at throat; limb 3–3.5(–8) mm in diameter; lobes obovate, 1.5–2(–4) mm long, 1–2(–3) mm wide. Capsule ovoid-cylindric, 4–5 mm long, 1.5–2.5 mm broad.

TANZANIA. Mbeya District: Mbeya airfield, 11 May 1956, *Milne-Redhead & Taylor* 10042!; Iringa District: Udzungwa Mts, 28 May 2002, *Luke et al.* 8430! & top of Lulanda escarpment, 1 Aug. 1999, *Kayombo & Kikoti* 2809!
DISTR. **T** 7; Angola, Zambia
HAB. Grassland and open miombo woodland; 1500–1650 m
CONSERVATION NOTES. Least Concern (LC)

NOTE. With his description of this species, Engler cited two syntypes from Pungo Adongo, *Welwitsch* 5832 and 5847. As the original material was destroyed in the World War II, I (D.P.) have chosen the iso-syntype of 5832 at Kew as the lectotype.

12. **Buchnera tenuifolia** *Philcox* in Bull. Soc. Brot. ser. 2, 64: 242 (1991). Type: Zambia, Mbala District: beyond Kawimbe, *Richards* 18131 (K!, holo.)

Annual herb, 30–50 cm tall, slender, much branched, erect, distichously minutely pubescent above to subglabrous; branches spreading-ascending, leafy throughout. Leaves opposite to subopposite, narrowly linear, 30–55 mm long, 0.5–2.3 mm wide, somewhat flexuous, sparsely short-hispid, 1-veined, frequently with smaller leaves arising in axils. Spikes slender with flowers opposite or alternate, closely appressed to axis, 4–9 (–18) cm long; bracts lanceolate-acuminate, 2.5–3(–3.5) mm long, ± 1 mm wide, ciliate; bracteoles linear, 1.5–1.8(–3.5) mm long, ciliate. Calyx 5-lobed, (5.5–)6.5–7(–9) mm long, 10-veined, minutely pubescent to subglabrous; lobes 0.5–1.5 mm long, narrowly triangular, ciliate. Corolla usually yellow, occasionally pinkish to mauve; tube 8–11 mm long, markedly arcuate following anthesis, glabrous; limb 3.5–4.5(–6) mm in diameter; lobes linear, 1.8–2.5(–3) mm long, 0.5 mm wide, obtuse, inrolled; throat densely pubescent. Capsule linear-elliptic, 9–10 mm long, ± 1 mm broad, slender, glabrous.

TANZANIA. Ufipa District: 6 km on Tatanda–Sumbawanga road, 26 Apr. 1997, *Bidgood et al.* 3493!
DISTR. **T** 4; Zambia
HAB. In well drained rocky soils; ± 1750 m
CONSERVATION NOTES. Data Deficient (DD) as known in the Flora area from the above collection only but possibly of Least Concern (LC) throughout its range

13. **Buchnera scabridula** *E.A.Bruce* in K.B. 1937: 426 (1937); U.K.W.F.: 259 (1994). Type: Kenya, Elgeyo District: Marakwet Hills near Moyben R., *Dale* 3429 (K!, holo.; EA, iso.)

Woody herb, 60–95 cm tall; stems erect, usually simple, rarely sparsely branched above, distichously minutely pubescent throughout but becoming subglabrous towards base. Leaves mostly alternate, occasionally subopposite, sessile, oblong-oblanceolate, 20–35 mm long, up to 6 mm wide, narrowing towards base, apex acute or obtuse, hispid-scabrid on both surfaces, more noticeably so on lower leaves, prominently 1-veined beneath; upper leaves similar but smaller gradually developing into bracts. Inflorescence terminal, densely spiciform with minutely pedicellate flowers, 4–10(–14) cm long, rarely becoming lax at fruiting, axis with few small bristly hairs present in addition to minute pubescence; bracts narrowly lanceolate, ± 8 mm long, 1 mm wide, acuminate, slightly keeled, scabrid on median vein and margins; bracteoles 2, narrowly linear-lanceolate, 5–5.5 mm long, 0.3–0.5 mm wide, hispid-scabrid. Calyx 5-lobed, 6.5–8 mm long, 10-veined, hispid-scabrid on veins; lobes narrowly lanceolate-acuminate, 2–2.5 mm long, hispid-ciliate. Corolla pink, pale purple to bluish-mauve, fragrant, up to 14 mm long; tube cylindric,

10–11 mm long, externally glabrous; limb 4–8.5 mm in diameter; lobes obovate, ± 5 mm long, 3 mm wide, rounded. Capsule ovoid, 6–6.5 mm long, 2.5 mm broad, apex truncate, shortly beaked.

UGANDA. Karamoja District: Mt Kadam [Dabasien], Jan. 1936, *Eggeling* 2766!
KENYA. West Suk/Elgeyo Districts: N of Chepkotet [Cheptoket], Cherangani Hills, Sondang Ridge, 8 Sept. 1969, *Mabberley & McCall* 303!; Kenya, Elgeyo District: Marakwet Hills near Moyben R., Apr. 1935, *Dale* 3429!
DISTR. **U** 1; **K** 2, 3; not known elsewhere
HAB. Alpine meadows and scrubland; 2600–3100 m
CONSERVATION NOTES. Known from a few specimens collected from a small area in northern Uganda and Kenya, it is assessed here as Vulnerable: VU B1 b(iii), with extent of occurrence less than 20,000 km² and a continuing decline in the quality of habitat due to grazing and developmental pressures

14. **Buchnera lastii** *Engl.* in P.O.A. C. 359 (1895); Skan in F.T.A. 4(2): 392 (1906); Philcox in F.Z. 8(2): 113 (1990); Norlindh & Weimarck in Bot. Not. 98: 121 (1951). Type: Mozambique, Namuli, Makua, *Last* s.n. (B†, holo; K!, iso.)

Perennial herb up to 30 cm or more tall (measurement based on 2 specimens from Flora area, up to 40 cm in F.Z. area); stems erect, simple or branched, several arising from a woody rootstock, ± angular, minutely pubescent to glabrous. Leaves mostly opposite, sessile, linear to linear-lanceolate, 11–24 mm long, 2–3 mm wide, apex acute, margins entire, glabrous or sometimes minutely pubescent, sparsely hispid-pubescent or glabrous beneath, especially on the veins, ciliate, 1-veined. Flowers in terminal spikes, slender, 6–30 cm long; lower flowers interrupted and arranged more laxly on spike; flowers imbricate until flowering; bracts lanceolate, 7–9 mm long, 2–3.5 mm wide, acuminate, ciliate or finely pubescent; bracteoles linear, 3–5 mm long, ± 1 mm wide, ciliate, ± pubescent. Calyx 5-lobed, 6–9 mm, 10-veined, glabrous; lobes unequal, lanceolate, 1–2 mm long, glabrous or ciliate. Corolla yellow to bluish purple; tube 7–8 mm long (up to 16 mm in F.Z. material), slender, glabrous or with a few minute hairs, somewhat curved; limb up to 5 mm in diameter (up to 8 mm in F.Z. material); lobes obovate, ± 3 mm long, thick with ± revolute margins. Capsule oblong, 4–5 mm long, ± 2 mm broad.

subsp. **lastii**

Corolla tube glabrous outside; flower spikes 6–10 cm long.

TANZANIA. Mpanda District: 12 km NW of Katuma, 13 Aug. 2005, *Mwangoka & Anderson* 4101!; Mbeya District: East of Muvwa & Mshewa villages, Itagano above Shokwa plateau, 27 Oct. 1990, *Lovett & Kayombo* 4841! & Chala Mt, 4 May 1997, *Bidgood et al.* 3695!
DISTR. **T** 4, 7; Zambia, Malawi, Mozambique, Zimbabwe
HAB. Montane wooded grassland and hillsides; 1700–2250 m
CONSERVATION NOTES. Data Deficient (DD) as known in the Flora area from the above two collections only (but see note under *B. scabridula*); possibly of Least Concern (LC) throughout its range

SYN. *B. similis* Skan in F.T.A. 4(2): 391 (1906). Types: Malawi, *Whyte* s.n. (K!, syn.) & Mt. Mulanji, *Whyte* s.n. (K!, syn; P!, isosyn.)
B. tuberosa Skan in F.T.A. 4(2): 391 (1906). Type: Malawi, Thuchila Plateau, Mt. Mulanje, *Purves* 21 (K!, holo.; E!, iso.)

NOTE. Subsp. *publiflora* Philcox found in Zambia and Malawi differs in its pubescent corolla tube and 18–30 cm long spikes.

15. **Buchnera leptostachya** *Benth.* in DC., Prodr. 10: 497 (1846); Skan in F.T.A. 4(2): 394 (1906), excl. *B. mossambicensis* var. *usafuensis* Engl.; Hepper in F.W.T.A. ed.2, 2: 369 (1963); Philcox in F.Z. 8(2): 115 (1990). Lectotype: West Africa, Senegal [Senegambia], 1838, *Guillemin* s.n. (K!, chosen here)

Annual or perennial, 40–50 cm tall; stems erect, simple or branched from middle or above, subglabrous to minutely and finely appressed, antrorse hispid-pubescent, sometimes scabrid; branches ascending, not widely spreading. Leaves opposite or occasionally alternate; upper linear-oblong to linear, 27–35(–50) mm long, 3–4(–8) mm wide, apex obtuse or acute, shortly hispid-scabrid to scabrid, 1-veined; lower and basal leaves elliptic, oblong or obovate, 20–75 mm long, 4–12(–25) mm wide, base mostly narrowing into a short petiole, margins entire to crenate, indumentum as for upper leaves, 1–3-veined. Inflorescence of long, slender, laxly-flowered spikes 7–25(–35) cm long; flowers almost opposite above, alternate below, subappressed, somewhat imbricate until fruiting; pedicels 0.5–1 mm long; bracts ovate to ovate-lanceolate, 2.5–3 mm long, ± 1 mm wide, concave, acute to acuminate, sparsely hairy to glabrous except on margins; bracteoles linear to linear-lanceolate, 1.5–3 mm long, 0.3–0.8 mm wide, indumentum as for bracts. Calyx 5-lobed, 4.5–6.5 mm long, 10-veined, sparsely short hispid-strigose on veins and lobes; lobes narrowly- to linear-triangular, 1–1.5 mm long, acute. Corolla pink, purplish-blue to white; tube 5.5–6.5 mm long, glabrous; limb up to 5.5 mm in diameter; lobes obovate, 1.5–3 mm long, 0.5–2 mm wide. Capsule narrowly oblong, 4.5–6 mm long, 1.8 mm broad, shortly apiculate.

UGANDA. Masaka District: near Lake Nabugabo, Aug. 1935, *Chandler* 1305! & Bale, 5 May 1969, *Lye* 2740!

KENYA. Kitui District: about 8 km N of Migwani, A.I. Mission. 5 May 1960, *Napper* 1603!; Kilifi District: Malindi road, head of salt creek, 6 Sept. 1936, *Swynnerton* 403! & Mida, Sept. 1929, *Graham* 2115!

TANZANIA. Tanga District: Nyamaku, 21 July 1957, *Faulkner* 2029!; Uzaramo District: Msasani, 26 May 1968, *Batty* 95! & July 1939, *Vaughan* 2843!

DISTR. **U** 4; **K** 4, 7; **T** 3, 6; ?**P**; Zambia, Malawi, Mozambique, Zimbabwe; widespread in West Africa

HAB. In wet grasslands, swamps and coastal bush, frequently with high salinity; 0–1700 m

CONSERVATION NOTES. Least Concern (LC)

SYN. *B. mossambicensis* Klotzsch in Peters, Reise Mossamb. Bot.: 224, t. 34 (1861). Type: Mozambique, Quirimba Island, *Peters* s.n. (B†, holo.)

NOTE: Two specimens are cited in the protologue: West Africa, Senegal [Senegambia], 1838, *Guillemin* s.n. & and Pemba, *Bojer* s.n.; I (S.A.G.) have not been able to trace the latter, and have therefore chosen the former housed at K as the lectotype.

16. **Buchnera hispida** *D.Don*, Prodr. Fl. Nepal.: 91 (1825); Benth. in DC., Prodr 10: 496 (1846); A. Rich., Tent. Fl. Abyss. 2: 128 (1850); Skan in F.T.A. 4(2): 397 (1906); Norlindh & Weimarck in Bot. Not. 98: 124 (1951); Hepper in F.W.T.A. ed. 2, 2: 369 (1963); Merxm. & Roessler, Prodr. Fl. SW Afr. 126: 20 (1967); Philcox in F.Z. 8(2): 119 (1990); U.K.W.F.: 259, pl. 111 (1994); Fischer in Fl. Ethiop. & Eritr. 5: 292 (2006). Type: Nepal, near Narainhetty, *Hamilton* s.n. (BM!, holo.)

Annual herb up to 1 m or more tall, erect; stems simple or much branched, branches widely spreading, usually long-pilose below, hispid-pilose to scabrid above. Leaves upper linear to linear-lanceolate, 15–25(–50) mm long, 1–2.5 mm wide, acute or obtuse, entire or occasionally remotely dentate, 1-veined; lower leaves lanceolate, ovate-lanceolate, to broadly or narrowly oblong, 45–54(–75) mm or more long, 9–12(–25) mm wide, narrowed gradually into a petiole-like base, acute or obtuse, margins entire to coarsely irregularly dentate, 3-veined with median vein usually pinnately divided; basal leaves where present broadly elliptic to subcircular, 22–35 mm long, 15–20 mm wide, entire, 3-veined with a pinnately divided median vein; all leaves appressed or antrorse more or less hispid-scabrid. Flowers in terminal spikes, initially crowded at apex but soon becoming lax, spikes (2–)6–25(–40) cm long; flowers sessile or with pedicels up to ± 0.5 mm long; bracts ovate to lanceolate, 4–5 mm long, 0.8–1(–3) mm wide, acute or acuminate, but bracts of flowers larger, more leaf-like, pilose, hispid or glabrous with ciliate margins; bracteoles linear-lanceolate to

FIG. 36. *BUCHNERA HISPIDA* — **1**, habit × ⅔; **1a**, habit with root; **2**, flower × 5; **3**, corolla opened with ovary and style × 6; **4**, anther × 20; **5**, style × 12; **6**, capsule with calyx and bract × 4; **7**, capsule dehisced × 4; **8**, seed × 60. 1 from *Faulkner* 2678; 1a, 2, 6 from *Milne-Redhead & Taylor* 10921; 3, 4, 5 from *Harley* 9165; 7, 8 from *Richards* 18336. Drawn by Juliet Williamson.

subulate, 2.8–4 mm long, 0.2–0.4(–1) mm wide. Calyx 4–8 mm long, clearly to obscurely 10-veined, sparsely to variously densely hispid to hispid-short pilose, hairs mostly spreading; lobes linear-lanceolate to narrowly deltoid, 1–2 mm long. Corolla blue, mauve to purple or white; tube cylindric, straight or rarely curved, 6–8 mm long, glabrous to sparsely or slightly densely pilose; limb 3–5 mm in diameter; lobes obovate to obovate-oblong, 2–3 mm long, 1.8–2 mm long, throat villous. Capsule ovoid, 5–6 mm long, 3 mm broad. Fig. 36, p. 135.

UGANDA. Teso District: Serere, Sept. 1932, *Chandler* 941!; Mbale District: Sebei, near Kabururoni–Bukwa road, 9.5 km ESE of Kabururoni, N slopes of Mt Elgon, 12 Oct. 1952, *Wood* 483!

KENYA. Northern Frontier District: Moyale, 10 Aug. 1952, *Gillett* 13707!; Ravine District: Nakuru-Eldama Ravine, 1 km S of railway crossing, 6 Oct. 1981, *Gilbert & Mesfin* 6381!; Kwale District: Tanga–Mombasa road, 0.8 km from Tanzania border, 14 Aug. 1953, *Drummond & Hemsley* 3741!

TANZANIA. Mbulu District: Tarangire National Park, Boundary Road, 29 Aug. 1971, *Richards* 27040!; Mpanda District: Sonta, Rukwa, 3 Nov. 1963, *Richards* 18336!; Tunduru District: 1.5 km E of Muhuwesi [Mawese], 19 Dec. 1955, *Milne-Redhead & Taylor* 7726!; Zanzibar: Uguga, 2 May 1960, *Faulkner* 2545!

DISTR. **U** 3; **K** 1–4, 6, 7; **T** 2–8; **Z**; **P**; widespread in tropical and subtropical Africa, from West through Central to East Africa; southwards from Oman and Yemen, through Sudan and Ethiopia, Angola, Namibia; Madagascar; India, Nepal

HAB. In well-drained grassland on sandy soils, frequently in *Brachystegia* and other open woodland; 0–1800 m

CONSERVATION NOTES. Least Concern (LC)

SYN. *Striga schimperiana* Hochst. in sched. Schimper Iter. Abyss. Sectio Prima No. 23 (1840), Sectio Tertia No. 1516 (1844) *nom. nud.*

Buchnera monocarpa Hochst. in sched. Schimper Iter Abyss., *nom. nud.*

B. longifolia Klotzsch in Peters, Reise Mossamb., Bot.: 225 (1861); Skan in F.T.A. 4(2): 398 (1906); Hutch. & Dalz., F.W.T.A. ed. 1, 2: 225 (1931), *non* Kunth (1821). Type: Mozambique, Rios de Sema, *Peters* s.n. (B†, holo.)

NOTE. As I (D.P.) commented in F.Z. (p. 119), there is great difficulty in distinguishing between material which has hitherto been identified as belonging either to *B. hispida* or *B. longifolia* Klotzsch. Seeing no reason to keep them separate and for reasons given in that work, they are again considered here under *B. hispida* Buch.-Ham. ex D.Don.

17. **Buchnera speciosa** *Skan* in F.T.A. 4(2): 394 (1906); Philcox in F.Z. 8(2): 120 (1990). Lectotype: Malawi, Masuku Plateau, July 1896, *Whyte* s.n. (K!, chosen here)

Perennial herb or undershrub, 60–200 cm or more tall; stems erect, rigid, branched, rarely simple, branches opposite to subopposite, spreading, terete to obscurely quadrangular, subglabrous to somewhat sparsely hispid with short, stout, patent white hairs, occasionally glabrous above, distichously hispid-scabrid pubescent below, at times glabrescent leaving only thickened scabrid bases of hairs. Leaves opposite to subopposite; upper linear to narrowly linear-lanceolate, (10–)15–40 mm long, 1–3 mm wide, narrowing at base, apex acute or obtuse, usually with small apiculus present, margins entire, hispid-scabrid, 1-veined; lower linear-lanceolate to broadly elliptic, (30–)40–80 mm long, 4–8.5(–25) mm wide, narrowing below into petiole-like base, apex obtuse or rarely acute, margins entire, 1–3-veined, hispid-scabrid; leaves often with clusters of small leaves or sterile branches in axils. Inflorescence of many long, spreading branches each of laxly-flowered spikes or racemes 6.5–18(–30) cm long or more; flowers opposite below, almost opposite to alternate towards apex; pedicels 0–2.5 mm long; bracts lanceolate-subacuminate to ovate-lanceolate, 2.5–6(–10) mm long, 0.5–1.3 mm wide, shortly acuminate, sparsely hairy on upper surface to glabrous, scabrid-ciliate; bracteoles linear to linear-lanceolate, 2.3–3(–5) mm long, 0.5–0.8 mm wide, scabrid-ciliate. Calyx 5-lobed, 6–12 mm long, 10-veined, externally glabrous or with few stiff hairs on veins; lobes lanceolate-oblong to ovate-lanceolate, 0.2–2(–4) mm long, acute to shortly acuminate, short-ciliate, spreading to slightly reflexed in fruit. Corolla white (see note below); tube cylindrical, 7–12(–18) mm long, externally glabrous, throat pilose; limb 6–12(–30) mm in diameter; lobes obovate to subcircular, 3–6(–14) mm long, 2–6(–12.5) mm wide. Capsule cylindric-ovoid to ellipsoid, 5.5–9 mm long, 2.5–3.8 mm broad, narrower above, slightly curved, shortly apiculate.

TANZANIA. Kigoma District: Mt Livandabe [Lubalisi], 31 May 1997, *Bidgood et al.* 4237!; Songea District: 0.8 km NW of Miyau, 27 May 1956, *Milne-Redhead & Taylor* 10533! & Matengo Hills, Lupembe Hill, 10 Jan. 1956, *Milne-Redhead & Taylor* 8185!

DISTR. **T** 4, 5, 7, 8; Zambia, Malawi, Mozambique, Zimbabwe

HAB. Grasslands (especially in the wetter areas), open woodland and swamp forests; 950–2000 m

CONSERVATION NOTES. Least Concern (LC)

SYN. *B. mossambicensis* Klotzsch var. *usafuensis* Engl. in E.J. 30: 404 (1901). Type: Tanzania, Mbeya/Chunya District: Usafwa [Usafua], *Goetze* 1053 (B†, holo.)

B. eylesii S.Moore in Journ. Bot. 46: 72 (1909). Type: Zimbabwe, Mazowe [Mazoe], Iron Mask Hill, *Eyles* 334 (BM!, holo.; SRGH, iso.)

B. usafuensis (Engl.) Melchior in N.B.G.B. 11: 680 (1932)

B. pruinosa Gilli in Ann. Naturhist. Mus. Wien 77: 44 (1973). Type: Tanzania, Njombe District: near Madunda, *Gilli* 503 (W, holo.; K!, iso.)

NOTE. Hitherto, authors, including me (D.P.), have kept *B. eylesii* S.Moore and *B. speciosa* Skan as distinct on the slenderest of evidence, that being little more than plant size along with calyx and corolla size. It is now felt that apart from the above, most other characters are similar and there appears to be no further valid reason for continued separation; they are therefore combined under the earlier name, *B. speciosa* Skan. It is of interest to note that to date, most collections of this species record the flower colour as 'white', this being one of its distinguishing characters. However, two collectors record the flower colour as 'red', a colour which is most unusual within the genus there being only one species with flowers of this colour, *B. rubriflora* Philcox, and that from South America. In every other way our plants match the description of *B. speciosa*. Also it is of interest that the two collections of this plant were made from exactly the same area of Tanzania, albeit two years apart in 1986 and 1988. As I do not feel that this sole character of colour warrants varietal status, the two collections are included under *B. speciosa*.

18. **Buchnera usuiensis** *Oliv.* in Trans. Linn. Soc. 29: 121, t. 122, fig. C (1872); Skan in F.T.A. 4(2): 391 (1906). Type: Tanzania, Biharamulo District: Usui, *Grant* 204 (K!, holo.)

Perennial herb, erect, up to 80 cm or more tall; stems sparingly branched, densely leafy, densely covered with mixture of small appressed white hairs and much longer, patent, stouter hairs, the stouter diminishing in number below. Leaves opposite or subopposite becoming alternate below, sessile, oblong to oblong-lanceolate, 15–27 mm long, 2.5–9 mm wide, apex acute or obtuse, margins entire, glabrous or with few hairs on prominent veins beneath, margins at times remotely hispid-scabrid, 1–3-veined. Inflorescence spicate, long and slender, 5–15(–20) cm long; flowers markedly imbricate; bracts leaf-like, ovate-lanceolate, 13–20 mm long, 4.5–5.5 mm wide, 1-obscurely 3-veined, glabrous, rather long-ciliate; bracteoles 2, lanceolate, 7–11 mm long, 1–1.5 mm wide, acuminate, glabrous, ciliate. Calyx greenish, 5-lobed, obscurely 10-veined, 10.5–12 mm long, glabrous; lobes unequal, lanceolate, 1.5–2.5 mm long, acuminate, ciliate. Corolla deep pink to magenta, 26–28 mm long; tube cylindric, up to 21 mm long, externally densely minutely pubescent, spreading, arcuate above midway; lobes obovate, 5–7 mm long, margins somewhat irregular, appearing slightly toothed, throat densely hirsute. Capsule oblong, ± 5.5 mm long, 2 mm broad, bisulcate, beaked.

TANZANIA. Biharamulo District: Biharamulo–Runazi, 6 Dec. 1956, *Gane* 98!; Ufipa District: 3 km on Muze road, from Sumbawanga–Mpanda road, 29 Apr. 1997, *Bidgood et al.* 3526! & Malonje Farm, Sumbawanga, 5 Mar. 1957, *Richards* 8466!

DISTR. **T** 1, 4; not known elsewhere

HAB. Grasslands; 1350–2100 m

CONSERVATION NOTES. Near Threatened (NT); known from a few collections and as far as is known at no threat but with grazing pressure in the area it is assessed as NT

SYN. *Striga senegalensis* T.Thoms. in Speke, Nile Append.: 642 (1863), *non* Benth. Type as above.

34. **STRIGA**

Lour., Fl. Cochinch.: 22 (1790); F.T.A. 4(4): 399 (1906)

Annual, rarely perennial herbs, semi-parasitic on roots, hirsute or scabrous, often turning black on drying. Root system greatly condensed, adventitious roots arising from subterranean scales, these fine roots terminated by small (1–2 mm diameter) haustoria, in some species a large (to 5 cm) primary haustorium present; stems stiffly erect, usually ridged. Leaves opposite or subopposite, sessile or subsessile, reduced to small scales near the base of the stem in most species. Inflorescence a spike, flowers in axils of bracts or in dense heads. Bracts leaf-like or reduced; bracteoles 2. Calyx tubular, 5-lobed or with 5 (rarely 4) teeth, in some species intercostal veins present. Corolla with a narrow tube and expanded 2-lipped limb; tube bent at right angles; upper lobes fused, lower 3 lobes spreading. Stamens 4, didynamous, inserted in the tube below the mouth; anthers monothecal, basifixed on short filaments. Style usually persistent; stigma bifid. Nectary present at base of ovary. Capsule loculicidal. Seeds minute ("dust seeds") with prominent encircling ridges.

A genus with 33 species in the Old World tropics and subtropics with occasional introduction into the warmer parts of N America. Semi-parasitic on roots of grasses, cereal crops and a wide range of other hosts; several species are serious cereal pests (especially of sorghum) in drier parts of West Africa.

1. Calyx 10- or more ribbed with at least 1 rib between each calyx tooth 2
   Calyx 4–5-ribbed, each rib ending in a calyx tooth 9
2. Plants glabrous; leaves reduced to scales; roots tuberous ... 15. *S. baumannii*
   Plants hairy; roots not tuberous 3
3. Calyx 10-ribbed (i.e. with 1 extra rib to sinus between each tooth) 4
   Calyx 15- or more ribbed (i.e. with 2 extra ribs between each calyx tooth) 7
4. Leaves 3–7(–10) mm wide, coarsely toothed; flowers salmon-pink .... 9. *S. forbesii*
   Leaves 1–2(–3) mm wide, more or less entire; flowers scarlet red 5
5. Plants usually well branched, slightly hairy; calyx glabrous with white bands between the ribs, lobes triangular, acute but not subulate .... 14. *S. fulgens*
   Plants usually unbranched and sparsely hairy; calyx lobes narrow and acute 6
6. Flowers in opposite pairs, many opening at the same time; corolla > 22 mm long .... 13. *S. elegans*
   Flowers alternate, usually 2 or 3 opening at the same time; corolla < 22 mm long .... 12. *S. asiatica*
7. Leaves 2–6 mm wide, margins with few coarse teeth; flowers bright orange-red .... 8. *S. latericea*
   Leaves 1–4 mm wide, margins entire; flowers cream, corolla tube green 8
8. Corolla tube ± 20 mm long, limb 15–20 mm across ...... 10. *S. pubiflora*
   Corolla tube 12–15 mm long, limb 10–12 mm across ..... 11. *S. angustifolia*

9. Plants grey-hairy, the hairs appressed to the stems and directed downwards .... 2. *S. linearifolia*
   Plants with erect or ascending hairs or glabrous 10
10. Plants tufted; stems yellow or reddish-purple, succulent, usually branching from the base, glabrous or slightly puberulous .... 7. *S. gesnerioides*
   Plants not tufted; stems simple or branching above, not succulent, conspicuously or sparsely pubescent to pilose or hispid or scabrid 11
11. Corolla tube ≥ 10 mm long, pubescent to glandular-pubescent externally 12
   Corolla tube ≤ 10 mm long; pubescent but never glandular 5. *S. gracillima*
12. Bracts shorter than calyx; flowers bright pink, showy; lower lobes of corolla 10–13 mm long .... 4. *S. hermonthica*
   Bracts as long as or longer than calyx; flowers pink to purple or white tinged purple, not showy; lower lobes of corolla < 10 mm long 13
13. Corolla 2-lipped, with the lobes of the lower lip fused to more than half their length; lobes ± 3 mm long, acute . 1. *S. bilabiata*
   Corolla not strongly 2-lipped, with the lobes of the lower lip fused to less than half their length; lobes > 3 mm long, rounded 14
14. Pubescence hispid with upward pointed hairs; leaves spreading; flowers alternate, in spikes shorter than vegetative stem; flowers pink .... 3. *S. aspera*
   Pubescence scabrid; leaves ascending; flowers opposite or alternate, in long lax spikes longer than vegetative stem, only two flowers at same node open per inflorescence branch; flowers white (rarely yellow or pinkish) .... 6. *S. passargei*

1. **Striga bilabiata** (*Thunb.*) *Kuntze*, Rev. Gen. Pl. 3, 2: 240 (1898). Type: South Africa, *Thunberg* s.n. (UPS, holo.)

Perennial herb; stems erect, terete or 4-angled, 10–30 cm high, simple or sparsely branched, branches arched, pilose or scabrid, hairs divergent. Leaves ascending or spreading linear to lanceolate, 2–25(–30) mm long, 1–2 mm wide, margins entire, pubescent, slightly scabrid. Flowers in spikes, alternate, the upper ones opposite, sometimes occupying much of the stem, as long as the vegetative part of the stem; bracts similar to the leaves; upper bracts 10–20 mm long, longer than the calyx; lower bracts lanceolate, shorter than the calyx. Calyx tubular, 5-ribbed, 6–12 mm long; calyx tube 2–5 mm long, 5-lobed, lobes unequal, lanceolate to triangular, 3–4 mm long, pubescent. Corolla pink to violet or whitish tinged purple with a pink spot inside lower lip, 2-lipped; tube 12–14 mm long, bent just above the calyx tube and expanding distally, pubescent to glandular-pubescent outside; lobes of the lower lip fused to more than half their length; lobes ± 3 mm long, acute, the upper lobes shorter. Capsule oblong, 4 mm long, 1.5 mm wide, usually included in the calyx, acute with persistent style or base of style.

SYN. *Buchnera bilabiata* Thunb., Prodr. Fl. Cap. 100 (1800)

subsp. **barteri** (*Engl.*) *Hepper* in K.B. 14: 414 (1960); Musselman & Hepper in K.B. 41: 211, fig. 2/B (1986); Mohamed et al. in Ann. Miss. Bot. Gard. 88: 70 (2001). Type: Nigeria, Nupe, *Barter* 1170 (K!, holo.)

Flowering spike longer than the vegetative stem; calyx lobes linear. Fig. 37, p. 140.

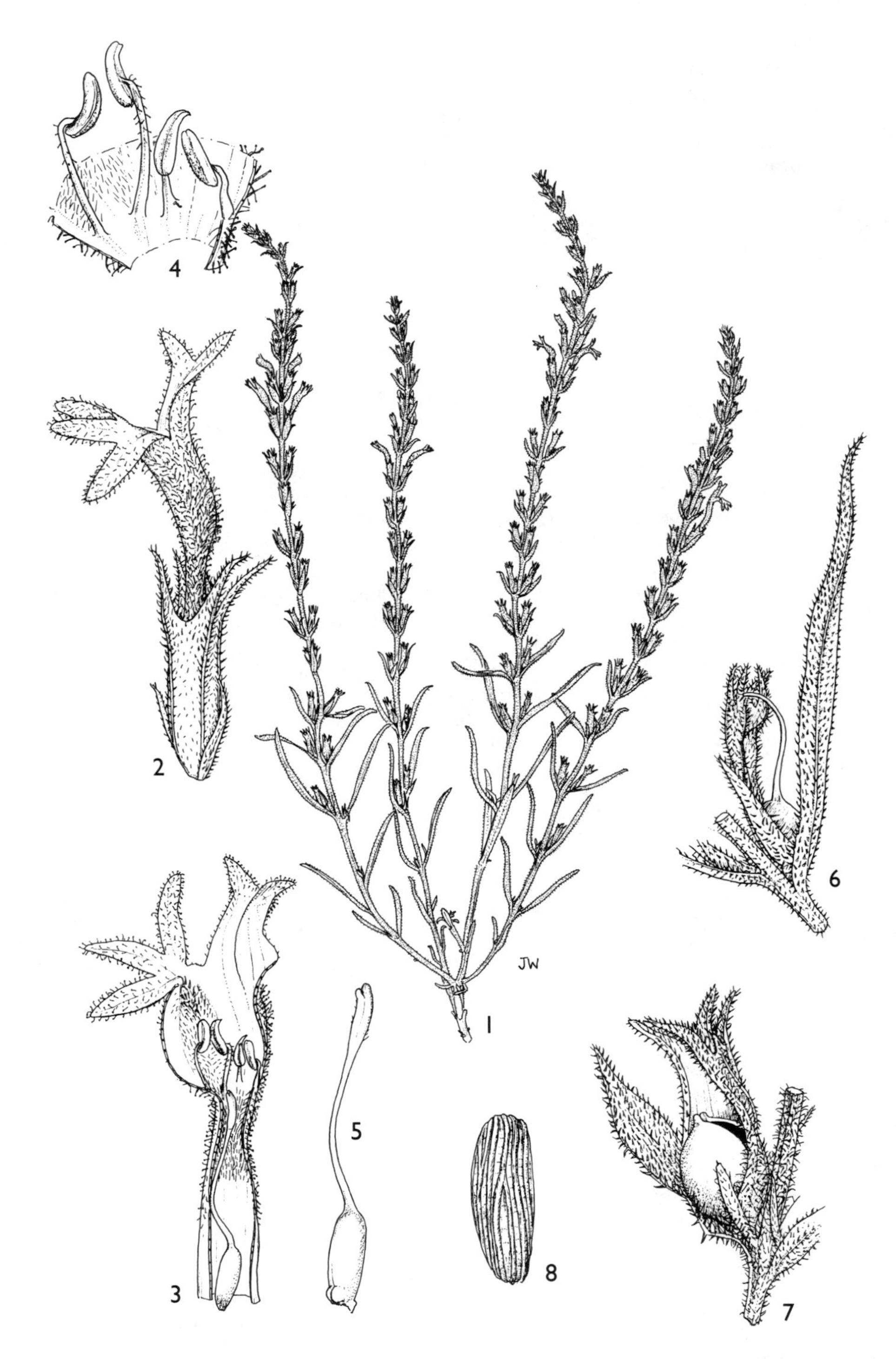

FIG. 37. *STRIGA BILABIATA* SUBSP. *BARTERI* — **1**, habit × ²⁄₃; **2**, flower × 4; **3**, corolla opened with stamens, ovary and style × 4; **4**, stamens detail × 4; **5**, young ovary and style × 6; **6**, young capsule × 5; **7**, mature capsule × 6; **8**, seed × 80. 1, 6, 7, 8 from *Lye* 2337; 2–5 from *Haarer* 2227. Drawn by Juliet Williamson.

UGANDA. Toro District: Kyaka, 1913, *Snowden* 27!; Teso District: Kapere Ferry, Soroti, 16 Sept. 1954, *Lind* 340!; Mubende District: Kikandwa, 16 Mar. 1969, *Lye* 2337!

TANZANIA. Ngara District: Mt Rukurazo, Bugufi, Aug. 1960, *Tanner* 5108A! & 5110!; Bukoba District: Ndama, Oct. 1931, *Haarer* 2227!

DISTR. **U** 2–4; **T** 1–4; West Africa, Rwanda, Burundi, Sudan

HAB. Seasonally wet grassland, semi-parasitic on grass roots; 1100–1700 m

USES. None recorded on specimens from our area

CONSERVATION NOTES. Least Concern (LC); common and widespread

SYN. *S. barteri* Engl. in E.J. 23: 514, t. 12, fig. E (1897); Skan in F.T.A. 4(2): 406 (1906); F.P.S. 3: 145 (1956)

S. glandulifera Engl. l.c., fig. H–J (1897); Brown & Massey, Fl. Sudan: 331 (1929). Type: Sudan, Niamniam, Mbala Ngia Mts, *Schweinfurth* 2931 (K!, iso.)

NOTE. *S. bilabiata* is a widespread and variable species. Hepper (in K.B. 14: 411–416 (1960)) and Mohamed et al. (op. cit.) have recognized up to 6 subspecies (all earlier recognised as separate species), two of which (according to Mohanmmed et al.) are found in the Flora area, the others restricted to southern, central and west Africa. Here, I (S.A.G.) have followed Hepper in recognising subsp. *barteri* but maintain subsp. *linearifolia* (Schum. & Thonn.) Mohamed et al. as a distinct species. Both taxa are widespread in the Flora area and are recognizably distinct. In subsp. *bilabiata* the flowering spike is as long as the vegetative stem, and calyx lobes subulate, differentiating it from subsp. *barteri*. The type subsp. is found in Congo-Kinshasa, Zambia, Angola, Zimbabwe, Botswana and South Africa.

2. **Striga linearifolia** (*Schum. & Thonn.*) *Hepper* in K.B. 14: 416 (1960) & in F.W.T.A. ed. 2, 2: 372 (1963) & in Proc. Third Symp. Parasitic Pl.: 265 (1984); Musselman & Hepper in K.B. 41: 218 fig. 4B (1986); U.K.W.F.: 261 (1994); Fischer in Fl. Ethiop. & Eritr. 5: 296 (2006). Type: Ghana, *Thonning* 284 (C, syn.; P-JU!, iso.)

Perennial herb; stems erect, 4-angled and -furrowed or -winged, 10–50 cm high, simple or sparsely branched, branches arched, pilose or scabrid, hairs grey, appressed, pointing downwards. Leaves ascending, linear to scale-like and reduced, 2–7 mm long, ± 1 mm wide, entire, pubescent, slightly scabrid. Flowers in spikes, shorter than the vegetative part of the stem, alternate, the upper ones opposite; bracts similar to the leaves, upper bracts 7–8 mm long, about as long as calyx; lower bracts shorter than the calyx. Calyx tubular, 5-ribbed, 7–10 mm long; calyx-tube 4–5 mm long, 5-lobed, lobes unequal, linear, 3–4 mm long, pubescent. Corolla pink to violet or whitish tinged purple with a pink spot inside lower lip, 2-lipped, 12–15 mm long; tube bent and expanding distally, pubescent to glandular-pubescent externally; lobes ± 3 mm long, acute, the upper lobes shorter. Capsule oblong, 4 mm long, 1.5 mm wide, usually included in the calyx, acute with persistent style or base of style.

UGANDA. Toro District: Kyaka, 1913, *Snowden* 27!; Teso District: Kapere Ferry, Soroti, 16 Sept. 1954, *Lind* 340!; Mubende District: Kikandwa, 16 Mar. 1969, *Lye* 2337!

TANZANIA. Ngara District: Mt Rukurazo, Bugufi, Aug. 1960, *Tanner* 5108A! & 5110!; Bukoba District: Ndama, Oct. 1931, *Haarer* 2227!

DISTR. **U** 2–4; **T** 1–4; West Africa, Rwanda, Burundi, Sudan

HAB. Seasonally wet grassland, semi-parasitic on grass roots; 1100–1700 m

USES. None recorded on specimens from our area

CONSERVATION NOTES. Least Concern (LC); common and widespread

SYN. *Buchnera linearifolia* Schum. & Thonn., Beskr. Guin. Pl.: 279 (1827)

Striga canescens Engl., P.O.A. C: 361 (1893) & in E.J. 23: 515, t. 12, figs. K, L (1897); Hiern in Cat. Afr. Pl. Welw. 1: 779 (1898); Moore in J.L.S. 37: 191 (1905); Skan in F.T.A. 4(2): 406 (1906). Type: Uganda, *Stuhlmann* s.n. (B†, holo.)

S. bilabiata (*Thunb.*) *Kuntze* subsp. *linearifolia* (Schum. & Thonn.) Mohamed et al. in Ann. Miss. Bot. Gard. 88: 70 (2001)

3. **Striga aspera** (*Willd.*) *Benth.* in Hook., Comp. Bot. Mag. 1: 362 (1836); DC., Prodr. 10: 501 (1846); Skan in F.T.A. 4(2): 403 (1906); F.P.S. 3: 145 (1956); Hepper in F.W.T.A. ed. 2, 2: 372 (1963) & in Proc. Third Symp. Parasitic Pl.: 263 (1984); Musselman & Hepper in K.B. 41: 209, fig. 2A (1986); Mohamed et al. in Ann. Miss. Bot. Gard. 88: 69 (2001); Fischer in Fl. Ethiop. & Eritr. 5: 297 (2006). Type: Ghana, Ada, *Isert* s.n. (B, Herb-Willd. 11182, holo.; C, iso., microfiche!)

Annual herb; stems erect, terete to slightly angled, to 30 cm high, branched or simple, hispid with upward pointed hairs. Leaves spreading, linear, usually up to 6 cm long, < 3 mm wide, scabrid. Flowers alternate, in spikes; bracts of lowest flowers ± 25 mm long, exceeding the calyx, bracts of upper flowers as long as or only slightly exceeding the calyx, 0.3 mm wide with hispidulous margins; bracteoles similar to bracts but greatly reduced, 3/4 length of calyx, 0.2 mm wide. Calyx 6–9 mm long, including subulate teeth, 5-ribbed; teeth 3–5 mm long. Corolla pink; tube 12–15 mm long, sparsely glandular-pubescent, bent and expanding above calyx; upper lip of corolla emarginate, slightly recurved, lower lobes spreading, 5–8 mm long. Capsule oblong, 4 mm long.

TANZANIA. Rungwe District: Kyimbila, 23 May 1912, *Stolz* 1304!; Songea District: Matengo Hills, Luiri Kitesa, 24 May 1956, *Milne-Redhead & Taylor* 10502!
DISTR. **T** 7, 8; Sudan and West Africa from Nigeria to Senegal
HAB. Grassland, parasitising the roots of wild grasses; 0–1300 m
USES. None recorded on specimens from our area
CONSERVATION NOTES. Least Concern (LC); common and widespread

SYN. *Euphrasia aspera* Willd., Sp. Pl. 3: 197 (1800)
Buchnera aspera (Willd.) Schum. & Thonn., Beskr. Guin. Pl.: 280 (1827)
Striga aspera (*Willd.*) *Benth.* var. *schweinfurthii* Skan in F.T.A. 4(2): 403 (1906); Brown & Massey, Fl. Sudan: 330 (1929). Type: Sudan, Djur, *Schweinfurth* 1992 (K!, holo.)

4. **Striga hermonthica** (*Delile*) *Benth.* in Hook., Comp. Bot. Mag. 1: 365 (1836); DC., Prodr. 10: 502 (1846); Engl., P.O.A. C: 361 (1895); Oliver in T.L.S. 29: 122 (1875); Skan in F.T.A. 4(2): 407 (1906); F.P.S. 3: 145, fig. 38 (1956); F.P.U.: 135 (1962); Hepper in F.W.T.A. ed. 2, 2: 372, fig. 290 (1963) & in Symp. Parasitic Weeds: 16, fig. p.15 (1973); Hepper in Proc. Third Symp. Parasitic Pl.: 264 (1984); Musselman & Hepper in K.B. 41: 216, fig. 1c (1986); U.K.W.F.: 261 (1994); Mohamed et al. in Ann. Miss. Bot. Gard. 88: 85 (2001); Fischer in Fl. Ethiop. & Eritr. 5: 295 (2006). Type: Egypt, *Delile* s.n. (MPU, holo.)

Stiffly erect, dark green annual, much branched, 20–50 cm high; stems 4-angled with a groove on each face, scabrous. Leaves opposite, linear to linear-lanceolate to lanceolate, 6–9 cm long, 1.1–1.5 cm wide, acute, margins entire, rather thick and leathery, both surfaces scabrous, margins with hispid hairs at regular intervals. Flowers opposite or subopposite, 8–10 open at same time per inflorescence branch, variable in size; bracts almost as long as the calyx with prominent hispid hairs on the margins, tip curved away from stem; bracteoles half the length of the calyx. Calyx 5-ribbed, 10–12 mm long, 2.5 mm wide, hispid on the ribs, intercostal portion whitish or pinkish and translucent, glabrous; lobes unequal, acuminate. Corolla bright pink (rarely white); tube 10–20 mm long, bent just above calyx teeth, outside with very scattered hairs, glandular-pubescent; upper lobe emarginate, erect, lower lobes 10–13 mm long, 8–10 mm wide. Capsule oblong, 12–15 mm long, 2–2.5 mm wide. Fig. 38: 4–6, p. 145.

UGANDA. Karamoja District: Iriri, June 1957, *J. Wilson* 349!; Ankole District: Bunyaruguru, Feb. 1939, *Purseglove* 560!; Mubende District: Bugandadzi, 1904, *Dawe* 120!
KENYA. South Kavirondo District: Kisiti, Sept. 1933, *Napier* 5343! & Homa Bay, 7 July 1958, *Bogdan* 4535!

TANZANIA. Musoma District: Mugango, 1 June 1959, *Tanner* 4290!; Shinyanga District: Mwanza to Shinyanga road, 1 May 1945, *Greenway* 7392!

DISTR. **U** 1–4; **K** 5; **T** 1; serious pest in Sudan, also across West Africa to Senegal, and parts of eastern Africa, Egypt; also Madagascar

HAB. Parasitic on the roots of maize (*Zea*), millet (*Pennisetum*), durra (*Sorghum*), sugar (*Saccharum*), finger millet (*Eleusine*) and wild grasses; 750–1550 m

USES. None recorded on specimens from our area

CONSERVATION NOTES. Least Concern (LC); common and widespread

SYN. *Buchnera hermonthica* Delile, Fl. Egypte: 245, t. 34, fig. 3 (1813)

NOTE. The size of the corolla varies immensely: it is typically large and decorative but smalll-flowered forms have been distinguished under the names *S. senegalensis* Benth. and *S. gracillima* Melchior, the latter having a slender inflorescence and the lower lip only 3 mm long. Although Melchior allied the latter species with *S. warneckei* Skan its affinities seem to lie with *S. hermonthica* owing to its broad bracts. Mohammed et al., op. cit. state that the corolla tube is never glandular-pubescent in *S. hermonthica.* I (S.A.G.) have examined many specimens of this taxon from the FTEA region and note the corolla tube to be more or less always glandular-pubescent, a character similar to that in *S. apsera. S. hermonthica* and *S. aspera* have been recorded to interbreed and it is possible that the hybrids have glandular-pubescent corollas.

5. **Striga gracillima** *Melch.* in N.B.G.B. 11: 681 (1932); Mohamed et al. in Ann. Miss. Bot. Gard. 88: 84 (2001). Type: Tanzania, Rungwe District: Kyimbila, *Stolz* 1304 (?B, holo.; BM!, K!, iso.)

Annual herb; stems erect, 25–50 cm high, simple or sparsely branched, slender, scabrid to pubescent, hairs pointing upwards. Leaves more or less thoughout the length of the stem, spreading, linear to linear-lanceolate, 14–50 mm long, 1–4 mm wide, entire, scabrid. Flowers alternate in terminal raceme; bracts similar to the leaves, 10–25 mm long, 1–2 mm wide; bracteoles as bracts, as long as the calyx; pedicels 1–2 mm. Calyx 5-ribbed, 5–7 mm long, 5-lobed, lobes unequal, linear to subulate, 1–2 mm, the adaxial < 1 mm long, shorter than the tube, pubescent on the ribs. Corolla purple to pinkish mauve, tube 6–8 mm, and expanding distally, hairy outside, never glandular; lobes obovate, 1–3 mm spreading; upper-lip obovate, 4–5 mm, emarginate. Capsule oblong, 4–5 mm, usually included in the calyx.

TANZANIA. Kigoma District: Mt Livandabe [Lubalisi], 31 May 1997, *Bidgood et al.* 4226!; Songea District: Matagoro Hills, 15 April 1960, *Wilsay* 91!; Luiri Kitesea, 24 May 1956, *Milne-Redhead & Taylor* 10502!

DISTR. **T** 4, 7, 8; Zambia, Malawi

HAB. In woodland and grassland, on slopes and on grey sandy soil; 1000–1800 m

USES. None recorded on specimens from our area

CONSERVATION NOTES. Least Concern (LC); not common but not at threat

6. **Striga passargei** *Engl.* in E.J. 23: 515, t. 12, figs. M, N (1897); Skan in F.T.A. 4(2): 403, note under *S. aspera* (1906); Hepper in F.W.T.A. ed. 2, 2: 372 (1963) & in Proc. Third Symp. Parasitic Pl.: 265 (1984); Musselman & Hepper in K.B. 41: 219, fig. 3B (1986); Mohamed et al. in Ann. Miss. Bot. Gard. 88: 93 (2001). Type: Cameroon, Jola, *Passarge* 48 (B†, holo.)

Annual herb, 10–40 cm high; stems simple or branched, somewhat succulent, slightly 4-angled, scabrid, drying black. Leaves few on lower half of stem, ascending, narrowly elliptic, 10–45 long, 1–3 mm wide. Flowers opposite or alternate, in long lax spikes, only two flowers at same node open per inflorescence branch; bracts much larger than leaves, linear, up to 5 cm long and 4 mm wide, often recurved; upper bracts as long as or longer than the calyx; bracteoles subulate, shorter than calyx-tube. Calyx in flower ± 8 mm long, 1.5 mm diameter, enlarging in fruit, 5-ribbed; teeth subulate, up to half as long as the tube and spreading in fruit, ribs scabrid, hyaline in between. Corolla white (rarely yellow or pinkish), tube 10–17 mm long, finely glandular-pubescent outside, throat hairy; upper lobe emarginate or bilobed, 2–5 mm long, 2–4 mm wide, reflexed; lower lobes spreading, 2–7 mm long, 1.5–4 mm wide. Capsule ovoid, 8–13 mm long, 2 mm wide.

TANZANIA. Mwanza District: Mwanza, 1926, *Davis* 255!; Mpanda District: Rukwa Valley, Tumba, 26 Feb. 1952, *Siame* 144!; Kilwa District: Kingupira Forest, 1 Mar. 1976, *Vollesen* in MRC 3311!

DISTR. **T** 1, 4, 8; Senegal, Ghana, Nigeria, Cameroon, Sudan; SW Arabia

HAB. Secondary *Acacia* woodland, by roadside, rocky places, parasitic on *Urochloa* and other grasses; 100–1450 m

USES. None recorded on specimens from our area

CONSERVATION NOTES. Least Concern (LC); common and widespread

7. **Striga gesnerioides** (*Willd.*) *Vatke* in Oesterr. Bot. Zeitschr. 25: 11 (1875) & in Linnaea 43: 310 (1882); Engl., P.O.A. C: 361 (1895) & in E.J. 30: 405 (1901); F.P.S. 3: 144, fig. 37 (1956); Hepper in F.W.T.A. ed. 2, 2: 373 (1963) & in Symp. Parasitic Weeds: 16, fig. p. 15 (1973); Hepper in Proc. Third Symp. Parasitic Pl.: 264 (1984); Musselman & Hepper in K.B. 41: 213, fig. 1B (1986); U.K.W.F.: 261, pl. 113 (1994); Mohamed et al. in Ann. Miss. Bot. Gard. 88: 80 (2001); Mohammed & Musselman in Thulin (ed.), Fl. Somal.: 3: 287 (2006); Fischer in Fl. Ethiop. & Eritr. 5: 296 (2006). Type: India, *Koenig* s.n. (B-W 11573, holo., photo!)

Tufted, greenish-yellow or reddish-purple, succulent annual or perennial, usually branching from the base, 11–25(–35) cm high, drying black; a single large primary haustorium ± 1–3 cm diameter is usually present on each plant when parasitizing; adventitious roots abundant from subterranean scales; stems quadrangular with obtuse angles, minutely puberulous. Leaves scale-like, appressed to the stem, 5–10 mm long, 2–3 mm wide; minutely puberulent with upward pointing hairs to almost glabrous. Flowers opposite or alternate, usually 2(–3) per node; bracts usually the same length and width as calyx, acuminate; bracteoles minute, $^{3}/_{4}$ length of calyx. Calyx 5-ribbed, 4–8 mm long (including the teeth), 2 mm wide; teeth linear, acuminate, shorter than the calyx-tube. Corolla light blue, pink or dark purple, 1.2–1.5 cm long; tube 7–11 mm, bent just below limb, glandular, and pubescent with small hairs, sometimes papillae-like; upper lobes 2–2.5 mm long, sharply recurved, lower 3–3.2 mm long. Capsule oblong, 1–2 cm long, 3 mm wide. Fig. 38: 3, p. 145.

UGANDA. Karamoja District: Kangole, July 1957, *J. Wilson* 373!; Busoga District: N of Nkondo, 9 July 1953, *G.H.S. Wood* 805!; Bunyoro District: Butiaba, Lake Albert, 1 Sept. 1943, *A.S. Thomas* 3753!

KENYA. Northern Frontier District: Dandu, 6 May 1952, *Gillett* 13104!; Machakos District: Ulu, 1 Oct. 1961, *Sangai* EAH 12546!; Tana River District: 46 km S of Garsen, 3 Oct. 1961, *Polhill & Paulo* 579!

TANZANIA. Pangani District: Pangani, 13 July 1953, *Drummond & Hemsley* 3319!; Iringa District: 19 km E of Ibuguziwa ferry, 13 Feb. 1972, *Bjørnstad* 1363!; Lindi District: Mlinguru, 26 Nov. 1935, *Schlieben* 5791!

DISTR. **U** 1–4; **K** 1–7; **T** 1–4, 6–8; tropical and South Africa; tropical Arabia, India, Sri Lanka

HAB. In permanent grassland parasitising grass roots (see note), and in cultivated ground; reported on many species including trees; on *Euphorbia, Indigofera, Tephrosia, Stylosanthes, Merremia, Sorghum* and wild grasses; 0–1700 m

USES. None recorded on specimens from our area

CONSERVATION NOTES. Least Concern (LC); common and widespread

SYN. *Buchnera gesnerioides* Willd., Sp. Pl. 3: 338 (1802)

B. orobanchoides R.Br. in Endl. Bot. Zeit. 2: 388, t. 2 (1832). Type: Ethiopia, without locality, *Salt* s.n. (BM!, holo.)

Striga orobanchoides (R.Br.) Benth. in Hook., Comp. Bot. Mag. 1: 361 (1836) & in DC., Prodr. 10: 501 (1846); Hiern in Cat. Afr. Pl. Welw. 1: 778 (1898) & in Fl. Cap. 4, 2: 380 (1904); Skan in F.T.A. 4(2): 402 (1906)

FIG. 38. *STRIGA ASIATICA* — **1**, habit × 1; **2**, calyx × 3. *STRIGA GESNERIOIDES* — **3**, habit of parasitic plant × 1. *STRIGA HERMONTHICA* – **4**, habit × 1; **5**, calyx × 2; **6**, corolla opened × $^3/_4$. Drawn by F.N. Hepper. Reproduced from K.B. 41(1): 1986.

S. orchidea Benth. in DC., Prodr. 10: 501 (1846). Type: Sudan, Kordofan, *Kotschy* 387 (BM!, iso.)

NOTE. Mohammed et al. (op. cit.) incorrectly state that this species is parasitic only on dicotyledons.

8. **Striga latericea** *Vatke* in Linnaea 43: 311 (1882); Skan in F.T.A. 4(2): 411 (1906); Hepper in Proc. Third Symp. Parasitic Pl.: 264 (1984); U.K.W.F.: 261 (1994); Mohamed et al. in Ann. Miss. Bot. Gard. 88: 898 (2001); Mohammed & Musselman in Thulin (ed.), Fl. Somal.: 3: 288 (2006); Fischer in Fl. Ethiop. & Eritr. 5: 295 (2006). Type: Kenya, Kitui, *Hildebrandt* 2752 (B†, holo.)

Perennial semi-parasitic herb, 15–65 cm high; stems usually unbranched, finely hispid, internodes 2–7 cm long. Leaves mostly opposite, lanceolate to linear-lanceolate, 20–80 mm long, 2–6 mm wide, margins usually with a few coarse teeth on each side, finely hispid. Inflorescences at first rather dense becoming laxer, 3–15 cm long; lower bracts leafy, upper shorter than calyx; bracteoles subulate, equalling calyx-tube. Calyx 15-ribbed, densely hispid-pubescent along veins, 10–21 mm long; teeth linear-lanceolate, nearly equalling tube especially in fruit. Corolla salmon-pink, hairy in throat; tube 20–30 mm long, inflated well below limb; upper lip broadly obovate, 6–8 mm long, emarginate; lower lip deeply 3-lobed; lobes obovate to obovate-oblong, central lobe 11–19 mm long, 7–12 mm broad. Capsule compressed-oblong, 9 mm long, acutely beaked.

KENYA. Northern Frontier District, June 1937, *Jex-Blake* in Mus. 6880!; Machakos District: Sultan Hamud, 22 Apr. 1902, *Kässner* 645!; Kwale District: Mwachi, 10 Sept. 1953, *Drummond & Hemsley* 4247!

TANZANIA. Arusha District: Lekuruki, 5 Dec. 1969, *Richards* 24897!; Pare District: Ngula, May 1928, *Haarer* 1244!; Lushoto District: Usambara Mts, 1893, *Holst* 2532!

DISTR. **K** 1, 4, 7; **T** 2, 3; Somalia

HAB. Seasonally wet places among grass; weed in sugar cane, maize, on *Chrysopogon* spp.; 100–1700 m

USES. None recorded on specimens from our area

CONSERVATION NOTES. Least Concern; (LC) common and widespread

SYN. *Rhamphicarpa stricta* Engl. in Abh. Preuss. Akad. Wiss. 24 (1894). Type: Tanzania, Usambara Mts, *Holst* 2532 (K!, isosyn.)

Cycnium strictum (Engl.) Engl. in P.O.A. C.: 361 (1895). Types: Tanzania, Tanga District: Duga, *Holst* 3201 (?B†, syn.); Lushoto District: Maschewa [Mascheua], *Holst* 8833 (?B†, syn.) & Hosiga, *Holst* 2532 (?B†, syn.)

9. **Striga forbesii** *Benth.* in Hook., Comp. Bot. Mag. 1: 364 (1836) & in DC., Prodr. 10: 503 (1846); Klotzsch in Peters, Reise Mossamb. Bot.: 227 (1862); Engl. in P.O.A. C: 361 (1895), excl. syn.; Hiern in Cat. Afr. Pl. Welw. 1: 780 (1898) & in Dyer, Fl. Cap. 4, 2: 384 (1904); Skan in F.T.A. 4(2): 410 (1906); U.O.P.Z.: 456 (1949); F.P.S. 3: 146 (1956); F.P.U.: 135, fig. 79 (1962); Hepper in F.W.T.A. ed. 2, 2: 371 (1963) & in Proc. Third Symp. Parasitic Pl.: 264 (1984); Musselman & Hepper in K.B. 41: 213, fig. 4C (1986); U.K.W.F.: 261, pl. 113 (1994); Mohamed et al. in Ann. Miss. Bot. Gard. 88: 78 (2001); Mohammed & Musselman in Thulin (ed.), Fl. Somal.: 3: 288 (2006); Fischer in Fl. Ethiop. & Eritr. 5: 295 (2006). Type: Mozambique, without locality, *Forbes* s.n. (K!, holo.)

Stiffly erect annual, dark green, up to 75 cm high, but usually much shorter; stems simple to sparsely branched, young parts glandular-pubescent, becoming scabrous with age. Leaves opposite, sessile, lanceolate, (1–)2–4(–7) cm long, 3–7(–10) mm wide, acute to obtuse, margins coarsely toothed, each tooth the termination of one of the secondary veins, margins recurved, scabrid-pubescent, 3-veined. Flowers opposite to alternate, in open racemes; bracts leaf-like, seldom more than 2 cm long; bracteoles linear, shorter than the calyx. Calyx 10-ribbed, tube 5–9 mm long, lobes lanceolate, as long as tube, veins scabrid hairy. Corolla salmon pink, no obvious scent; tube ± 2 cm long, bent near top, glandular-pubescent; middle lower lobe obovate, 7–9(–13) mm long, 4–7(–11) mm wide, upper lobes smaller, emarginate. Capsule oblong, flattened, shorter than calyx-tube, rounded at apex.

UGANDA. Karamoja District: Nonyili Ridge, Mar. 1960, *J. Wilson* 880!; Teso District: Serere, Dec. 1931, *Chandler* 383!; Mengo District: Busana Bugerere, Apr. 1930, *Liebenberg* 1532!

KENYA. Nandi District: Kaimosi, near Yala Bridge, 3 June 1933, *Gilbert Rogers* 709!; Masai District: Mara Plains, 15 Sept. 1960, *R.M. Stewart* 327!; Tana River District: Wema, 15 July 1972, *Gilbert & Kibuwa* 19932!

TANZANIA. Tanga District: Kange Estate, near Mawmi, 14 Dec. 1965, *Faulkner* 3271!; Mpanda District: Tumba, 7 Mar. 1951, *Bullock* 3748!; Morogoro District: 10 Apr. 1925, *Ritchie* 3!; Zanzibar, Masingini, 5 Oct. 1931, *Vaughan* 1613!

DISTR. **U** 1–4; **K** 3, 4, 6, 7; **T** 1, 3, 4, 6–8; **Z**; **P**; widespread in tropical Africa; Madagascar

HAB. Low lying grassy places with cracking clay in dry season; 10–1700 m

USES. None recorded on specimens from our area

CONSERVATION NOTES. Least Concern (LC); common and widespread

SYN. *Buchnera forbesii* (Benth.) D.Dietr., Syn. Pl. 3: 526 (1843)
Cycnium pratense Engl. in P.O.A. C: 301 (1895). Type: Tanzania, Usambara Mts, Lutindi, *Holst* 3269 (B†, holo.)

10. **Striga pubiflora** *Klotzsch* in Peters, Reise Mossamb. Bot.: 227 (1861); Engl. in P.O.A. C: 361 (1895); Skan in F.T.A. 4(2): 412 (1906); Moriarty, Wild Flow. Malawi: 77, t. 39, 3 (1975); Hepper in Proc. Third Symp. Parasitic Pl.: 265 (1984); Mohamed et al. in Ann. Miss. Bot. Gard. 88: 94 (2001). Type: Mozambique, *Peter* s.n. (B†, holo.)

Erect perennial herb with pale purple root tubers; stems slender, 30–65 cm high, usually several simple stems from a woody base, sometimes branched above, shortly pubescent. Leaves opposite below, alternate above, held vertically, linear, 15–38 mm long, ± 2 mm wide, acute, shortly scabrid-pubescent. Flowers 4–14, alternate, in loose terminal inflorescences; pedicels ± 2 mm; bracts and bracteoles subulate, usually shorter than calyx. Calyx tubular, strongly 15-ribbed, tube 8–11 mm long, teeth 5, linear, 3–8 mm long. Corolla white with green tube; tube ± 20 mm long, sharply bent above, densely pubescent outside; upper lip broadly truncate-ovate, undivided; lower lip broadly 3-lobed, lobes finely pubescent on margins and outside, central lobe obovate, up to 15 mm long, 10 mm broad. Capsule ovoid, 8 mm long, 2.5 mm wide, valves recurved at apex after dehiscence.

KENYA. Lamu District: near Lamu, Dec. 1875, *Hildebrandt* 1907!; Kwale District: NE of Lunga Lunga, 23 Sept. 1958, *Moomaw* 953! & Shimba Hills, 13 Apr. 1968, *Magogo & Glover* 849!

TANZANIA. Tanga District: Muva, June 1893, *Holst* 2989!; Uzaramo District: Pugu Forest Reserve, June 1954, *Semsei* 1720!; Songea District: 5 km east of Gumbiro, Jan. 1956, *Milne-Redhead & Taylor* 8421!

DISTR. **K** 7; **T** 3, 4, 6–8; Malawi, Mozambique

HAB. Woodland among grass in damp sandy soil and in salt marshes, on roots of grasses; 100–1100 m

USES. None recorded on specimens from our area

CONSERVATION NOTES. Least Concern (LC); common and widespread

SYN. *Striga zanzibarensis* Vatke in Linnaea 43: 310 (1882). Type: Kenya, Lamu, "Zanzibar coast", *Hildebrandt* 1907 (B†, holo.; K!, iso.)
S. pubiflora Klotzsch var. *zanzibarensis* (Vatke) Engl. in P.O.A. C: 361 (1893) as "*sansibarensis*"; Skan in F.T.A. 4(2): 412 (1906)

11. **Striga angustifolia** (*D.Don*) *Saldanha* in Bull. Bot. Surv. India 57, 1: 70 (1963) & in Fl. Hassan Distr.: 526 (1976); Mohamed et al. in Ann. Miss. Bot. Gard. 88: 66 (2001). Type: Nepal, *Wallich* s.n. (K!, holo.)

Erect annual or biennial herb, 15–50 cm high, with usually erect, simple branches; stems stiff, ± pubescent to densely covered with very short hispid hairs. Leaves opposite to subopposite, alternate above, linear to linear-lanceolate, 10–20 mm long, 1–4 mm wide, acute, entire, finely pubescent. Flowers alternate, axillary, solitary, forming long lax terminal spikes; lower bracts leaf-like, the upper linear to subulate; bracteoles subulate, ± 4 mm long. Calyx tubular, 10–12 mm long, very prominently 15-ribbed, 5-toothed, finely pubescent; teeth linear-lanceolate, 3–5 mm long, elongating in fruit. Corolla white or cream with greenish tube, pubescent outside; tube 12–15 mm long, abruptly curved just above calyx teeth and inflated above; upper lip obovate, emarginate or truncate; lobes of lower lip obovate, 6–8 mm long, 5 mm wide, rounded. Capsule ovoid 4–5 mm long, ± 3 mm in diameter, apiculate, valves recurved sharply after dehiscence.

TANZANIA. Songea District: 3 km W of Gumbiro, 9 May 1956, *Milne-Redhead & Taylor* 10024! & S of Hanga R., 64 km from Songea on Njombe road, 16 Nov. 1966, *Gillett* 17889!; Tunduru District: E of Songea district boundary, 5 May 1956, *Milne-Redhead & Taylor* 10577!

DISTR. **T** 8; Zambia, Malawi, Mozambique, Zimbabwe; India, Nepal, Sri Lanka

HAB. Wet grassland, abandoned cultivations, *Brachystegia–Uapaca* woodland; 800–900 m

USES. None recorded on specimens from our area

CONSERVATION NOTES. Least Concern (LC); common and widespread

SYN. *Buchnera angustifolia* D.Don, Prodr. Fl. Nepal: 91 (1825)
Buchnera euphrasioides auctt. *non* Vahl; Wight, Ic. Pl., t. 855 (1844–45)
Striga euphrasioides sensu Benth. in Hook., Comp. Bot. Mag. 1: 364 (1836) & in DC., Prodr. 10: 503 (1846); Skan in F.T.A. 4(2): 412 (1906); Hepper in Proc. Third Symp. Parasitic Pl.: 264 (1984), *non* Vahl

12. **Striga asiatica** (*L.*) *Kuntze*, Rev. Gen. Pl. 2: 466 (1891); U.O.P.Z.: 456 (1949); F.P.S. 3: 145 (1956); F.P.U.: 135 (1962); Hepper in F.W.T.A. ed. 2, 2: 372 (1963) & in Rhodora 76: 45 (1974); Blundell, Wild Flow. Kenya: 100 (1982); Hepper in Proc. Third Symp. Parasitic Pl.: 262 (1984); Musselman & Hepper in K.B. 41: 207, fig. 1A (1986); U.K.W.F.: 261, pl. 113 (1994); Mohamed et al. in Ann. Miss. Bot. Gard. 88: 67 (2001); Fischer in Fl. Ethiop. & Eritr. 5: 295 (2006). Lectotype: "Habitat in Zeylona, China", *Torén*, Herb. Linn. No. 790.10, branched specimen (LINN) designated by Hepper (1974)

Erect annual; stems square, sparsely branched or if branched then from above the middle, 15–20 cm high, densely hispid. Leaves ascending, linear to narrowly elliptic, 8–16 mm long, ± 1 mm wide, acute, entire, hispid. Flowers opposite or alternate, only 2 flowers per inflorescence branch; bracts 9–12 mm long, longer than the calyx; bracteoles 2 mm or less long. Calyx 10-ribbed, 6 mm long, including teeth, 2.5 mm wide; teeth 2 mm long. Corolla scarlet red, outside yellowish, occasionally entirely yellow; tube 11–14 mm long, bent near limb, sparsely to densely pubescent; limb with the upper lobe emarginate, 4 mm wide; lower lobes spreading, 4 mm long, 1.5 mm wide; whole limb positioned at right angles to the tube; style persistent, usually with white pollen mass on the stigma. Capsule oblong, 7 mm long, 2 mm wide. Fig. 38: 1–2, p. 145.

UGANDA. West Nile District: Koboko, Sept. 1940, *Purseglove* 1048!; Kigezi District: Kambuga, May 1947, *Purseglove* 2424!; Mubende District: Katera, 16 Mar. 1969, *Lye, Lester & Morrison* 2296!

KENYA. Nakuru District: S of Lake Naivasha, 4 Sept. 1971, *Gilbert* 4840!; Nyeri District: Zawadi Estate, 19 May 1974, *R.B. & A.J. Faden & Evans* 74/572!; Kwale District: Buda Mafisini Forest, 17 Aug. 1953, *Drummond & Hemsley* 3839!

TANZANIA. Musoma District: Handajega, 21 Feb. 1968, *Greenway, Kanuri & Turner* 13322!; Mpanda District: Tumba, 13 Mar. 1951, *Bullock* 3770!; Kilosa District: R. Ruaha, 140 km E of Iringa, 2 May 1975, *Hepper, Field & Mhoro* 5218!; Zanzibar: Chaani, 26 Jan. 1929, *Greenway* 1204!

DISTR. **U** 1–4; **K** 1–7; **T** 1–4, 6–8; **Z**; **P**; widespread in tropical Africa, tropical Asia, South Africa, Mascarenes, introduced into U.S.A.

HAB. Semi-parasitic on roots of wild grasses and cultivated maize (*Zea*); 0–1900 m (–2480 m fide U.K.W.F.)

USES. None recorded on specimens from our area

CONSERVATION NOTES. Least Concern (LC); common and widespread

SYN. *Buchnera asiatica* L., Sp. Pl.: 630 (1753)

Striga lutea Lour., Fl. Cochinch.: 22 (1790); F.T.A. 4(2): 409 (1906). Type: China, Canton, *Loureiro* s.n. (P, holo., photo.!)

Buchnera hirsuta Benth., Scroph. Ind.: 41 (1835). Type: India, *Wallich* 3869 (K, holo.)

Striga hirsuta (Benth.) Benth. in DC., Prodr. 10: 502 (1846)

NOTE. An extremely variable species, simple or branched, leafy or almost leafless, densely pubescent (*S. hirsuta*) to almost glabrous (but strigose), flowers brilliant scarlet or yellow (*S. lutea*). Tall, slender plants with white flowers from Tanzania (Tunduru District, *Richards* 17912, 18024) may be separable as a distinct species.

13. **Striga elegans** *Benth.* in Hook., Comp. Bot. Mag. 1: 363 (1836); DC., Prodr. 10: 502 (1846); Engl., Hochgebirgsfl. Trop. Afr.: 382 (1892) & in P.O.A. C: 361 (1895); Hiern in Cat. Afr. Pl. Welw. 1: 779 (1898) & in Dyer, Fl. Cap. 4, 2: 282 (1904); Engl. in E.J. 28: 480 (1900); S. Moore in J.L.S. 37: 190 (1905); Skan in F.T.A. 4(2): 408 (1906); Hepper in Proc. Third Parasitic Pl. Symp.: 263 (1984); U.K.W.F.: 261 (1994); Mohamed et al. in Ann. Miss. Bot. Gard. 88: 77 (2001). Type: South Africa, 1836, *Drege* 3591 (K!, holo.)

Erect annual 11–36(–40) cm high; stems usually simple, obtusely square, densely strigose, when fresh yellowish green, drying greenish. Leaves mostly opposite, erect, linear, 8–30(–45) mm long, 1–3 mm wide, entire, densely strigose. Inflorescences terminal, 3–11(–19) cm long, usually rather dense or interrupted with several opposite pairs of flowers out at the same time; lower bracts longer than calyx, upper ones shorter; bracteoles subulate shorter than calyx. Calyx 10-ribbed, 8–9 mm long, including teeth, strigose; tube 6–7(–9) mm, lobes linear to lanceolate, 2–5 mm long. Corolla bright scarlet, yellowish outside, fragrant; tube 14–16(–18) mm long, bent and expanding above calyx, densely glandular-pubescent; upper lip bilobed, lower lip very deeply 3-lobed, lobes ovate, 8–10 mm long. Capsule oblong, ± 7 mm long (young).

KENYA. Laikipia District: near Colchecio Lodge, 5 Nov. 1978, *Hepper & Jaeger* 6618!; Fort Hall District: Murang'a [Fort Hall], 1925, *McDonald* 903!; Masai District: Laitokitok, 26 Feb. 1933, *Gilbert Rogers* 560!

TANZANIA. Musoma District: Manjira, Ikoma, 22 Nov. 1959, *Tanner* 4461!; Moshi District: Rombo to Useri, E slopes of Mt Kilimanjaro, 28 June 1946, *Greenway* 7840!; Iringa District: Image Mts, Mar. 1954, *Carmichael* 390!

DISTR. **K** 3, 4, 6; **T** 1, 2, 4, 7; Angola, Zambia, Malawi, Zimbabwe, South Africa

HAB. Semi-parasitic on grass roots in upland grassland; 900–1900 m

USES. None recorded on specimens from our area

CONSERVATION NOTES. Least Concern (LC); widespread

SYN. *Buchnera elegans* (Benth.) D.Dietr., Syn. Pl. 3: 525 (1843)

14. **Striga fulgens** (*Engl.*) *Hepper* in K.B. 38: 598 (1984). Type: Tanzania, Iringa District, Tengulinyi, Uhehe, *Goetze* 697 (B†, holo.; K!, iso.)

Erect perennial 25–50 cm high; stems simple or usually diffusely branched above, hispid with short ascending hairs. Leaves opposite or the upper subopposite, sessile, narrowly linear to linear-lanceolate, 5–21 mm long, 1–2 mm wide, acute, entire, hispid. Inflorescence lax, few-flowered; flowers nearly sessile, alternate, opposite or subopposite; bracts broadly lanceolate, ± 2 mm long, ciliate; bracteoles linear, 1–2 mm long. Calyx cylindrical, 7–9 mm long, strongly 10-ribbed, with white bands present between the ribs, glabrous except the triangular teeth 1–2 mm long. Corolla bright scarlet or orange-red, tube ± 10 mm long, curved above, throat densely hairy; lobes obovate, upper pair ± 4 mm long, lower 3 ± 8 mm long, 4–5 mm wide. Capsule oblong, 6 mm long.

TANZANIA. Iringa District: Mufindi area, N of Mafinga on Madabera road, 25 Nov. 1986, *Goldblatt et al.* 8270! & Ifunda, Nov. 1928, *Haarer* 1643!; Njombe District: N of Lake Nyasa, *Thomson* s.n.!
DISTR. **T** 7; not known elsewhere
HAB. Montane grassland; 1600–1800 m
USES. None recorded on specimens from our area
CONSERVATION NOTES. Known only from a few collections; here assessed as Data Deficient (DD)

SYN. *Buchnera fulgens* Engl. in E.J. 28: 478 (1900); Skan in F.T.A. 4(2): 392 (1906)

NOTE. A. Raynal in Bull. Mus. Natl. Hist. Nat., B, Adansonia 3 (1993) pointed out that this species is apparently a member of the Verbenaceae, possibly belonging in the genus *Chascanum* on the basis of its free lobes of the upper corolla lip, glabrous calyx, white bands between the ribs and short calyx lobes.

15. **Striga baumannii** *Engl.* in E.J. 23: 515, t. 12, figs. O–T (1897); Skan in F.T.A. 4(2): 414 (1906); Hepper in F.W.T.A. ed. 2, 2: 371 (1963) & in Proc. Third Parasitic Pl. symp.: 263 (1984); Raynal-Roques in Bull. Mus. Nat. Hist. Nat. Paris B, Adansonia 7(2): 123–133 (1985); U.K.W.F.: 261 (1994). Type: Sierra Leone, without locality, *Scott-Elliot* 5085 (K!, lecto.)

Erect, glabrous perennial with stout base and fleshy roots; stems 30–90 cm high, erect, branched above, glaucous. Leaves reduced to scales clasping stems, nodes 1–4 cm apart. Inflorescences slender, up to 20 cm long with numerous pairs of flowers appressed to stem; bracts lanceolate, shorter than calyx, bracteoles subulate. Calyx narrowly tubular, 15–veined, ± 7 mm long; teeth 2 mm long, glabrous, ligneous. Corolla dull yellow (maroon fide U.K.W.F.), glabrous; tube slightly longer than calyx, curved, expanding above, 2-lipped, 5-lobed; lobes 1–2 mm long, recurved (in dry state), acute; style 3 mm long, pubescent. Capsule oblong, ± 6 mm long, ± 1.5 mm wide.

UGANDA. Mengo District: Nakasongola, 19 Jan. 1956, *Langdale-Brown* 1827!
KENYA. Uasin Gishu District: Kipkarren, Mar. 1932, *Brodhurst Hill* 755!
DISTR. **U** 4 ; **K** 3; Mali to Sierra Leone, Cameroon, Congo-Kinshasa
HAB. Among *Hyparrhenia* in *Combretum* woodland and moist grassland; ± 1650 m

NOTE. A specimen from Tanzania, Moshi, Iringa District: *H. horsburgh-Porter* s.n. (BM) is similar but with broader leaves (5 mm) and larger flowers (23 mm in diameter). Mohamed et al. (op. cit. p. 95) exclude *S. baumannii* from *Striga* on the basis of its tuberous roots and ligneous calyx and suggest that it could belong to a new genus.

35. PARASTRIGA

Mildbr. in J. Arn. Arb. 11: 52 (1930)

Parasitic herbs with simple unbranched stems, glandular, drying black. Leaves opposite, sessile or shortly petiolate. Flowers solitary in the axils of leaves. Calyx campanulate, inflated, unequally 5-lobed, 10-veined, persistent. Corolla tubular, somewhat oblique; tube bent and widening below throat; limb unequally 5-lobed. Stamens 4, somewhat didynamous, inserted at the distal part of the corolla tube, included; anthers monothecal, slightly curved, attached transverse to the filament. Style glabrous, included; stigma bent apically, flattened. Capsule ovoid. Seeds cylindrical, reticulate.

Monotypic.

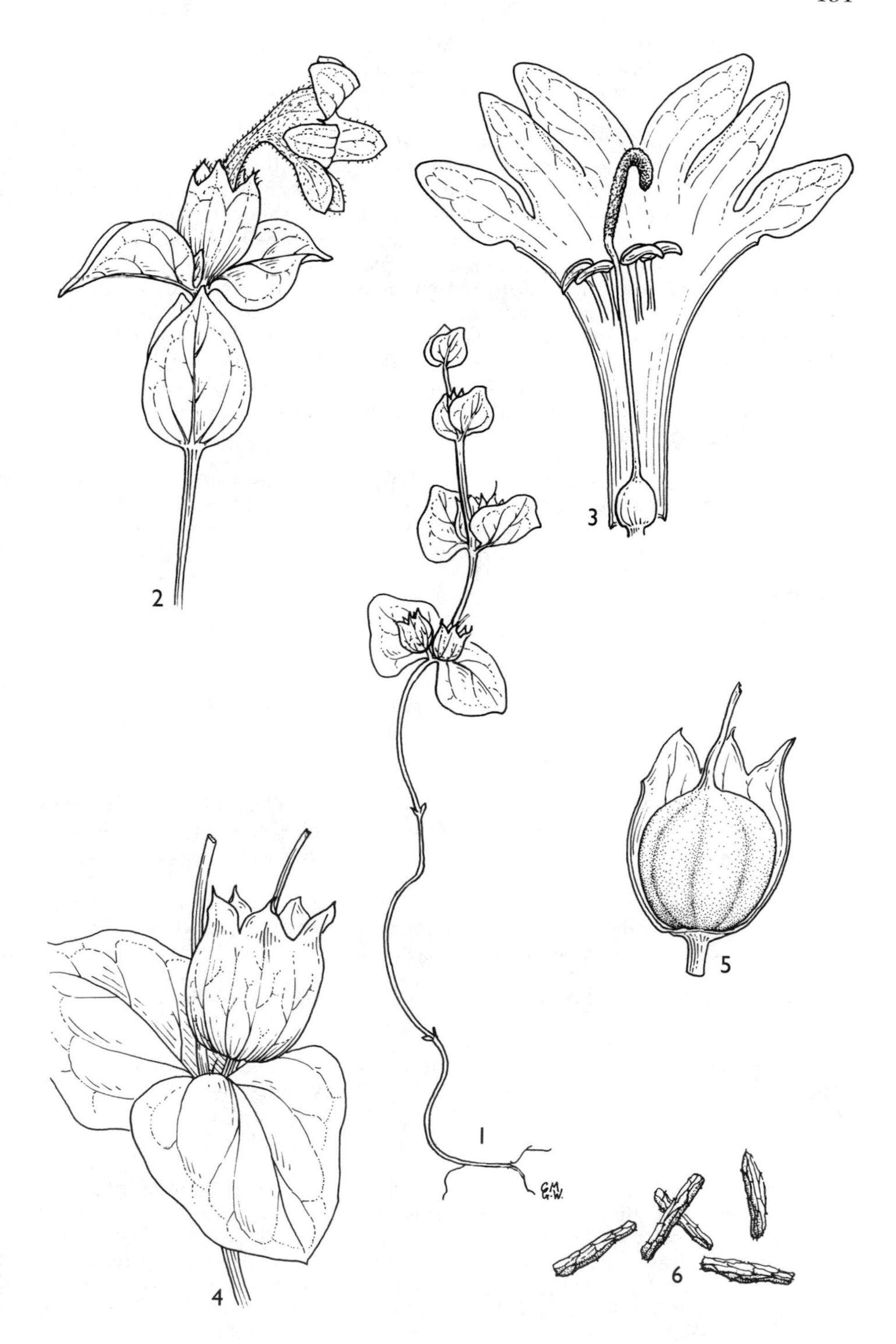

FIG. 39. *PARASTRIGA ALECTROIDES* — **1**, habit × 1; **2**, flower × 2; **3**, flower opened × 4; **4**, fruiting node × 3; **5**, capsule × 4; **6**, seeds × 18. 1 from *Purseglove* 2123; 2, 3, from *Purseglove* 3635; 4–6 from *Gillett* 14673. Drawn by F.N. Hepper.

Parastriga alectroides *Mildbr.* in J. Arn. Arb. 11: 52 (1930); Fischer in Fl. Ethiop. & Eritr. 5: 298 (2006). Type: Congo-Kinshasa, volcanic region near Kivu Lake, SW slope of Mt Mikeno, *D.H. Linder* 2428 (GH, holo.; B†, iso.)

Parasitic herb, up to 20 cm tall; stems simple, unbranched, glandular-villous, drying black. Leaves opposite, sessile or shortly petiolate, broadly ovate, 6–15 mm long, 6–14 mm wide, base truncate to cordate, apex acute to obtuse, margins entire, glabrous to sparsely glandular on the midrib on the abaxial side. Flowers solitary in the axils of leaves; pedicels up to 2 mm long. Calyx inflated, 8–9 mm long, unequally 5-lobed, 10-veined; lobes ovate, ± 3 mm, acute, sparsely ciliate with long ± glandular hairs. Corolla pink, tubular, 5-lobed, somewhat oblique; tube curved, 10–12 mm long, expanding above; lobes subequal, ovate-oblong, 5–6 mm long, obtuse, ?reflexed. Stamens inserted at the distal part of the corolla tube, included; upper filaments ± 3 mm long, lower 2.5 mm long; anthers ± 1.5 mm long, slightly curved, attached transverse to the filament. Ovary glabrous; style glabrous, included; stigma bent apically, flattened. Capsule ovoid, ± 5 mm in diameter, glabrous. Seeds cylindric, ± 0.8 mm long, reticulate. Fig. 39, p. 151.

UGANDA. Kigezi District: Lake Bunyonyi, 12 Aug. 2001, *Lye & Namaganda* 25175! & Bukimbiri swamp, Sept. 1946, *Purseglove* 2123! & Kachwekano farm, June 1951, *Purseglove* 3635!
TANZANIA. Lushoto District: Mkussu, 15 Feb. 1959, *Faulkner* 2235!; Magamba forest, 17 Jan. 1970, *Archbold* 1196!
DISTR. **U** 2; **T** 3; Congo-Kinshasa, Ethiopia
HAB. Swamps and damp places near streams; 1500–2000 m
USES. None recorded on specimens from our area
CONSERVATION NOTES. Least Concern (LC); localised but not uncommon

36. CYCNIOPSIS

Engl. in E.J. 36: 233 (1905)

Annual or perennial herbs. Leaves opposite. Bracts similar to leaves; bracteoles absent. Flowers solitary, axillary or not. Calyx tubular, lobed to about half its length, straight or relexing in fruit. Corolla tubular, 4-lobed above, tube bent in the upper part, extended or gibbous near the apex. Stamens 4, didynamous; filaments attached at the same level about 1/3 from the apex of the corolla tube; anthers monothecal. Ovary globose; style shorter than the stamens, included; stigma clavate. Fruit a septicidal capsule. Seeds numerous, black, cylindrical, reticulate.

A small genus of 3 species distributed in tropical E and NE Africa, extending to Yemen.

Cycniopsis humifusa (*Forssk.*) *Engl.* in E.J. 36: 233 (1905); Schwartz in Mitt. Instit. Allgem. Bot. Hamb. 246 (1939); U.K.W.F.: 261 (1994); Wood, Handb. Fl. Yemen: 266 (1997); Fischer in Fl. Ethiop. & Eritr. 5: 298 (2006). Type: Yemen: Al Hadiyah, *Forsskål* 401 (C, lecto. & isolecto.)

Annual or perennial mat-forming prostrate herb; stems creeping, unbranched or sparsely branced, rooting at the nodes; stems and leaves strigose, ± glandular, hairs white with a bulbous base. Leaves ovate to ovate-oblong to elliptic-oblong, 4–10 long, 2.5–7 mm wide, base attenuate, apex rounded to subacute, margins entire. Bracts similar to leaves, 4–5 mm long. Flowers solitary, axillary or not; pedicels 0.5–2 mm long. Calyx tubular, 9–10 mm, lobed to about half its length; lobes oblong, 5, acute, strigose on the veins and margins, straight or relexing in fruit. Corolla mauve, lilac, pink or pale pink; tube 16–25 mm long, bent in the upper part, extended or gibbous near the apex, glandular outside, sometimes sparsely so, hairy within; lobes 4, obovate, 10–13 mm long, 4–6(–7) mm wide, lip 9.5–10 mm wide, emarginate.

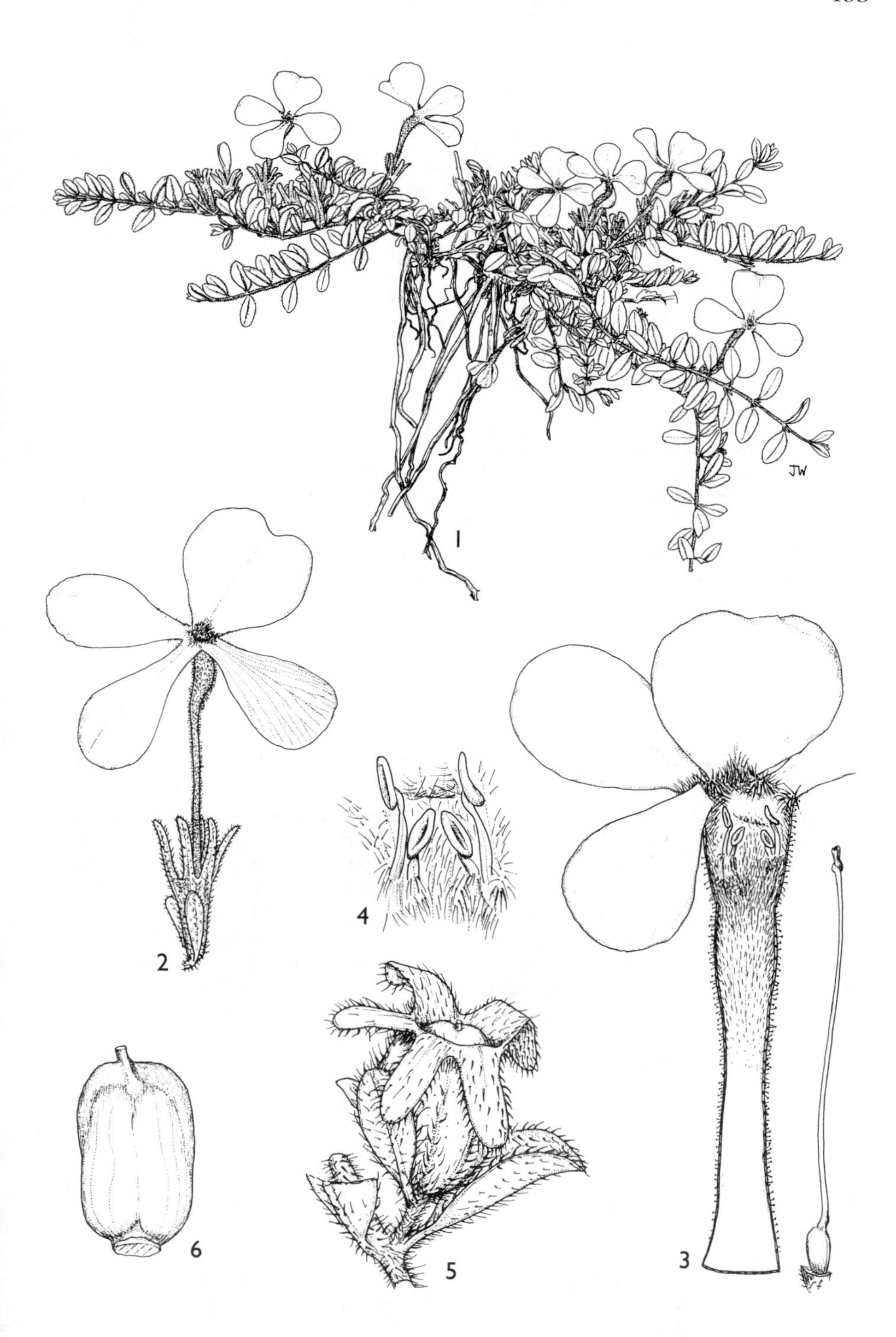

FIG. 40. *CYCNIOPSIS HUMIFUSA* — **1**, habit × $^{2}/_{3}$; **2**, flower × 2; **3**, corolla opened with ovary and style × 3; **4**, stamens detail × 8; **5**, capsule with calyx × 6; **6**, capsule × 6. 1–4 from *Greenway & Kanuri* 13889; 5, 6 from *Glover, Gwynne, Samuel & Tucker* 2508. Drawn by Juliet Williamson.

Filaments of the longer pair of stamens 1.5–2 mm, those of the shorter pair 0.4–1 mm long, filaments attached about 1/3 from the apex of the corolla tube. Ovary subglobose; style 15–20 mm long, included; stigma clavate. Capsule oblong, 4.4–5 mm long, 2–3 mm wide. Fig. 40, p. 153.

UGANDA. Karamoja District: Kidepo valley, *J. Wilson* 928!; Mt Elgon, 1905, *Evan James* s.n.!;
KENYA. Naivasha District: Longonot, near Lake Naivasha, July 1952, *Verdcourt* 692!; Nairobi District: Nairobi, July 1913, *Battiscombe* 716!; Masai District: N slopes of Ngong Hills, Mar. 1957, *Greenway* 9163!
TANZANIA. Masai District: Loliondo, Kingarana Forest Reserve, 24 Mar. 1995, *Congdon* 428!
DISTR. **U** 1, 3; **K** 2–4, 6; **T** 2; Ethiopia; Yemen
HAB. Woodland, secondary forest, grassland, also on black cotton soil; 1000–2500 m
USES. None recorded on specimens from our area.
CONSERVATION NOTES. Least Concern (LC); common and widespread

SYN. *Browallia humifusa* Forssk., Fl. Aegypt.-Arab. 112 (1775); Christensen, Dansk Bot. Ark. 4, 3: 22 (1922)
Buchnera humifusa (Forssk.) Vahl, Symb. Bot. 1793: 81
Striga humifusa (Forssk.) Benth. in Hook., Comp. Bot. Mag. 1: 362 (1835)
Cycnium humifusum (Forssk.) Benth. & Hook.f., Gen. Pl. 2: 969 (1876)
Cycniopsis obtusifolia Skan in F.T.A. 4(2): 416 (1906). Type: Kenya, Nandi District, *Johnston* s.n. (K!, lecto.)

NOTE. This species is quite variable in its pubescence from being densely strigose to almost glabrous especially in Ethiopia.

37. RHAMPHICARPA

Benth. emend Engl., P.O.A. C.: 361 (1895); O.J. Hansen in Dansk. Bot. Tidsskr. 70:103–125 (1976)

Rhamphicarpa Benth. in Hooker's Comp. Bot. Mag. 1:368 (1835)

Annual herbs, turning black on drying; stems erect, terete or quadrangular, glabrous. Leaves opposite or subopposite, filiform or pinnatisect with filiform segments. Flowers axillary, solitary in upper axils or slightly supra-axillary, pedicellate. Calyx 5-lobed, valvate in bud. Corolla tubular with a long, slender tube; lobes equal or subequal, imbricate in bud, spreading in flower. Stamens 4, inserted in the slightly distended upper portion of corolla tube; filaments bearded above base; anthers monothecal; dehiscence introrse by longitudinal slit. Ovary 2-locular, globose or ovoid; style long and slender. Capsules distinctly and usually obliquely beaked, sometimes winged along the sutures. Seeds numerous, reticulate.

A small genus of 6 species, occurring mostly in tropical Africa and extending to India, New Guinea and Australia. The genus as interpreted by O.J. Hansen, following Engler, has fewer recognised species than hitherto owing to the transference of many to *Cycnium*.

Rhamphicarpa fistulosa (*Hochst.*) *Benth.* in DC., Prodr. 10: 504 (1846), excl. syn. *Macrosiphon elongatus* Hochst.; Engl. in Abh. Kön. Akad. Wiss. Berlin 1891: 382 (1892); Hiern in Fl. Cap. 4(2): 399 (1904); Skan in F.T.A. 4(2): 420 (1906); Hepper in F.W.T.A. ed. 2, 2: 370 (1963); O.J. Hansen in Dansk Bot. Tidsskr. 70: 117, fig. 11 a–d (1976); Philcox in F.Z. 8(2): 135 (1990); U.K.W.F.: 261 (1994); Fischer in Fl. Ethiop. & Eritr. 5: 304 (2006). Type: Sudan, Cordofan, *Kotschy* 77b (W, holo.; BM, K!, iso.)

Annual herb 5–50 cm or more high, simple or with a few lateral branches, mainly from near the base. Leaves pinnatisect, (0.5–)1–6(–10) cm long; segments 2–7, filiform, 3–9 mm long, less than 1 mm broad. Flowers axillary, solitary; bracteoles 1–5(–7) mm long, inserted near the middle of the pedicel; pedicels (5–)9–20(–30) mm

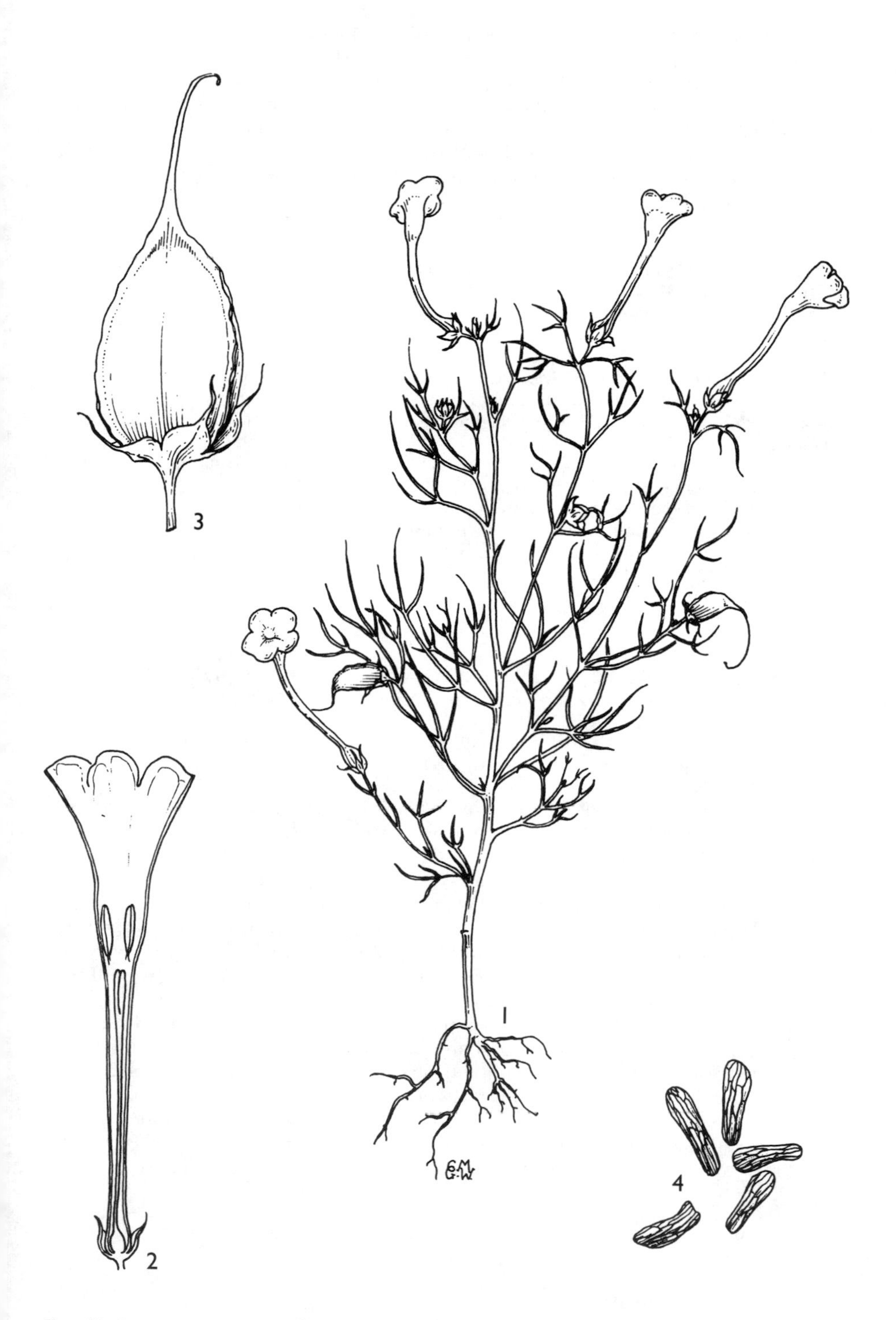

FIG. 41. *RHAMPHICARPA FISTULOSA* — **1**, habit × $^{2}/_{3}$; **2**, flower dissected × 2; **3**, capsule × 4; **4**, seeds × 48. All from *Biegel* 3021. Drawn by Christine Grey-Wilson. Reproduced with permission from F.Z.

long. Calyx tube 0.5–2 mm long, 1–4 mm broad; lobes ovate with filiform apex. Corolla white or pale pink, pale mauve or greenish outside, fragrant, stipitate-glandular outside; tube (22–)25–30(–35) mm long, ± 1 mm in diameter, upper portion below lobes up to 4 mm in diameter; lobes ± equal, suborbicular to spatulate, 6–9 mm long, 5–8 mm broad, rounded. Filaments 1.5–2 mm long; anthers 2–3.5 mm long. Ovary ovoid, 3–4 mm long, 1.5–2 mm wide; style 17–30 mm long; stigma tongue-shaped, 3–5 mm long. Capsules often covered by persistent corolla remnants, slightly obliquely beaked, 7–15 mm long from the base to the tip of the beak, slightly winged along the sutures. Fig. 41, p. 155.

UGANDA. West Nile District: Maracha, Dec. 1937, *Hazel* 403!; Busoga District: Lake Nakuwa, 28 Jan. 1951, *G.H.S. Wood* 217!; Masaka District: Kalengu county, 2–3 km S of West Mengo border, 11 July 1971, *Lye & Katende* 6447!

KENYA. Uasin Gishu District: Kipkarren, 1931, *Mrs. Brodhurst-Hill* 29!

TANZANIA. Ufipa District: Sumbawanga, 18 Mar. 1957, *Richards* 8786!; Ulanga District: Ifakara, Dec. 1959, *Haerdi* 393!; Songea District: R. Luhira, near Mshangano, 24 Apr. 1956, *Milne-Redhead & Taylor* 9804!; Pemba, Makongwe, 15 Aug. 1929, *Vaughan* 486!

DISTR. **U** 1, 3, 4; **K** 3; **T** 4, 6–8; **P**; widespread in tropical Africa and in Madagascar; also in New Guinea and Australia

HAB. Open wet places and among grass, in pools on rock outcrops, and sometimes as a weed of cultivation in rice fields; flowers opening in the evening; 0–2500 m

USES. None recorded on specimens from our area

CONSERVATION NOTES. Least Concern (LC); common and widespread

SYN. *Macrosiphon fistulosa* Hochst. in Flora 24(1): 373 (1841)

38. **CYCNIUM**

Benth. emend. Engl., P.O.A. C: 360 (1895); Skan in F.T.A. 4(2): 430 (1906), pro parte; Hansen in Dansk Bot. Arkiv 32(3): 23 (1978)

Rhamphicarpa Benth., Comp. Bot. Mag. 1: 368 (1835); Skan in F.T.A. 4(2): 418 (1906), pro parte

Annual or perennial herbs or small shrubs with a tuberous or elongated rhizome, often blackening on drying. Leaves opposite or subopposite, occasionally alternate or in whorls of three, bi- or pinnatisect with variously incised or entire margins, sessile or very shortly petiolate. Flowers axillary or supra-axillary, solitary or in terminal spikes or (sometimes interrupted) racemes. Bracts leafy. Calyx campanulate or tubular, enlarged or at least broadened in fruiting stage; tube with (4–)5–11 ribs, rarely with a ventral split; lobes 4 or 5, equal or unequal. Corolla salver-shaped (hypocrateriform), showy; tube straight or curved, sometimes pouched and constricted on one side (gibbous of Hansen) above the anthers; throat and inner side of tube bearded; limb bilabiate; upper lip 2-lobed or emarginate, lower lip deeply 3-lobed. Stamens 4, didynamous; filaments inserted in the corolla tube, unilaterally bearded; anthers monothecal, dorsifixed, dehiscence introrse by a longitudinal slit. Ovary 2-locular, glabrous; style terminal, never exceeding the lower pair of anthers; stigma simple, terete or compressed. Fruits either a capsule or a berry, very variable in shape, sometimes with an oblique beak, sometimes apiculate by the persistent base of the style. Seeds numerous, testa variously reticulate or smooth.

A genus of 15 species distributed in East Africa and southwards to South Africa

1. Leaves bi- or pinnatisect 2
   Leaves entire or variously toothed, never bi- or pinnatisect 3
2. Annual; leaves very narrow (0.5–1.5(–2)) mm wide; corolla tube 8–18 mm long; limb 6–10 mm in diameter .... 1. *C. recurvum*

Perennial; leaves broader (up to 2 mm wide); corolla tube 13–20 mm long; limb 12–30 mm in diameter . . . 2. *C. tenuisectum*
3. Fruit indehiscent, a fleshy berry . . . . 12. *C. adonense*
Fruit dehiscent, capsule 4
4. Calyx narrow tubular, in fruit urceolate; calyx tube with a deep ventral slit . . . . 13. *C. erectum*
Calyx campanulate; calyx tube without a ventral split 5
5. Calyx lobes unequal 6
Calyx lobes equal 7
6. Corolla white or pale-pink, tube glandular, not gibbous; limb 18–22 mm long; anthers apiculate . . . . 6. *C. ajugifolium*
Corolla pink to pink-purple, tube gibbous in the upper half, adpressed hairy with short glandular hairs; limb 10–15 mm long; anthers acute to shortly apiculate . . . 7. *C. brevifolium*
7. Capsule with a straight beak or beak absent 8
Capsule with an oblique beak 9
8. Flowers white with dark purple throat, mostly solitary; corolla tube cylindrical throughout, not gibbous or sometimes gibbous . . . . 11. *C. herzfeldianum*
Flowers pink to red, rarely white, in racemes; corolla tube gibbous . . . . 8. *C. cameronianum*
9. Corolla tube gibbous above the middle 10
Corolla tube straight, not gibbous 12
10. Flowers white or cream; corolla tube ≤ 22 mm long . . . . 3. *C. filicalyx*
Flowers pink, red or white; corolla tube ≥ 22 mm long 11
11. Leaves lanceolate, elliptic or ovate 8–16 mm wide; margins coarsely and regularly or irregularly toothed or lobed . . . . 9. *C. veronicifolium*
Leaves linear, 0.5–2 mm wide, lower ones sometimes rudimentary; margins entire or with a few distant teeth . . . . 10. *C. jamesii*
12. Perennial; leaves lanceolate, oblong to elliptic, 2–11 (–15) mm wide; flowers pink . . . . 5. *C. volkensii*
Annual or perennial; leaves linear to ovate, 0.5–4 mm wide; flowers pink, cream or white 13
13. Flowers white or pink, solitary (plants then decumbent) or in racemes; pedicel 2–50(–90) mm long; corolla limb 15–70 mm in diameter . . . . 4. *C. tubulosum*
Flowers cream or white, in spikes or racemes; pedicel 0–1 mm; corolla limb 8–16 mm in diameter . . . . 3. *C. filicalyx*

1. **Cycnium recurvum** (*Oliv.*) *Engl.*, P.O.A. C: 361 (1895); Hansen. in Dansk Bot. Arkiv 32(3): 25, figs 7, 8 a–d (1978); Philcox in F.Z. (2): 137 (1990); U.K.W.F.: 260, pl. 112 (1994). Type: Uganda, Bukoba District: Karagwe [Karague] *Grant* 402 (as 642 in Speke's Journ.) (K!, holo)

Annual, 20–80 cm high, erect; stems unbranched to much-branched with ascending, usually opposite branches, ± strigose, sometimes glandular as well. Leaves bi- or pinnatisect, 1–7 cm long; segments and rhachis 0.5–1.5(–2) mm wide; venation of simple veins in rhachis and segments, ± strigose. Inflorescence a spike or a raceme; bracteoles subulate to linear, 2.5–8 mm long, adnate to the base of the calyx tube; pedicels 0–2 mm long. Calyx campanulate, tube 2–5 mm long, ribbed; lobes 5, triangular to narrowly triangular or rarely linear, reflexed or spreading, erect in fruit, 1–8 mm long, the upper one often reduced in size; densely strigose to glabrous outside, inner side of lobes with shorter ± appressed hairs intermingled with sessile glands. Corolla white or pink; tube curved, slightly extended at the middle and ±

gibbous above the middle, 8–18 mm long, glandular outside, limb 6–10 mm in diameter. Filaments of the shorter pair of stamens 2–3 mm long, those of the longer pair 3–7 mm long; anthers 1.5–2 mm long; tip acutely apiculate. Ovary laterally compressed, subglobose, 1–2 mm long and broad; style 4–6 mm long, including the 1–3 mm long compressed stigma. Capsule ± globose, obliquely beaked, 4–8 mm in diameter, distal suture of beak 7–14 mm long; winged and dehiscent along upper-distal suture only. Seeds 0.5–1 mm long, reticulate or reticulate-tuberculate.

UGANDA. Acholi District: Chua, *Eggeling* 2376!; Teso District: Serere, Kyere Rock, July 1926, *Maitland* 1305!; Busoga District: Namaiera Hill, 9 July 1953, *G.H.S. Wood* 800!

KENYA. Turkana District: Lodwar, 27 Sept. 1963, *Paulo* 105!; NE Elgon, Nov. 1957, *Tweedie* 1477!; North Kavirondo, Nov. 1931, *Jack* 141!

TANZANIA. Shinyanga District: Shinyanga, Nov. 1938, *Koritschoner* 209!; Masai District; Loliondo, 6 July 1959, *Sangai* 699!; Iringa District: Nyangolo Scarp, 17 Apr. 1962, *Polhill & Paulo* 2042!

DISTR. **U** 1, 3, 4; **K** 2, 3, 5; **T** 1, 2, 4, 5, 7; S Sudan to Malawi

HAB. Shallow soil on rocky outcrops, on sandy soil, and in disturbed wet or dry grassland; flowering most of the year except December and January; 400–2200 m

USES. None recorded on specimens from our area

CONSERVATION NOTES. Least Concern (LC); widespread and common

SYN. *Rhamphicarpa recurva* Oliv. in Trans. Linn. Soc. 29: 122 (1875); Wettstein in E. & P. Pf.: 95 (1891); Skan in F.T.A. 4(2): 420 (1906); Staner in B.J.B.B. 15: 149 (1938)

NOTE. Hansen (op. cit.) draws attention to *Tanner* 585 and *Rounce* 301 from Tanzania that have 5 equal and almost linear calyx lobes and exceptionally dense foliage. He retains within *C. recurvum* several collections from Kenya and Tanzania (e.g. *Tanner* s.n, 10 April 1959 & 4125, *Perdue* 460) that have characteristic capsules with a very long and less compressed beaks, and rather wide leaf segments.

2. **Cycnium tenuisectum** (*Standl.*) *O.J.Hansen* in Dansk Bot. Arkiv 32(3): 28, figs 10, 11 a–e (1978); U.K.W.F.: 260, pl. 112 (1994); Fischer in Fl. Ethiop. & Eritr. 5: 303 (2006). Type: Kenya, Mt Kenya, *Mearns* 2338 (US, holo.)

Perennial herb with tuberous rhizome; stems decumbent or ascending, up to 40 cm long, unbranched or with a few branches; stems and leaves glabrescent or shortly strigose often with scattered stipitate glands and/or hyaline, erect hairs intermingled with shorter hairs. Leaves bi- or pinnatisect, 0.5–3(–5) cm long; segments and rhachis filiform or flat, up to 2 mm wide; venation of simple veins in segments and rhachis. Inflorescence a spike or raceme; bracteoles linear, 5–9 mm long, adnate to the base of the calyx-tube; pedicels 0–4(–7) mm long. Calyx shortly tubular, campanulate at fruiting stage; tube 5–9 mm long, ribbed; lobes 5, triangular to narrowly triangular, 4–10 mm long, subequal to unequal; strigose outside especially along margins and ribs. Corolla pink or mauve-pink; tube curved, gibbous above the middle, 13–20 mm long, glandular outside; limb 12–30 mm in diameter. Filaments of the shorter pair of stamens 3–4 mm long, those of the longer pair 6–8 mm long; anthers 2–2.5 mm long, with rounded tip. Ovary compressed, ovate to elliptic, 2–3 mm long; style ± 5 mm long, including the 2 mm long compressed stigma. Capsule ovoid, oblique or with a very short obliquely set beak, 8–16 mm long; winged and dehiscent mainly along upper-distal suture. Seeds 0.8–1 mm long, tuberculate.

UGANDA. Karemoja District: Kadam Mt, Apr. 1959, *J. Wilson* 790! & Mt Moroto, 11 June 1970, *Lye* 5624!; Mt Elgon, Jan. 1918, *Dummer* 3366!

KENYA. Northern Frontrier District: Mt Kulal, Nov. 1978, *Hepper & Jaeger* 7038!; Naivasha District: Longonot, 28 July 1952, *Verdcourt* 693!; Ravine District: Timboroa, 11 Sept. 1958, *Napper* 757!

TANZANIA. Arusha District: Mt Meru crater, 18 Dec. 1966, *Richards* 21715!; Masai District: Oldeani Mt, Ngorongoro, 19 Nov. 1957, *Tanner* 3808!; Mbulu District: Mt Hanang, 8 Feb. 1946, *Greenway* 7657!

DISTR. **U** 1, 3; **K** 1–6; **T** 2; Sudan, Ethiopia

HAB. Open montane vegetation such as grassland, moorland, clearings in bushland, roadsides, evidently avoiding permanently wet soil, and tolerating fires; (1800–)2000–3000(–3500) m
USES. None recorded on specimens from our area
CONSERVATION NOTES. Least Concern (LC); widespread and common

SYN. *Rhamphicarpa meyeri-johannis* Engl. in Abh. Königl. Akad. Wiss. Berlin 1891: 392 (1892); Skan in F.T.A. 4(2): 421 (1906); R.E. Fries in Acta Horti Berg. 8: 62 (1924); Cufodontis in B.J.B.B. 33: 908 (1963). Type: Tanzania, Kilimanjaro, *Meyer* 266 (B†, syn.), *Höhnel* 139 (B†, syn)
Cycnium meyeri-johannis (Engl.) Engl., P.O.A. C: 361 (1895), *non C. meyeri-johannis* Engl. (1892) (which is *C. herzfeldianum* (Vatke) Engl.)
Rhamphicarpa tenuisecta Standl. in Smithsonian Misc. Collect. 65(5): 17 (1917); Pole Evans in Botanical Survey Memoirs 22: 294 (1948)

NOTE. The wide range of flower size gave rise to the recognition of the large flowered plants in Tanzania as *Rhamphicarpa meyeri-johannis*, (e.g. *Grimshaw* 93/1248, *Geilinger* 3655, *Chuwa* 2712) in Tanzania, but Hansen has indicated that in Kenya they merge with small-flowered forms without a discontinuity.

3. **Cycnium filicalyx** (*E.A.Bruce*) *O.J.Hansen* in Dansk Bot. Archiv 32(3): 29, figs 11 f–h, 12 (1978); Philcox in F.Z. (2): 138 (1990). Type: Tanzania, Iringa District: plain of the little Ruaha R., *Lynes* P.g. 70 (K!, lecto.)

Annual 10–50 cm high, erect; stems simple or much branched, glabrescent or strigose to hispid (often hairy in longitudinal stripes), and ± verrucose. Leaves linear, 1–8 cm long, 0.5–1 mm wide, the lower ones rudimentary; margins usually entire or rarely with a few distant teeth; venation inconspicuous or a single midrib present in well developed leaves, strigose to hispid or glabrescent. Inflorescence a spike or raceme; bracteoles usually absent, if present filiform, 1–3 mm long, variously inserted on the pedicels; pedicels 0–1 mm long. Calyx campanulate, broadly so in fruit; tube 2–3 mm long, ribbed, sparsely hispid or in fruiting stage glabrescent; lobes (4–)5, narrowly triangular to linear, equal, 11–23 mm long, scabrous along the margins and midribs. Corolla white or cream; tube straight, 15–22 mm long, gibbous or not, glandular outside; limb 8–16 mm in diameter. Filaments of the shorter pair of stamens 1.5–2 mm long, those of the longer pair 3–4 mm long; anthers 1.8–2 mm long, apiculate. Ovary compressed, ovoid, 2–3 mm long; style 5–7 mm long, including the 2–3 mm long compressed stigma. Capsule ± ovoid, obliquely beaked, 3–6 mm long; distal suture of beak 7–14 mm long; winged and dehiscent along upper suture only. Seeds 0.7–1 mm long, reticulate or reticulate-tuberculate.

TANZANIA. Ufipa District: Lake Mbugu, 19 Mar. 1950, *Bullock* 2672!; Singida District: Kiomboi, Iramba Plateau, 28 Apr. 1962, *Polhill & Paulo* 2223!; Mbeya District: Igawa, 2 Apr. 1962, *Polhill & Paulo* 1976!
DISTR. **T** 4, 5, 7; Zambia, Zimbabwe, Namibia
HAB. In scrub on sandy soil, in open wet grassland with sandy or clayey soil, on shallow soil over rock, and in open waterlogged depressions in miombo woodland; 1200–1650 m
USES. None recorded on specimens from our area
CONSERVATION NOTES. Least Concern (LC); probably at its northernmost limit in the Flora area

SYN. *Rhamphicarpa filicalyx* E.A.Bruce in K.B. 1933: 475 (1933)

4. **Cycnium tubulosum** (*L.f.*) *Engl.*, P.O.A. C: 361 (1895); Philcox in F.Z. (2): 138 (1990); U.K.W.F.: 260, pl. 112 (1994); Fischer in Fl. Ethiop. & Eritr. 5: 301 (2006). Type: South Africa, *Thunberg* (UPS!, IDC-microfiche, holo.)

Perennial or rarely annual herb with woody tuberous or elongated rhizome; stems decumbent, ascending, erect or straggling, 5–40(–70) cm long; stems, leaves and calyx glabrescent, pubescent or hispid, smooth or verrucose. Leaves linear to ovate, 1–7(–10) cm long, 2–4 mm wide, the lower ones sometimes rudimentary; apex acute or obtuse, margins entire or irregularly toothed or lobed; venation inconspicuous or of 1–3 ± parallel veins in narrow leaves or pinnate in broader leaves. Flowers solitary (in decumbent plants) or in racemes; bracteoles absent or present, variable in shape, 0–10(–20) mm long, variously inserted on the pedicels or adnate to the base of the calyx tube; pedicels 2–50(–90) mm long, axillary or supra-axillary. Calyx campanulate or shortly tubular, in fruiting state campanulate to broadly campanulate; tube 3–10(–23) mm long, ribbed or not; lobes (4–)5, variable in shape from ovate to linear-filiform, 2–12(–18) mm long, equal, erect, reflexed or recoiled. Corolla white or pink; tube curved or straight, 14–55 mm long, glandular outside; limb 15–70 mm in diameter. Filaments of the shorter pair of stamens 3–6 mm long, those of the longer pair 5–10 mm long; anthers 2–4.5 mm long, apiculate or rounded at tip. Ovary compressed, ± obliquely ovate, 2.5–4 mm long; style 4–11 mm long, including the 2–4 mm long terete or compressed stigma. Capsule obliquely beaked, globose or obliquely ovoid, 4–8 mm long; distal suture of beak 6–15 mm long; winged and dehiscent mainly along the upper-distal suture. Seeds 0.7–1 mm long, testa reticulate.

This is a polymorphic species in which Hansen recognised two subspecies characterised by differences in indumentum, habit, size and arrangenent of the flowers. There are several specimens from Kenya and Tanzania that show characters intermediate between the two subspecies (e.g. *Richards* 8586, *Glover, Gwyne* & *Samuel* 877, *Kirika* 179, *Gillett* 12795) and are also sympatric in distribution with the two subspecies. I (S.A.G.) would be more tempted to recognize *C. tubulosum* as one polymorphic species for the FTEA region, but here follow Hansen's treatment as it is based on his studies throughout the range of the taxa:

Plants ascending, straggling or erect; calyx (in the flowering state) glabrous except for minute hairs (less than 0.2 mm long on margins of lobes and along the ribs); flowers in racemes . . . . subsp. *tubulosum*

Plants prostrate or ascending, if erect then calyx conspicuously hairy; calyx (in the flowering state) usually with spreading hyaline hairs, some of which are 0.5 mm long or more; flowers solitary or in poorly defined racemes . . . . subsp. *montanum*

subsp. **tubulosum;** Hansen in Dansk Bot. Arkiv 32(3): 32 (1978); Thulin, Fl. Somal.: 3: 289 (2006); Fischer in Fl. Ethiop. & Eritr. 5: 301 (2006). Fig. 42: 1–4, p. 161.

UGANDA. Mbale District: Budadiri Bugishu, Jan. 1932, *Chandler* 484!; Masaka District: Katera, 2 Oct. 1953, *Drummond & Hemsley* 4575! & Kakuto, Aug. 1945, *Purseglove* 1776!

KENYA. Northern Frontier District: Moyale, 3 May 1958, *Everard* in EA 11447!; Kiambu District: Nairobi, 12 Mar. 1930, *Napier* 93!; Kisumu-Londiani District: road to Kisumu, 25 Apr. 1975, *Friis & Hansen* 2603!

TANZANIA. Tanga District, Kirindomi, 25 Mar. 1966, *Faulkner* 3788!; Ufipa District; Lake Kwela, 11 Mar. 1959, *Richards* 11145!; Mbeya District: Njerenje, 5 May, 1975, *Hepper, Field & Mhoro* 5294!

DISTR. **U** 1, 3, 4; **K** 1–6; **T** 1, 3, 4, 6, 7; Nigeria, Congo-Kinshasa, Rwanda, Sudan, Angola, Zambia, Mozambique, Zimbabwe, Botswana, Namibia, South Africa

HAB. Wet places such as swamps and riversides and rice-fields, reported on a wide range of soils; flowering all the year round; 0–2150 m

USES. None recorded on specimens from our area

CONSERVATION NOTES. Least Concern (LC); widespread and common

SYN. *Gerardia tubulosa* L.f., Suppl. P1.: 279 (1781)

Rhamphicarpa tubulosa (L.f.) Benth. in Hook., Comp. Bot. Mag. 1: 368 (1835); Oliver, J.L.S. 15: 91 (1876); Skan in F.T.A. 4(2): 428 (1906); F.P.U.: 134 (1962); Hepper in F.W.T.A. 2: 270 (1963)

R. curviflora Benth., Comp. Bot. Mag. 1: 368 (1835). Type: Mozambique, *Forbes* s.n. (K!, holo.)

?*R. serrata* Klotzsch in Peters, Reise Mossamb. Bot. 228 (1861); Skan in F.T.A. 4(2): 429 (1906); Cufodontis in B.J.B.B. 33: 909 (1963). Type: Mozambique, *Peters* s.n. (B†, holo.)

Fig. 42. *CYCNIUM TUBULOSUM* subsp. *TUBULOSUM* — **1**, flowering branch × $^2/_3$; **2**, basal leaf × 1; **3**, flower dissected × 1. Subsp. *MONTANUM* — **4**, habit of erect plant; **5**, habit of procumbent plant (reduced). 1, 3 from *Allen* 34; 2 from *Jackson* 2091; 4, 5 after *Hansen*, fig. 13. Drawn by Christine Grey-Wilson. Reproduced with permission from F.Z.

Cycnium serratum (Klotzch) Engl., P.O.A. C: 360 (1895)
C. aquaticum Engl. in E.J. 28: 479 (1900). Type: Tanzania, Iringa District: Ifweme swamp near Tengulinye *Goetze* 690 (B†, holo.; BM!, lecto., K!, drawings)
C. hamatum Engl. & Gilg in Warburg, Kunene-Sambesi-Exped.: 368 (1903). Type: Namibia, *Baum* 403 (B†, holo.; BM, lecto.; COI, HBG, K!, W, isolecto.)
C. questieauxianum De Wild. in Ann. Mus. Congo Bot., sér. 4: 124 (1903). Type: Congo-Kinshasa, Shaba [Katanga], *Verdick* 165 (BR, holo.)
Rhamphicarpa aquatica (Engl.) Skan in F.T.A. 4(2): 429 (1906)
R. hamata (Engl. & Gilg) Skan in F.T.A. 4(2): 429 (1906)

subsp. **montanum** (*N.E.Br.*) *O.J.Hansen* in Dansk Bot. Arkiv 32(3): 35, figs 13d–g, 14 (1978); Fischer in Fl. Ethiop. & Eritr. 5: 301 (2006). Type: Zimbabwe [Rhodesia], *Elliot* s.n. (K!, lecto.). Fig. 42: 5, p. 161.

UGANDA. Karamoja District: Moroto, May 1956, *J. Wilson* 245!; Ankole District: Rubaare, 8 Dec. 1968, *Lye & Lester* 585!; Teso District: Mt Alekilek, 16 Nov. 1968, *Lye & Lester* 402!
KENYA. Northern Frontier District: Maralal, Nov. 1978, *Hepper & Jaeger* 6693!; Naivasha District: Njoroa Gorge, 31 Dec. 1959, *Verdcourt* 2601!; N Kavirondo: Broderick Falls, Apr. 1938, *Tweedie* 437!
TANZANIA. Arusha District: Momella, 24 Apr. 1975, *Hepper & Field* 5111!; Lushoto District: Lushoto to Mombo road, 16 June 1953, *Drummond & Hemsley* 2933!; Morogoro District: NE of Turiani, 26 Mar. 1953, *Drummond & Hemsley* 1818!
DISTR. **U** 1–4; **K** 1–6; **T** 1–8; Congo-Kinshasa, Rwanda, Burundi, Ethiopia, Mozambique, Zimbabwe, South Africa
HAB. In grassland (damp or dry), reported from a wide range of soils; the commonest *Cycnium* in East Africa, flowering throughout the year; (0–)200–2600(–3000) m
USES. None recorded on specimens from our area
CONSERVATION NOTES. Least Concern (LC); widespread and common

SYN. *Rhamphicarpa heuglinii* Schweinf., Beitr. Fl. Aethiopiens: 100 (1867). Type: Sudan, "Khartum in Sennar", 1854, *Heuglin* s.n. (B†, holo.)
?*Cycnium serratum* (Klotzsch) Engl. forma *elatum* Engl., P.O.A. C: 360 (1895). Types: E Africa, *Peter* s.n. (B†, syn.), *Böhm* 127 (B†, syn.)
C. serratum (Klotzsch) Engl. forma *longipedicellatum* Engl., P.O.A. C: 361 (1895). Type: Tanzania, Meru, *Volkens* 1612 (B†, holo.)
C. serratum (Klotzsch) Engl. forma *paucidentatum* Engl., P.O.A. C: 360 (1895). Types: East Africa, *Stuhlmann* 6751 (B†, syn.), *Holst* 3942 (B†, syn.) & *Volkens* 2385 (B†, syn.)
C. serratum (Klotzsch) Engl. forma *subintegra* Engl., P.O.A. C: 360 (1895). Type: Tanzania, Uzaramo District: Usaramo, *Hildebrandt* 1129 (B†, holo.; L, lecto.)
C. paucidentatum (Engl.) Engl. in Ann. Ist. Bot. Roma 7: 29 (1897)
C. rubrifolium Engl. in E.J. 30: 405 (1901). Type: Tanzania, Mbeya District: Mpagara, Mbowu R., *Goetze* 1402 (B†, holo.; BR, lecto.)
Rhamphicarpa montana N.E. Br. in K B. 1901: 129 (1901); Skan in F.T.A. 4(2): 427 (1906); Troupin, Syllabus fl. Rwanda: 124 (1971); F.P.U.: 134 (1964); U.K.W.F.: 564 (1974)
R. multicaulis Skan in F.T.A. 4(2): 427 (1906). Type: Tanzania, Lushoto District: Usambara, Umba Valley, *Smith* s.n. (K!, holo.)
R. paucidentata (Engl.) Fiori in Nuovo Giorn. Bot. Ital. 47: 37 (1940)
R. neghellensis Fiori in Nuovo Giorn. Bot. Ital. 47: 36 (1940). Type: Ethiopia, Neghelli, *Saccardo* s.n. (FT, holo.)

5. **Cycnium volkensii** *Engl.*, P.O.A. C: 360 (1895); Hansen in Dansk Bot. Arkiv 32(3): 36, figs 16, 17a–d (1978); U.K.W.F.: 260 (1994); Fischer in Fl. Ethiop. & Eritr. 5: 302 (2006). Type: Tanzania, between Meru and Kilimanjaro, *Volkens* 366 (B†, holo.; BM!, lecto.)

Perennial (15–)30–120 cm high, erect, somewhat woody, with a ± distinct tuber; stems, leaves, and calyx ± strigose. Leaves lanceolate, oblong or elliptic, 20–35 mm long, 2–11(–15) mm wide; base cuneate; margins coarsely toothed, lobed or in small leaves entire; venation pinnate, inconspicuous in small leaves. Inflorescence a spike or raceme; bracteoles linear to lanceolate, 3–7 mm long inserted at the uppermost part of the pedicel or adnate to the base of the calyx tube; pedicels 0–3 mm long. Calyx campanulate to shortly tubular, in fruiting state campanulate; tube 4–8 mm long, ± distinctly ribbed; lobes (4–)5, triangular to narrowly triangular, 3–4 mm long, equal or the upper ones reduced in size or absent, erect or spreading, strigose. Corolla pink; tube straight or slightly curved, 13–20(–26) mm long, glandular outside; limb 14–26 mm in diameter. Filaments of the shorter pair of stamens 2.5–4 mm long, those of the longer pair 4–7 mm long; anthers 1.8–3 mm long, tip rounded to shortly apiculate; upper pair well below the throat or rarely slightly exceeding the throat. Ovary ± obliquely ovoid or globose; style 3–7 mm long, including the 1.5–3 mm long compressed stigma. Capsule ovoid, oblique or with a short poorly defined obliquely set beak, 4–10 mm long; winged and dehiscent along the upper-distal suture only. Seeds 0.5–1 mm long, reticulate.

KENYA. Northern Frontier District: Furroli, 15 Sept. 1952, *Gillett* 13896!; Machakos District: Chyulu Mts above Kibwezi, 4 May 1975, *Friis & Hansen* 2711!; Masai District: Narok to Nairobi, 14 July 1962, *Glover & Samuel* 3118!

TANZANIA. Arusha District: Ngurdoto National Park, 13 Oct. 1965, *Greenway & Kanuri* 12130! & Usa R. to Egari Nanyuki, 27 Oct. 1959, *Greenway* 9587!; Kilimanjaro, 23 Feb. 1953, *Drummond & Hemsley* 1283!

DISTR. **K** 1, 3, 4, 6; **T** 2, 3; Ethiopia

HAB. Open grassland avoiding shade and permanently wet soil, also in open thickets and rock outcrops; flowering throughout the year; 0–2450 m

USES. None recorded on specimens from our area

CONSERVATION NOTES. Least Concern (LC)

SYN. *C. gallaense* Engl. in Ann. Reale Ist. Bot. Roma 7: 29 (1897). Type: Ethiopia, Galla, *Riva* 1257 (FT, holo.)

C. bricchettii Engl. in Ann. Reale Ist. Bot. Roma 7: 30 (1897); Skan in F.T.A. 4(2): 435 (1906); Cufodontis in B.J.B.B. 33: 906 (1963). Type: Ethiopia, Harar, *Bricchetti* s.n. 1889 (FT, holo.)

C. asperrimum Engl. in E.J. 36: 231 (1905). Type: Ethiopia, Arusi-Galla, *Ellenbeck in Erlanger* 1947 (B†, holo.)

C. ellenbeckii Engl. in E.J. 36: 231 (1905). Type: Ethiopia, Arusi-Galla, *Ellenbeck* 1967 (B†, holo.)

Rhamphicarpa asperrima (Engl.) Skan in F.T.A. 4(2): 424 (1906)

R. volkensii (Engl.) Skan in F.T.A. 4(2): 425 (1906)

R. ellenbeckii (Engl.) Skan in F.T.A. 4(2): 425 (1906)

R. volkensii (Engl.) Skan var. *keniensis* R.E.Fries in Acta Horti Berg. 8: 62 (1924). Type: Kenya, Mt Kenya, *Fries* 1599 (S, holo.; BR, K!, iso.)

R. gallaensis (Engl.) Cufodontis in B.J.B.B. 33: 908 (1963)

6. **Cycnium ajugifolium** *Engl.* in E.J. 18: 74 (1893) & in P.O.A. C: 360 (1895); Hansen in Dansk Bot. Arkiv 32(3): 39, figs 17e–g, 18 (1978); U.K.W.F.: 260 (1994). Type: Tanzania, without locality, *Fischer* 1/104 (HBG, holo.)

Perenial herb up to 25 cm long, with woody tuber; stems decumbent or ascending, densely strigose or hispid. Leaves elliptic, ovate or obovate, 5–12(–32) mm long, 3–8 mm wide; base cuneate, apex acute, margins coarsely few-toothed or lobed, obtuse or rounded; venation pinnate, sometimes with the lower pair of secondary veins reaching half way to the tip of the leaves. Inflorescence a spike or a raceme; pedicels 0–3 mm long; bracteoles linear or lanceolate, 5–8 mm long, adnate to the base of the calyx tube. Calyx campanulate to shortly tubular, broadly campanulate in fruit; tube 2.5–8 mm long, obliquely cut, ± distinctly ribbed; lobes 5, narrowly triangular to almost linear, 3–7 mm long, unequal, the upper one reaching the middle of the lateral ones. Corolla white or pale pink; tube straight or curved, 18–22 mm long, glandular outside and sometimes also with a few scattered, white multicellular hairs; limb 18–25 mm in diameter. Filaments of the shorter pair of stamens 3–4 mm long, those of the longer pair 5–6 mm long; anthers 2–2.5 mm long, apiculate, the upper pair inserted in the throat. Ovary compressed, ± obliquely ovate,

2–4 mm long; style ± 5 mm long including the ± 2.5 mm long compressed stigma. Fruit a capsule, oblique or with a short poorly defined obliquely set beak, ovoid, 5–8 mm long; winged and dehiscent along upper-distal suture only. Seeds ± 1 mm long, reticulate.

KENYA. Nairobi District: Nairobi National Park, 25 Dec. 1968, *Greenway* 13546!; Machakos District: Kilima Kui, Dec. 1933, *Jex-Blake* 5904!; Masai District: Ngong Hills, 26 Nov. 1950, *Greenway* 8472!

TANZANIA. Masai District: Ngaserai Plain, 13 Dec. 1969, *Richards* 24937!; Arusha District, Arusha to Sanya Junction, 20 Jan. 1969, *Richards* 23783A!; Mbulu District; N of Mto wa Umbu, 24 Jan. 1965, *Greenway & Kanuri* 12067!

DISTR. **K** 4, 6; **T** 2; not found elsewhere

HAB. Dry grassland and *Balanites–Acacia–Commiphora* grassland, surviving annual fires, on black cotton soil and volcanic loam; flowering mainly November to June; 900–1850 m

USES. None recorded on specimens from our area

CONSERVATION NOTES. Least Concern (LC), but may qualify as Near Threatened (NT) due to rapid degradation of vegetation where it is found

SYN. *Rhamphicarpa ajugifolia* (Engl.) Skan in F.T.A. 4(2): 426 (1906); Cufodontis in B.J.B.B. 33: 907 (1963)

7. **Cycnium breviflorum** *Ghaz.* in K.B. 60(3): 461 (2005). Type: Tanzania, Arusha District, Oldoinyo Sambu, *Bally* 12003 (K!, holo.)

Perennial herb up to 30 cm tall with a woody rhizome; stems dark brown to blackish, erect or ascending, simple or branched, often with many stems arising from the base, strigose to hispid, often densely so. Leaves sessile, elliptic to ovate to ovate-lanceolate, 4–12(–22) mm long, 1.5–3(–6) mm wide, base cuneate, apex acute or obtuse, margins coarsely and irregularly few-toothed or lobed, venation pinnate, sometimes with the lower pair of secondary veins reaching half way to the tip of the leaf, hispid to strigose, often verrucose, dull green, drying blackish. Inflorescence a spike or a raceme; bracteoles linear, 4–7 mm long, ± 5 mm wide, adnate to the base of the calyx tube; pedicels 0–3 mm long, strigose. Calyx campanulate to shortly tubular, campanulate in fruit; calyx tube obliquely cut, 5-lobed, 3–5 mm long, ribs distinct or obscure, strigose; lobes triangular to narrowly triangular, unequal, 3–5 mm long, the upper one reaching the middle of the lateral ones, erect, not reflexed in fruit. Corolla creamy pink to pink or purple pink; tube straight or slightly curved, gibbous in the upper half near the apex, 10–15 mm long, densely appressed hairy with short scattered glandular hairs; limb 5–6 mm long, 8–16 mm in diameter; lobes rounded, undulate, sparsely hairy especially on the veins on the outside, glabrous within. Filaments of the shorter pair of stamens 2–3 mm long, those of the larger pair ± 4 mm long, the upper pair inserted in the throat; anthers 1.5–2 mm long, tip acute to shortly apiculate. Ovary compressed, ovoid, 1.5–2 mm long; style 7–8 mm long, including the 2–3 mm long compressed stigma. Capsule ovoid, 3–4 mm in diameter, obliquely beaked; beak ± 1.5 mm long. Seed ± 1 mm long, testa reticulate.

KENYA. Naivasha/Masai District: Suswa, steamjet ridge near Stafford camp, 1 Jan. 1963, *Glover et al.* PEG 3432!

TANZANIA. Arusha District: Engari Nanyuki, 6 Apr. 1948, *Greenway & Kanuri* 13434! & 9 Apr. 1965, *Richards* 20124!; Loonguru-m-yu road, off road on the north side of Mt Meru, 26 Dec. 1968, *Richards* 23484!; Kilimanjaro, Loloval Hill, NAFCO estates, 17 Dec. 1993, *Grimshaw* 93/1193!

DISTR. **K** 6; **T** 2; very local, not found elsewhere

HAB. Grassland and wooded grassland, noted to be semi-parasitic on *Barleria prionites* (cf. *Greenway & Kanuri* 13434); 1200–1500 m

USES: Children chew the flowers to blacken tongue, lips and gums

CONSERVATION NOTES. Least Concern (LC); localised but common in grassland

SYN. *Cycnium* sp. nov. aff. *ajugifolium* Engl., O.J. Hansen in Dansk Bot. Arkiv 32(3): 40, figs 20a, b (1978)

8. **Cycnium cameronianum** (*Oliv.*) *Engl.*, P.O.A. C: 361 (1895); Hansen in Dansk Bot. Arkiv 32(3): 40, figs. 20c–e (1978); U.K.W.F.: 260 (1994). Type: Tanzania, District unclear, Lake Tanganyika, *Cameron* s.n. (K!, holo.)

Annual (10–)20–60(–90) cm high, erect; stems sparingly branched, hispid or strigose, sometimes glabrescent and ± verrucose. Leaves lanceolate to elliptic, (10–)30–110 mm long, 4–25 mm wide; base cuneate, apex acute to acuminate, margins dentate, serrate, or coarsely and ± irregularly toothed or lobed, hispid or strigose, venation pinnate. Inflorescence racemose; bracteoles linear to filiform, 2–10 mm long, inserted on the pedicel close to the calyx tube; pedicels 1–4 mm long. Calyx campanulate, broadly so in fruit; tube 3–5 mm long, ribbed; lobes 5, triangular to narrowly triangular, equal, 3–9 mm long, erect or spreading, ribs and margins of the lobes hispid with shorter appressed hairs intermingled, rarely glabrous. Corolla pink or red, occasionally white; tube straight or slightly curved, gibbous above the middle, 24–32 mm long, glandular outside; limb 20–50 mm in diameter. Filaments of the shorter pair of stamens 3–4 mm long, those of the longer pair 6–8 mm long; anthers ± 3 mm long, tip acute. Ovary compressed, suborbicular in lateral view, 2–2.5 mm in diameter; style 5–7 mm long, including the 2–5 mm long terete or compressed stigma. Capsule laterally compressed, suborbicular in lateral view, 7–10 mm long, 9–12 mm broad, sometimes apiculate with the persistent base of the style, winged and dehiscent along both sutures. Seeds 0.7–1 mm long, reticulate or reticulate-tuberculate.

KENYA. Machakos District: Kibwezi, 28 Jan. 1904, *Scheffler* 60!; Kitui District: 20 km N of Mutomo, 25 Sept. 1958, *Gillett* 18552!; Masai District: Olorgesaile, 12 July 1954, *Bally* 9794!
TANZANIA. Musoma District: Kilimfeza, 15 Feb. 1968, *Greenway & Kanuri* 13184!; Lushoto District: Soni to Mombo, 24 June 1960, *Drummond & Hemsley* 2980!; Iringa District: Ruaha National Park, 29 Jan. 1966, *Richards* 21073!
DISTR. **K** 4, 6; **T** 1–7; not found elsewhere
HAB. Grassland and *Acacia-Commiphora*, or *Combretum* wooded grassland; also in grassy wasteland and in neglected cultivation, and on rocky outcrops; 600–1850 m
USES. None recorded on specimens from our area
CONSERVATION NOTES. Least Concern (LC)

SYN. *Rhamphicarpa cameroniana* Oliv. in J.L.S. 15: 95 (1876); Skan in F.T.A. 4(2): 423 (1906)
[*R. veronicifolia* sensu Agnew, U.K.W.F.: 564 (1974)]

NOTE. This species may be semi-parasitic on *Sorghum* according to evidence by Fuggles-Couchman in E. Afr. Agric. J. Kenya 1: 145, 147 (1935) cited by Hansen in Dansk Bot. Arkiv 32(3): 23, 42 (1978). The flowers are fragrant at night.

9. **Cycnium veronicifolium** (*Vatke*) *Engl.*, P.O.A.C: 361 (1895); Hansen in Dansk Bot. Arkiv 32(3): 42, figs. 22–24 (1978); U.K.W.F.: 260 (1994). Type: Kenya, *Hildebrandt* 2753 (B†, holo.); Kenya, Kwale District: Mariakani, *Polhill & Paulo* 903 (K!, neo.; BR, EA, FT, P!, S, isoneo.)

Annual; stems erect (10–)20–120 cm tall, sparingly branched, hispid or strigose, glabrescent and ± verrucose. Leaves lanceolate, elliptic or ovate, (7–)15–110 mm long, 8–16 mm wide, base cuneate, apex ± acute, margins coarsely and regularly or irregularly toothed or lobed, proximal teeth larger than the distal ones, strigose, glabrescent and ± verrucose, venation pinnate. Inflorescence a raceme; bracteoles absent or present and up to 10(–15) mm long, subulate, filiform, linear or lanceolate, variously inserted on the pedicels; pedicels 1–5 mm long. Calyx shortly tubular or campanulate, broadly campanulate in fruit; tube 2–8 mm long, ± distinctly ribbed; lobes 5, equal, triangular to narrowly triangular or ovate, 2–5(–9) mm long, erect or spreading, glabrous or minutely strigose on margins and ribs. Corolla pink or red; tube straight or slightly curved, gibbous above the middle, 22–30 mm long, glandular outside, sometimes with a few hyaline hairs intermingled with the glands;

limb 15–40 mm in diameter. Filaments of shorter pair of stamens 3–4 mm long, those of longer pair 6–8 mm long; anthers 2.5–3 mm long, apex rounded, acute or shortly apiculate. Ovary compressed, suborbicular or ovoid, 2–3 mm long and broad; style 5–9 mm long including the 2(–5) mm long, terete or compressed stigma. Capsule ovoid to subglobose, obliquely beaked, 5–8 mm long; distal suture of beak 10–20 mm long; winged and dehiscent mainly along the upper distal suture. Seeds 0.7–1 mm long, reticulate-tuberculate.

KENYA. Masai District: Olorgesaile, July 1957, *Quelea Control* s.n.!; Kwale District: Mariakani, 8 Dec. 1961, *Polhill & Paulo* 903!; Kilifi District: Rabai Jatoni, *C.W. Elliot* in F.D. 3373!

TANZANIA. Tanga District: Duga, July 1893, *Holst* 3217!; Lushoto District: Magila, Sept 1877, *Kirk* s.n.!; Iringa District: Udzungwa Mt National Park, 18 Mar. 2003, *Luke* 9375!

DISTR. **K** 6, 7; **T** 3, 6, 7; not found elsewhere

HAB. Grassland, wooded grassland, bushland, open hillsides and rock outcrops, and in degraded forest whether wet or dry; also occurring as a weed in rice fields and disturbed ground; flowering almost througout the year; semi-parasitic; 0–1700 m

USES. None recorded on specimens from our area

CONSERVATION NOTES. Least Concern (LC); apparently common

SYN. *C. suffruticosum* Engl. in E.J. 28: 479 (1900). Type: Tanzania, Morogoro District: S Uluguru Mts, *Goetze* 231 (B†, holo.), *Schlieben* 2789 (K!, BR, isoneo.)
Rhamphicarpa suffruticosa (Engl.) Skan in F.T.A. 4(2): 423 (1906)
Cycnium veronicifolium (Vatke) Engl. subsp. *suffruticosum* (Engl.) O.J.Hansen in Dansk Bot. Arkiv 32(3): 45, figs 22e, 24 (1978)

NOTE. Hansen (op. cit.) recognizes subspecies *suffruticosum*, earlier recognised as a separate species, but also acknowledges that there are intermediates, especially in the western Usambaras, that are difficult to name. The two subspecies are found in different habitats (subsp. *veronicifolium* on open hillsides and rocky outcrops from 1000–1200 m in Tanzania, confined to two small areas (**T** 3, 6)); in my opinion (S.A.G.) this is an ecotype of a single widespread species.

10. **Cycnium jamesii** (*Skan*) *O.J.Hansen* in Dansk Bot. Arkiv 32(3): 45, figs 25, 26a–c (1978); U.K.W.F.: 260, pl. 112 (1994). Type: Kenya, Kavirondo, *Scott Elliot* 7030 (BM!, lecto.; K!, isolecto.)

Erect perennial 40–100 cm tall with a woody tuber; stems usually several, sparingly branched; stems, leaves, and calyx glabrous and smooth or verrucose, drying glossy and black. Leaves linear, 3–11 cm long, 0.5–2 mm wide, lower ones sometimes rudimentary; apex acute to obtuse, margins entire or with a few distant teeth; venation of 1 or 3 ± parallel veins. Inflorescence a spike or a raceme; bracteoles absent; pedicels 1–3(–10) mm long, axillary, often adnate to the stems. Calyx shortly tubular or campanulate, broadly so in fruit; tube 4–8 mm long, ± distinctly ribbed; lobes 5, triangular or lanceolate, 3–8 mm long, equal, erect, often 3-veined. Corolla white or pink, tube straight or curved, gibbous above the middle, 22–36 mm long, glandular outside; limb 25–50 mm in diameter. Filaments of the shorter pair of stamens 3–5 mm long, those of the larger 6–8 mm long; anthers 2.5–4 mm long, apex rounded, acute or shortly apiculate. Ovary compressed, ± obliquely ovoid or suborbicular, 2–3.5 mm long; style 6–9 mm long, including the 3–4.5 mm long, terete stigma. Capsule ovoid, obliquely beaked, 6–10 mm long, distal suture of beak 13–19 mm long. Seeds 0.7–1 mm long, almost smooth.

UGANDA. Karamoja District: Nakipiripirit, July 1965, *J. Wilson* 1694!; Mt Elgon, 21 Mar. 1924, *Snowden* 867!; Mengo District: Kagoye, Buruli, 16 May 1941, *A.S. Thomas* 3915!

KENYA. Trans-Nzoia District: Kitale, Mar. 1935, *Thorold* 3207!; Kericho District: Kericho-Lumbwa, 24 Aug. 1965, *S.F. Polhill* 127!; Masai District: Lolgorien, Sept. 1933, *Napier* 5348!

TANZANIA. Mwanza, 1926, *R.L. Davis* 272!; Kahama District: 85 km S of Mbugwe, 31 July 1950, *Bullock* 3054!

DISTR. **U** 1–4; **K** 3, 5, 6; **T** 1, 4; Burundi, Ethiopia; confined to the eastern and western rift valleys

HAB. Wet grassland, permanent swamps, and open water-logged depressions in wooded grassland on black cotton soil, flowering most of the year; 950–2500 m
USES. None recorded on specimens from our area
CONSERVATION NOTES. Least Concern (LC)

SYN. *Rhamphicarpa jamesii* Skan in F.T.A. 4(2): 423 (1906)

11. **Cycnium herzfeldianum** (*Vatke*) *Engl.* in E.J. 18: 74 (1893); Hansen in Dansk Bot. Arkiv 32(3): 46, fig. 26d–g, 27 (1978); U.K.W.F.: 260, pl. 112 (1994); Fischer in Fl. Ethiop. & Eritr. 5: 301 (2006). Type: Kenya, Kitui District: Kitui, *Hildebrandt* 2736 (B†, holo.; K!, lecto.; M, isolecto.)

Perennial herb with ± distinct woody tuber; stems weak, all decumbent or the younger ones ascending, up to 35 cm long; stems, leaves, and calyx sparingly hispid, glabrescent or ± verrucose, sometimes also stipitate-glandular. Leaves lanceolate to ovate, 20–60 mm long, 8–12 mm wide, the lower ones often rudimentary, base cuneate, apex acute, obtuse, or rounded, margins serrate-crenate, the proximal teeth often distinctly larger than the distal ones; venation pinnate, sometimes with the lower pair of secondary veins reaching halfway to the tip of the leaves. Flowers axillary and usually solitary, but the arrangement of flowers may approach that of a raceme; bracteoles filiform to linear-lanceolate, 4–6(–13) mm long, usually inserted on the pedicel close to the calyx tube; pedicels slender, sinuous, flattened, (8–)20–90 mm long. Calyx campanulate, in fruiting stage broadly campanulate; tube 5–12 mm long, ribbed; lobes 5, very variable in shape from broadly ovate to lanceolate and triangular, equal, 7–17 mm long, 3-veined. Corolla white with a dark purple or red throat and tube; tube sinuous, 16–25 mm long, gibbous or not, glandular on the outside; limb 10–25 mm in diameter. Filaments of shorter pair of stamens 2.5–0.5 mm long, those of longer pair 5–7 mm long; anthers 2.5–3.5 mm long, acute to shortly apiculate. Ovary compressed, ± obliquely ovoid, 2–3.5 mm long; style 5–7 mm long, including the 3–4 mm long, compressed stigma. Capsule (?fleshy), subglobose or slightly laterally compressed, 7–12 mm long, rarely with a short, straight beak, but often apiculate by the persistent base of the style; winged and dehiscent along both sutures. Seeds 0.7–1 mm long, finely reticulate to almost smooth.

UGANDA. Toro District: Nyakasura, 18 Feb. 1932, *Shillito* 63!; Mbale District: Budadiri, Jan. 1932, *Chandler* 480!; Masaka District: Malabigambo forest, 2 Oct. 1953, *Drummond & Hemsley* 4585!
KENYA. Mt Elgon, 24 June 1951, *Irwin* 72!; Kiambu District: Muguga forest, 4 Sept. 1965, *Kokwaro & Kabuye* 331!; Teita District: Wundanyi to Mwatate, 1 May 1975, *Friis & Hansen* 2643!
TANZANIA. Musoma District: Bologonja River, 18 Aug. 1962, *Greenway* 10758!; Lushoto District: Mazumbai Forest Reserve, 8 Jan. 1976, *Cribb & Grey-Wilson* 10109! & Mkuzi, 8 Apr. 1953 *Drummond & Hemsley* 2051!
DISTR. **U** 1, 3, 4; **K** 2–4, 6, 7; **T** 1–3; Congo-Kinshasa, Rwanda, Ethiopia
HAB. Wet grassland and forest margins with seepage on a wide range of soils; flowering throughout the year; 700–2500 m
USES. None recorded on specimens from our area
CONSERVATION NOTES. Least Concern (LC)

SYN. *Rhamphicarpa herzfeldiana* Vatke in Linnaea 43: 311 (1882); Skan in F.T.A. 4(2): 426 (1906); F.P.U. 134 (1962); U.K.W.F.: 564 (1974)
R. herzfeldiana Vatke var. *subauriculata* Vatke in Linnaea 43: 311 (1882). Type: Kenya, Teita, *Hildebrandt* 2440 (B†, holo.)
Cycnium meyeri-johannis Engl. in Abh. Königl. Akad. Wiss. Berlin 1891: 383 (1892), *non* Engl. (1895). Types: Tanzania, Pare District: Ugweno, *Meyer* 191 (B†, syn.); Kilimanjaro, *Meyer* 58 (B†, syn.)
C. herzfeldianum (Vatke) Engl. var. *subauriculatum* (Vatke) Engl., P.O.A. C: 360 (1895)
C. herzfeldianum (Vatke) Engl. forma *holstii* Engl., P.O.A. C: 360 (1895); Skan in F.T.A. 4(2): 426 (1906). Type: description
C. bequaertii De Wild. in Rev. Zool. Bot. Afr. 8: B44 (1920), and in Pl. Bequaert. 1, 2: 277 (1922). Type: Congo-Kinshasa, Ruwenzori, *Bequaert* s.n. (BR, lecto.)

12. **Cycnium adonense** *Benth.* in Comp. Bot. Mag. 1: 368 (1835), Skan in F.T.A. 4(2): 431 (1906); F.P.U. 134 (1962); Hansen in Dansk Bot. Arkiv 32(3): 48 (1978); Cribb & Leedal, Mount. Fl. S Tanz.: 119, pl. 29A (1982); Philcox in F.Z. (2): 140 (1990); U.K.W.F.: 260, pl. 112 (1994); Fischer in Fl. Ethiop. & Eritr. 5: 303 (2006). Type: South Africa, *Drège* 2295a (K!, lecto.)

Perennial with woody tuber; stems prostrate, decumbent, ascending, or erect (rarely straggling), 5–35 cm long; stems, leaves, and calyx hispid, scabrous, or pilose, and sometimes stipitate-glandular as well. Leaves oblanceolate, lanceolate, elliptic, oblong, or ovate, up to 80(–120) mm long, 10–38 mm wide, base cuneate, apex or obtuse, rarely acuminate or rarely rounded, margins variously toothed, venation pinnate. Flowers solitary or in racemes; bracteoles subulate, filiform, linear, or lanceolate, (1–)2–15 mm long, inserted on the pedicel close to the calyx or (sometimes) adnate to the base of the calyx tube; pedicels 2–30(–50) mm long, axillary or supra-axillary. Calyx shortly to long tubular, or (rarely) campanulate, in fruiting stage campanulate to urceolate; tube 5–60 mm long, 3–13 mm broad, ribbed and sometimes almost plicate between the ribs; lobes 5, narrowly triangular, triangular, lanceolate, or ovate, 3–15 mm long, equal or unequal, 1- or 3-veined. Corolla white or pink; tube straight sometimes indistinctly gibbous above the middle, 30–95 mm long, glandular on the outside; limb–25–70(–90) mm in diameter. Filaments of shorter pair of stamens 3–13 mm long, those of longer pair 6–17 mm long; anthers 4–7 mm long, tip rounded, acute or apiculate. Ovary ovoid, ± compressed, 2–5 mm long; style 6–30 mm long, including the 2.5–16 mm long, compressed stigma. Fruit a berry, globose to ovoid, 12–14 mm long and 10–12 mm broad, apiculate by the peristent base of the style. Seeds 0.8–1.2 mn long, finely reticulate.

NOTE. A widely distributed species with considerable variation in habit and corolla size etc. which has been carefully analysed by Hansen and presented as histograms (op. cit. pp. 50–52).

Plants usually prostrate or decumbent; calyx including the lobes more than 20 mm long in flowering stage; corolla tube 50 mm or more long .... subsp. *adonense*
Plants usually erect or ascending; calyx including the lobes up to 30 mm long in flowering stage; corolla tube less than 55 mm long .... subsp. *camporum*

subsp. **adonense** *O.J.Hansen* in Dansk Bot. Arkiv 32(3): 52, fig. 28a–b, 33 (1978)

Stems usually prostrate, rarely ascending or erect. Calyx tube (15–)20–60 mm long and 5–13 mm broad. Corolla tube (45–)50–95 mm long; limb 40–70(–90) mm in diameter; lower lobes 17–26 mm broad. Filaments of shorter pair of stamens 3–13 mm long, those of longer pair 7–17 mm long; anthers 5–7 mm long. Ovary 3–5 mm long; style 17–27(–30) mm long, including the 5–16 mm long stigma.

UGANDA. West Nile District: W of Oleiba, 2 Aug. 1953, *Chancellor* 88!; Bunyoro District: Kijanjubwa, Feb. 1943, *Purseglove* 1276!; Mengo District: Busana, Apr. 1930, *Liebenberg* 1526!
KENYA. W Suk District: Kongoli road, July 1961, *Lucas* 194!; North Kavirondo District: Kakamega Forest, 11 Apr. 1973, *Hansen* 913!; Kwale District: Shimba Hills, 9 Feb. 1953, *Drummond* & *Hemsley* 1192!
TANZANIA. Ufipa District: S of Sumbawanga, 30 Dec. 1961, *Robinson* 4825!; Dodoma District: S of Itigi Station, 20 Apr. 1964, *Greenway* & *Polhill* 11646!; Songea District: W of Songea, 1 Jan. 1956, *Milne-Redhead* & *Taylor* 8014!
DISTR. **U** 1–4; **K** 2–7; **T** 1, 4–8; W Africa to Guinea, Gabon, Central African Republic, Burundi, W Sudan, Angola, Zambia, Zimbabwe, South Africa
HAB. Grassland, and grassland derived from woodland, subject to frequent fires; 0–2500 m

USES. None recorded on specimens from our area
CONSERVATION NOTES. Least Concern (LC); widespread and common

subsp. **camporum** *O.J.Hansen* in Dansk Bot. Arkiv 32(3): 53, figs 28c–d, 34 (1978). Type: Sudan, Djur country, Wau R., *Schweinfurth* 1645 (K!, lecto.; P, S, isolecto.)

Stems usually erect or ascending, rarely prostrate. Calyx tube 5–25(–30) mm long and 4–10 mm broad. Corolla tube 30–55(–60) mm long; limb 25–50 mm in diameter. Filaments of shorter pair of stamens 3–8 mm long, those of longer pair 6–13 mm long. Ovary 2–4 mm long; style 6–13(–24) mm long, including the 2.5–12 mm long stigma.

UGANDA. Ankole District: Bugambe, Feb. 1943, *Purseglove* 1244!; Busoga District: Kiumu Hill, N of Jinja, 4 Sept. 1952, *G.H.S. Wood* 361!; Mengo District: Mabira Forest, Chagwe, (fl.) Dec. (fr.) Mar., *Ussher* 17!
KENYA. N Kavirondo District: Kakamega Forest, 17 Apr. 1965, *Gillett* 16676! & 11 Apr. 1973, *Hansen* 912!; S Kavirondo District: Utembe, 9 Sept. 1933, *Napier* 5350!
TANZANIA. Mpanda District: Silkcub Highlands, 4 Apr. 1956, *Richards* 7143!; Ufipa District: Mbizi Forest, 26 Nov. 1958, *Napper* 1053!; Rungwe District: Mbosi, 14 Nov. 1932, *Davies* 671!
DISTR. **U** 1–4; **K** 5; **T** 4, 7; Guinea to Congo-Brazzaville, Central African Republic, Congo-Kinshasa, Burundi, Ethiopia, Angola, Zambia, Malawi, Zimbabwe
HAB. Grassland, and grassland derived from woodland, subject to frequent fires; 0–2500 m
USES. None recorded on specimens from our area
CONSERVATION NOTES. Least Concern (LC); widespread and common

SYN. *Cycnium camporum* Engl. in E.J. 18: 73 (1843); Skan in F.T.A. 4(2): 432 (1906); Cufodontis in B.J.B.B. 33: 906 (1963); Hepper in F.W.T.A., ed. 2, 2: 373 (1963)
Cycnium dewevrei De Wild. & Durand in Compt. Rend. Soc. Bot. Belge 38: 129 (1899). Type: Congo-Kinshasa(?), *Dewèvre* 401 (BR, holo.)

13. **Cycnium erectum** *Rendle* in J.B. 34: 128 (1896); Skan in F.T.A. 4(2): 434 (1906); Cufodontis in B.J.B.B. 33: 906 (1963); Hansen in Dansk Bot. Arkiv 32(3): 56, figs. 37, 38 (1978); U.K.W.F.: 260 (1994); Fischer in Fl. Ethiop. & Eritr. 5: 301 (2006). Type: Ethiopia, Sheik-Hussein, *Donaldson Smith* s.n. (BM!, holo.)

Erect, shrubby perennial, up to 2 m tall, with a woody tuber; stems, leaves, and calyx tomentose, becoming verrucose with age. Leaves lanceolate to ovate, up to 40(–70) mm long, 5–20 mm wide, base cuneate, apex acute to obtuse, margins crenate-serrate, entire towards the base of the blade, venation pinnate, prominent on the lower surface. Inflorescence a raceme; bracteoles linear to filiform, 5–8 mm long, adnate to the base of the calyx tube; pedicels 3–15 mm long. Calyx tubular, in fruiting stage narrowly urceolate; tube 9–31 mm long and 3–9 mm broad, with a deep split on the lower side, lobes 4–6, lanceolate to linear, 1–3 mm long, equal to unequal, erect or variously bent. Corolla white, tube curved or straight, somewhat widening upwards, 25–60 mm long, densely glandular on the outside; limb 30–60 mm in diameter. Filaments of the shorter pair 3–6 mm long, those of longer pair 7–12 mm long; anthers 5–8 mm long, sometimes with a ± 1 mm long apiculum. Ovary compressed, quadrangular to rectangular, 2–3 mm long and broad; style abruptly set from the ovary, 12–15 mm long, including the 4–6 mm long, compressed stigma. Capsule straight with woody walls, oblong or ovoid (rarely globose), 7–15 mm long; apex truncate, sometimes oblique; winged and dehiscent mainly along the apical parts of both sutures. Seeds 2–3 mm long, smooth.

UGANDA. Acholi District: Agoro, 15 Nov. 1945, *A.S. Thomas* 4383!; Mbale District: Sabei, Mt Elgon, 20 July 1924, *Snowden* 935! & Bukwa, 4 July 1971, *Katende* 1127!
KENYA. Trans-Nzoia District: Kimilili forest, 23 Apr. 1975, *Friis & Hansen* 2571!; Mt Elgon, Oct. 1958, *Tweedie* 1724! & June 1933, *Dale* 3094
DISTR. **U** 1, 3; **K** 3; Sudan (Imatong Mts), southern Ethiopia
HAB. In degraded forest patches and grassland; flowering throughout the year; (1200–) 1500–2400 m

USES. None recorded on specimens from our area
CONSERVATION NOTES. Least Concern (LC)

SYN. *Cycnium tomentosum* Engl. in Ann. Reale Ist. Bot. Roma 7: 29 (1897); Skan in F.T.A. 4(2): 434 (1906); F.P.U.: 134 (1962). Type: Ethiopia, Biddume [Biddimo], *Riva* 1219 (Fl, holo.)

39. **BUTTONIA**

Benth. in Hook., Ic. Pl. 11: 63, t. 1080 (1871); Hemsley & Skan in F.T.A. 4(2): 438 (1906)

Slender perennial climbing shrubs. Leaves opposite, pinnatisect or lobed, petiolate. Flowers solitary, axillary, pedicellate, bracteate; bracts deciduous. Calyx campanulate, 5(–4)-lobed, inflated in fruit. Corolla 5-lobed; lobes rounded, subequal; tube funnel-shaped with wide throat, somewhat curved. Stamens 4, didynamous, included; filaments linear; anthers connivent in pairs, bithecal with one empty theca or empty theca absent on lower stamens. Ovary bilocular; ovules numerous. Capsule enclosed in the persistent calyx. Seeds numerous, conical-oblong, truncate at both ends.

A genus of 2 species native to tropical East and South Africa.

Buttonia natalensis *Benth.* in Hook., Ic. Pl. 11: 63, t. 1080 (1871); Hiern, Fl. Cap. 4(2): 385 (1904); Philcox in F.Z.: 8(2): 141 (1990); U.K.W.F.: 258 (1994); K.T.S.L.: 589, fig., map (1994). Type: South Africa, Natal, *Button* 2 (K!, holo.)

Climbing stems ± woody, many-branched, glabrous or occasionally glabrescent with few minute hairs. Leaves toothed or 3-lobed, 3–6.5 cm long, 1.5–4 cm wide, lower lobes ovate to broadly elliptic, narrowed at base, obtuse, margins entire to 1- or 2-toothed, central lobe broadly lanceolate to rhombic, variously toothed or laciniate, glabrous to rarely shortly pubescent; petioles narrowly winged. Flowers supra-axillary, solitary, ± in short racemes, fragrant; bracteoles 2, opposite, suborbicular to reniform, 8.5–16 mm in diameter, subsessile immediately below calyx, glabrous or sparsely furfuraceous, deciduous; pedicels 8–25 mm long, stout, spreading, glabrous, becoming recurved in fruit. Calyx campanulate to urceolate, 15–20 mm long in flower, longer in fruit, shortly 4–5 lobed, 10-veined; lobes reticulate-veined inside. Corolla carmine, rose-lilac to violet, throat darker; tube ± 30 mm long, upward curving, inflated above middle; limb 3–7 cm in diameter, spreading; lobes 15–20 mm long, 20–30 mm wide. Stamens glabrous. Capsule globose, 15–22 mm in diameter. Seeds 2–3.3 mm long. Fig. 43, p. 171.

KENYA. Machakos District: Garabani Hill, 26 Mar. 1940, *van Someren* 219!; Masai District: Oldonyo Orok, 29 Mar. 1948, *Gardner* in *Bally* 6172!; Teita District: Kasigau Mt, 1 June 1969, *Gillett* 18740!

TANZANIA. Musoma District: Ushashi, 15 June 1959, *Tanner* 4361!; Mbulu District: Lake Manyara National Park, 26 Nov. 1963, *Greenway & Kirrika* 11082!; Handeni District: Kwa Mkono, 30 km E of Hudeni, 16 Oct. 1973, *Archbold* 2238!

DISTR. **K** 4, 6, 7; **T** 1–4; Zimbabwe, Mozambique, South Africa
HAB. Dry rocky places in woodland and on riverbanks; 100–1900 m
USES. None recorded on specimens from our area
CONSERVATION NOTES. Least Concern (LC)

SYN. *B. hildebrandtii* Engl. in E.J. 23: 509 (1897); Hemsley & Skan in F.T.A. 4(2): 439 (1906). Type: Kenya, Teita District: Voi River, *Hildebrandt* 2493 (B†, holo.)

NOTE. There is a wide range of variation in this species with the South African plants described as *B. natalensis* with distinctly pinnatisect leaves and larger flowers than the East African *B. hildebrandtii*; the latter may have been described from a poorly grown plant.

FIG. 43. *BUTTONIA NATALENSIS* – **1**, flowering branch × $^2/_3$; **2**, flower dissected × 1; **3**, stamens × 2; **4**, fruiting calyx × $^2/_3$; **5**, capsule with part calyx removed × $^2/_3$. 1–3 from *Balsinhas* 689; 4, 5 from *Chase* 8271. Drawn by Christine Grey-Wilson. Reproduced with permission from F.Z.

40. **PSEUDOSOPUBIA**

Engl. in Ann. Ist. Bot. Roma 7: 28 (1897) & E.J. 23: 511, t. 13/L–Q (1897)

Perennial herbs or undershrubs, erect, trailing or ascending, branched or not, glabrous to scabrid or strigose. Leaves opposite. Inflorescence terminal, racemose; flowers solitary, axillary, bracteate, pedicellate or not, 2-bracteolate. Calyx tubular to campanulate, 5-lobed; lobes valvate, ± equal. Corolla tube short, expanding and inflated above; limb oblique, 2-lipped; the lower lip overlapping the upper in bud, 3-lobed; the upper lip obovate, shortly bilobed. Stamens 4, didynamous, included, upper pair of anthers without appendage, lower pair with connective, the upper branch of which bearing one anther, the lower curving upwards and ending in a rounded appendage; anthers monothecal, dehiscing by an apical pore. Ovary 2-locular. Capsule 2-valved. Seeds numerous, angular, pitted.

A small genus of 3 or 4 species, endemic to tropical East and NE Africa.

Annual or perennial herb; calyx and corolla hairy; corolla 10–18 mm . . . . 1. *P. hildebrandtii*
Shrub; calyx and corolla glabrous; corolla 25–30 mm . . . . . . 2. *P. delamerei*

1. **Pseudosopubia hildebrandtii** (*Vatke*) *Engl.* in Ann. Ist. Bot. Roma 7: 28 (1897) & in E.J. 23: 511 (1897); Hemsley & Skan in F.T.A. 4(2): 442 (1906); Blundell, Wild Fl. E. Afr.: 377, pl. 787 (1987); U.K.W.F.: 259, pl. 111 (1994); Fischer in Thulin (ed.), Fl. Somal.: 3: 285 (2006) & in Fl. Ethiop. & Eritr. 5: 290 (2006). Type: Kenya, Kwale/Kilifi District: Maji ya Chumvi [Tchamtéi] in Duruma, *Hildebrandt* 2314b (B†, holo.). Neotype: Kenya, Northern Frontier District: Moyale, *Gillett* 12843 (K!, chosen here)

Annual or perennial herb, up to 60 cm; stems erect to ascending, branched, glabrescent to scabrid or strigose, hairs retrorse often with tubercled base. Leaves ovate to ovate-lanceolate to lanceolate to linear-lanceolate, (2.5–)10–35 mm long, (1.5–)4–8 mm wide, base cuneate, apex acute, margins entire, strigose, often black-dotted and verrucose, often drying black. Flowers axillary or in the axils of bracts, scented (recorded on one sheet only); bracts lanceolate, 4–5 mm; bracteoles linear to lanceolate, placed close under the calyx, shorter than the calyx tube; pedicels 4–12 mm. Calyx 4–6 mm, distinctly veined, glabrescent to strigose with short stipitate-glandular hairs, persistent; lobes triangular, 3.5–4 mm, strigose on the margins. Corolla mauve to pink or white flushed with pink, oblique, 10–18 mm long; tube 8–10 mm long, ± 8 mm in diameter, strigose with or without short stipitate glandular hairs, often with stipitate glands only outside, glabrous inside; limb 2-lipped, 5–6 mm strigose, sometimes very sparsely so or with hairs on the veins only, with or without short stipitate-glandular hairs, sparsely hairy inside. Larger anthers ± 5 mm, smaller ± 3 mm, orange, connective white. Ovary ovoid, strigose. Capsule subglobose, laterally compressed, 5–5.5 mm in diameter, glabrescent to strigose. Fig. 44, p. 173.

UGANDA. Karamoja District: Upe, near Kisei, June 1955, *M.S. Philip* 694! & 20 km S of Moroto, 13 Sept. 1956, *Bally* 10808!; Nakiloro, 16 June 1963, *Kertland* 472!

KENYA. Baringo District: Kampi ya Samaki, Baringo lodge, 11 June 1977, *Gilbert* 4730!; Machakos District: Garabani valley, 30 Mar. 1940, *van Someren* 259!; Kwale District: Tanga–Mombasa road, 14 Aug. 1953, *Drummond & Hemsley* 3748!

TANZANIA. Lushoto District: Usambaras, Umba Valley, Dec. 1892, *Smith* s.n.!

DISTR. **U** 1, 3; **K** 1–4, 7; **T** 3; Ethiopia, Somalia

HAB. Grassland, grassland with trees, rocky and sandy scree, river and swamp margins, also on black cotton soil; 0–1500 m

FIG. 44. *PSEUDOSOPUBIA HILDEBRANDTII* — **1**, habit × ²⁄₃; **2**, leaf underside detail × 2; **3**, flower × 1; **4**, flower opened × 1½; **5**, calyx with style × 2; **6**, calyx opened with gynoecium × 2; **7**, stamens × 4; **8**, capsule × 4; **9**, seed × 14. 1 from *Gillett* 18975; 2–7 from *Bally & Smith* B14747; 8, 9, from *Gilbert, Kanuri, Kitui & Mungai* 5477. Drawn by Juliet Williamson.

USES. A cold fusion of leaves left standing for 5 days is used for bathing babies suffering from malaria (recorded on sheet from Kenya, *Bally* 5843)
CONSERVATION NOTES. Least Concern (LC); widespread and common

SYN. *Sopubia hildebrandtii* Vatke in Linnaea 43: 314 (1882)
S. kituiensis Vatke in Linnaea 43: 313 (1882). Type: Kenya, Kitui, *Hildebrandt* 2757 (K!, holo.)
Pseudosopubia obtusifolia Engl. in Ann. Ist. Bot. Roma 7: 28 (1897) & in E.J. 23: 511, t. 13l–q (1897); Hemsley & Skan in F.T.A. 4(2): 444 (1906). Type: Ethiopia, between Karena and Daua, *Riva* 950 (FT, holo.)
P. kituiensis (Vatke) Engl. in Ann. Ist. Bot. Roma 7: 28 (1897) & in E.J. 23: 511 (1897); Hemsley & Skan in F.T.A. 4(2): 442 (1906)
P. ambigua Hemsl. & Skan in F.T.A. 4(2): 443 (1906). Type: Tanzania, Lushoto District: Usambaras, Umba Valley, *Smith* s.n. (K!, holo.)
P. elata Hemsl. & Skan in F.T.A. 4(2): 443 (1906). Type: Kenya, Lamu District: Witu, *A.S. Thomas* 9 (K!, holo.)
P. polemonioides Chiov., Fl. Somal. 2: 338 (1932). Type: Somalia, Lake Dera, *Balladelli* 287 (FT, holo.)

NOTE. This species is very variable in its habit, pubescence of stem, leaf and calyx, and the size and shape of leaves. Sparsely strigose to glabrescent plants with a straggling habit, and generally narrower leaves have been separated as *P. kituiensis*. I (S.A.G.) have looked at most of the material of this and that identified as *P. kituiensis*, and note that variation in characters noted above is present throughout the range of the species. It seems to me that *P. kituiensis* is a herbaceous and glabrescent form of *S. hildebrandtii* localized in one area from where the type has been collected. Other material from the same area identified as *P. kituiensis*, (**K** 4, Machakos/Kitui District: e.g. *Napper* 1640, *Bally* 8065, *Gardner* 3695) do not show the same characters as the type. Based on my observations it is difficult to give *P. kituiensis* a separate status.

2. **Pseudosupobia delamerei** *S.Moore* in J.B. 1901: 261 (1901); Hemsley & Skan in F.T.A. 4(2): 441 (1906); Fischer in Thulin (ed.), Fl. Somal.: 3: 286 (2006). Type: Kenya, Dadaro, 1898, *Delamere* s.n. (BM!, holo.)

Shrub to 2 m; stems spreading, branched, bark of the older stems white and splitting, scabrid with retrorse tubercled based hairs. Leaves in fascicles of 4–5 on the main stem and branches, opposite on the flowering stems, linear-oblancelate to oblanceolate, 8–30 mm long, 1–3 mm wide, base cuneate, apex obtuse, margins entire, strigose to ± glabrous, drying black to blackish-green. Flowers solitary on the stems; bracts similar to leaves; bracteoles linear, 3–4 mm, placed 2/3 up on the pedicel; pedicels 5–8 mm, sparsely scabrid but almost glabrous above the bracteoles. Calyx campanulate, 7–9 mm, distinctly 5-veined, glabrous, with or without whitish tubercles on the lobes, persistent; lobes broadly triangular, 3–4 mm. Corolla pink to lilac, 25–30 mm long, oblique; tube 18–20 mm long, expanding to ± 20 mm in diameter above, glabrous; limb ± 10 mm, 2-lipped, the upper shallowly 2-lobed, the lower 3-lobed; lobes rounded. Larger anthers ± 8 mm, smaller ± 6 mm, white with the connective orange. Ovary glabrous; style ± 30 mm long, curved at the top, included. Capsule not seen.

KENYA. Northern Frontier District: SW of Damassa Pan, 26 May 1952, *Gillett* 13330! & below W flank of Danissa [Gari] Hills, 15 Dec. 1971, *Bally & A.R. Smith* B14626! & Dadaro, 1898, *Delamere* s.n.!
DISTR. **K** 1; Somalia
HAB. *Acacia–Commiphora* bushland; 400–600 m
USES. None recorded on specimens from our area
CONSERVATION NOTES. Least Concern (LC)

41. **SOPUBIA**

D.Don, Prodr. Fl. Nep.: 88 (1825)

Annual or perennial herbs or undershrubs, usually erect, branched. Leaves opposite or verticillate, or upper alternate. Inflorescence terminal, racemose or spicate. Flowers bracteate; bi-bracteolate; pedicellate or not. Calyx campanulate, 5-lobed; lobes valvate. Corolla tube usually short, at times exserted, enlarged at throat; limb 5-lobed. Stamens 4, didynamous, slightly included; anthers bithecal, all coherent or coherent in pairs; one cell of each anther perfect, the other cell much smaller, quite empty or nearly so. Style elongated, thickened or flattened at apex. Capsule often compressed above, loculicidal. Seeds numerous, oblong to obovoid or at times narrowly cylindric.

A genus of about 40 species, from tropical and southern Africa, and from the Himalayas to Indo-China and Hong Kong; a single species in Australia.

1. Calyx glabrous to subglabrous to scabrid-pubescent or only hairy along the margins of the calyx lobes; not densely woolly 2
   Calyx densely woolly on the outside 7
2. Leaves undivided 3
   Leaves trifid 5
3. Stem densely pubescent with pubescence usually in 3 or 4 rows alternating with leaf insertions; pedicels 1.5–4.5 mm long, elongating to 6.5 mm in fruit .... 4. *S. ramosa*
   Stems glabrous to minutely sparingly pubescent or scabrid throughout; pedicels 4–26 mm 4
4. Stems many-ribbed; leaves few, appressed to the stems, the plants thus appearing almost leafless .... 3. *S. simplex*
   Stems not ribbed, leafy; leaves closely or loosely arranged, often with leafy branches in axils .... 2. *S. mannii*
5. Inflorescence lax; corolla yellow or white with dark purple centre; calyx sparsely to more or less densely orange to black glandular-punctate when dried .... 1. *S. eminii*
   Inflorescence compact; corolla pink, mauve or lilac; calyx glabrous to scabrid-pubescent, not glandular-punctate 6
6. Leaves always trifid; inflorescence short, densely spicate, 15–30 mm long; calyx glabrous, sparsely hairy on margins of lobes .... 5. *S. ugandensis*
   Leaves rarely trifid; inflorescence many-flowered, lax, up to 20(–35) cm long; calyx scabrid-pubescent .... 4. *S. ramosa*
7. Inflorescence lax, the rhachis being visible between the whorls or pairs of flowers at least in the lower part 8
   Inflorescence compact, the rachis not visible between the flowers 9
8. Plant below inflorescence only sparsely woolly; leaves simple or 3- or 5-fid, leaves or segments ± 1 mm broad; calyx lobes triangular; ovary glabrous .... 6. *S. karaguensis*
   Plant covered with a dense woolly or appressed silvery grey indumentum; retaining its grey colour when dry; leaves always simple, 2–10 mm broad; calyx lobes linear-triangular; ovary densely pubescent .... 7. *S. lanata*
9. Plant covered with a densely woolly or appressed silvery grey indumentum, retaining its grey colour when dry; leaves always simple; capsule sparsely pilose .... 7. *S. lanata*
   Plant only sparsely woolly, not silvery; black, brown or green when dry; leaves simple or trifid; capsule glabrous .... 8. *S. conferta*

1. **Sopubia eminii** *Engl.*, P.O.A. C: 359 (1895); Skan in F.T.A. 4(2): 447 (1906); Hansen in K.B. 30: 544 (1975); Philcox in F.Z. 8(2): 146 (1990); U.K.W.F.: 259, pl. 111 (1994); Fischer in Fl. Ethiop. & Eritr. 5: 289 (2006). Types: Tanzania, *Stuhlmann* 3334, 3582 and 3848 (B†, syn.). Neotype: Kenya, Nandi–Mumias, *Whyte* s.n. (K! chosen by Hansen, 1975)

Annual, erect to slightly erect, scabrid herb, (20–)30–60 cm tall; stems simple to much branched especially near base, obscurely angled, branches spreading to slightly erect, scabrid, frequently with minute, retrorse, bristly hairs. Leaves up to 4 cm long, opposite to almost opposite, occasionally in whorls of 3, usually trifid particularly below with narrowly linear lobes, or occasionally 5-fid, becoming entire above, strict to somewhat flexuous when dry, bracteate within inflorescence, often subtending small axillary leafy branches. Inflorescence lax, terminal, racemose; flowers opposite; bracteoles 3–4 mm long, 0.4–0.6 mm wide, narrowly linear to linear-lanceolate, scabrid, arising towards apex of pedicel; pedicels 5–10(–25) mm long, slender, scabrid. Calyx 3.5–4.5 mm long, subglabrous to minutely scabrid on the outside, subglabrous within, not densely woolly; lobes broadly triangular, ± 1.3 mm long, slightly to barely tomentose at margins, sparsely to more or less densely orange to black glandular-punctate when dried. Corolla yellow or white with dark purple centre, 6.5–7 mm long, limb 8–10 mm in diameter. Fertile anther thecae cylindric-ellipsoid, ± 2 mm long. Capsule broadly cylindric-ellipsoid, 3.5–5 mm long, ± 3 mm in diameter, glabrous or sparsely reddish glandular-punctate when dried.

UGANDA. Mbale District: Sukulu, 7 km SW of Tororo, 17 Nov. 1965, *J. Stewart* 790!; Masaka District: Dumu, Lake Victoria, 18 May 1971, *Lye* 6121!

KENYA. S Elgon, Sept. 1946, *Tweedie* 691!; Nairobi District: Thika Road House, 8 July 1951, *Verdcourt* 542!; Kisumu–Londiani District: Londiani–Fort Ternan road, shoulder of Limutet [Limutit], 26 Sept. 1953, *Drummond & Hemsley* 4462!

TANZANIA. Bukoba District: Maruku, Feb. 1932, *Haarer* 2495!; Dodoma District: W of Itigi Station, 14 Apr. 1064, *Greenway & Polhill* 11565!; Songea District: Matengo Highlands, WSW from Songea, 4 Mar. 1936, *Zerny* 498!

DISTR. **U** 3–4; **K** 3–5; **T** 1–2, 4–8; Zambia, Malawi, Zimbabwe

HAB. Woodland and grassy places and hillsides; 1000–2300 m (–2790 m fide U.K.W.F.)

USES. None recorded from our area

CONSERVATION NOTES. Least Concern (LC); widespread

SYN. *Sopubia trifida* sensu Skan in F.T.A. 4(2): 446 (1906), *non* Buch.-Ham. ex D.Don (1825)

2. **Sopubia mannii** *Skan* in F.T.A. 4(2): 450(1906); Hansen in K.B. 30: 545 (1975); Philcox in F.Z. 8(2): 144(1990); U.K.W.F.: 259 (1994); Fischer in Fl. Ethiop. & Eritr. 5: 290 (2006). Type: Cameroon, Mt Cameroon, *Mann* 1281 (K!, lecto., see Hepper, K.B. 14(3): 409 (1960))

Erect, rigid, scabrid, branched herb, 20–150 cm tall; stems very leafy, numerous arising from a thick root-stock. Leaves opposite or verticillate, undivided, closely or loosely arranged, often with leafy branches in axils, filiform or narrowly linear, 15–55 mm long, 0.3–1(–1.5) mm wide, apex acute, subscabrid on margins, margins reflexed or not, midvein indistinct or not. Inflorescences of usually terminal racemes, up to 10(–25) cm long. Flowers usually numerous, opposite or in whorls of 3; bracteoles 1.5–4 mm long, 0.2–0.8 mm wide, arising immediately beneath or up to 4 mm below calyx; pedicels 4–26 mm long. Calyx 3.5–7 mm long, variously hirsute inside and outside. Corolla pink, mauve-pink to mauve or purple (?yellow, see below); limb 10–22 mm in diameter. Anthers 2–3.5 mm long, 0.7–1 mm wide. Capsule ovoid, 3.5–8 mm long, 2.5–4.5 mm broad.

1. Leaves linear or filiform, 0.9–1(–1.3) mm wide, mostly congested on stem; pedicels less than 10 mm long; calyx 3–5 mm long; corolla limb rarely exceeding 12 mm in diameter . . . . a. var. *mannii*
   Leaves linear or filiform, more loosely arranged on stem; pedicels usually more than 10 mm long; corolla limb usually more than 12 mm in diameter 2
2. Leaves filiform, 0.3–0.5(–0.8) mm wide; midrib mostly indistinct; pedicels 12–18(–26) mm long; corolla limb (12–)14–18(–22) mm in diameter . . . . b. var. *tenuifolia*
   Leaves linear (flat), some up to 1.5 mm wide; midrib distinct beneath; pedicels 10–15 mm long; corolla limb 16–18(–20) mm in diameter . . . . c. var. *linearifolia*

a. var. **mannii** *Hepper* in K.B. 14: 409 (1960) & F.W.T.A. ed. 2, 2: 369 (1963); Hansen in K.B. 30: 545 (1975); Philcox in F.Z. 8(2): 144 (1990)

Perennial herbs. Leaves undivided, 0.8–1.3 mm wide, mostly congested on stem. Flowers closely arranged on inflorescence axis; pedicels 4–8.5(–10) mm long. Calyx 3.5–5.5 mm long, smooth, glabrous or subglabrous, tomentose on margins. Corolla limb 10–12(–13.5) mm in diameter.

UGANDA. Masaka District: 1.4 km on Katera–Kyebe [Kiebbe], 1 Oct. 1953, *Drummond & Hemsley* 4519!

TANZANIA. Ngara District: Kirushya, Bugufi, 5 Apr. 1960, *Tanner* 4825!; Ufipa District: Sumbawanga, Mbisi [Mbesi] Forest, 13 Mar. 1957, *Richards* 8694!; Njombe/Mbeya District: Kitulo [Elton] Plateau, 2 Mar. 1986, *Bidgood & Congdon* 135!

DISTR. **U** 4; **T** 1, 4, 7; Ivory Coast, Cameroon, Zambia, Malawi, Mozambique, Zimbabwe

HAB. Grassland and wooded grassland; 1500–2600 m

USES. None recorded from our area

CONSERVATION NOTES. Least Concern (LC)

b. var. **tenuifolia** (*Engl. & Gilg*) *Hepper* in K.B. 14: 410 (1960) & F.W.T.A. ed. 2, 2: 369 (1963); Hansen in K.B. 30: 546 (1975); Philcox in F.Z. 8(2): 145 (1990); U.K.W.F.: 259 (1994). Type: Angola, Habungu, *Baum* 475 (K!, lecto.)

Perennial herbs. Leaves undivided, 0.3–0.5(–0.8) mm wide, loosely arranged on stem. Flowers loosely arranged on inflorescence axis; pedicels (10–)12–18(–26) mm long. Calyx 4–5.5(–7) mm long, smooth, glabrous outside, tomentose on margins. Corolla limb (12–)14–18(–22) mm in diameter.

UGANDA. Busoga District: Bukoli, Igwe, 16 km S of Bugiri, 17 Dec. 1952, *Wood* 528!; Masaka District, Lake Nabugabo, 6 Oct. 1953, *Drummond & Hemsley* 4656!

KENYA. Nairobi District: Thika road, near Nairobi, 5 July 1952, *Kirika* 205! [fls reported as 'white turning scarlet'] (see note after next variety)

TANZANIA. Bukoba District: Ngono swamp, Jan. 1931, *Haarer* 2474!; Mpanda District: Illembo Plain, Illembo–Mpanda road, 5 July 1968, *Sanane* 202!; Iringa District: Dabaga Highlands, Kibengu, 29 km S of Dabaga, 17 Feb. 1962, *Polhill & Paulo* 1523!

DISTR. **U** 3, 4; **K** 4; **T** 1, 4, 7; Sierra Leone, Angola, Zambia, Malawi, Zimbabwe, Mozambique, Botswana, Namibia (Caprivi Strip)

HAB. In wet and dry grassland, bordering rivers and among rocks; 1000–2500 m

USES. None recorded from our area

CONSERVATION NOTES. Least Concern (LC); widespread

SYN. *Sopubia dregeana* Benth. var. *tenuifolia* Engl. & Gilg in Warb., Kunene-Samb. Exped. Baum: 365 (1903)

Sopubia trifida D.Don forma *humilis* Engl. & Gilg in Warb., Kunene-Samb. Exped. Baum: 365 (1903) excl. *S. decumbens* Hiern. Type: Zimbabwe, Kassinga, n.d. Kunbango, Oct. 1899, *Baum* 233 (K!, iso.)

c. var. **linearifolia** *O.J.Hansen* in K.B. 30: 546, fig. 1/e (1975). Type: Tanzania, Mbeya District, Kitulo [Elton] Plateau, *Richards* 14183 (EA, holo.; K! iso.)

NOTE. O.J. Hansen (op. cit.) with his description of this variety, cites the EA specimen as the holotype with that at K as the isotype. However, the K material is annotated in his own hand as the holotype.

Perennial herbs. Leaves opposite or verticillate, linear (flat), up to 18–25 mm long, 1 mm or more wide, somewhat crowded in stem; midvein clearly visible. Flowers many, regularly arranged on inflorescence axis, not crowded; pedicels 10–24 mm long, slender. Calyx 4–8 mm long, smooth, glabrous; lobes marginally lanuginose. Corolla limb 16–18(–20) mm in diameter.

TANZANIA. Iringa District: Mt Image, 80 km NE of Iringa, 4 Mar. 1962, *Polhill & Paulo* 1682!; Mbeya District: Mbeya Mt, NE side, 15 May 1956, *Milne-Redhead & Taylor* 10202!; Njombe District: Kipengere Mts, 10 Jan. 1957, *Richards* 7649b! [flowers yellow, see note below]
DISTR. **T** 7; known only from this area of Tanzania
HAB. Wooded grassland; 2000–2550 m
USES. None recorded from our area
CONSERVATION NOTES. Least Concern (LC)

This species has been considered to have simple, undivided or 3–5-multifid leaves. I have been unable to find any evidence of divided leaves and consider that what was thought to be such leaves, were in fact young axillary branches that are frequently quite common. As described, the flowers of this species vary in colour through shades of pink to mauve. However, *Richards* 7649b describes the flowers of her collection of var. *linearifolia* as "lemon-yellow". Although unusual, this cannot be accepted as unique since a similarly coloured plant was collected by *Philips* 1374 from the Nyika Plateau in Malawi, an area similar in altitude and vegetation to the *Richards* 7649b habitat on the other side of the Rift Valley, the two localities being separated only by Lake Malawi. This is not the only questionable record of flower colour, as a specimen of var. *tenuifolia, Kiddikan* 205 from Kenya was recorded as having flowers "white turning scarlet". I suspect this is possibly due to the specimen being accompanied by an incorrect and unrelated label.

3. **Sopubia simplex** (*Hochst.*) *Hochst.* in Flora 27: 27 (1844); Skan in F.T.A. 4(2): 450 (1906); Hepper in F.W.T.A. ed. 2, 2: 369(1963); F.P.S. 3: 142 (1956); Hansen in K.B. 30: 550 (1975); Philcox in F.Z. 8(2): 144(1990); U.K.W.F.: 259, pl. 111 (1994); Fischer in Fl. Ethiop. & Eritr. 5: 290 (2006). Type: South Africa, Port Natal, *Krauss* 400 (K!, holo.)

Perennial herb, erect, up to 65 cm tall; stems several arising from a woody base, simple or branched above, markedly angled, many-ribbed, glabrous to minutely sparingly pubescent, sparsely leafy. Leaves alternate to verticillate, undivided, narrowly linear, (5.5–)12–20 mm long, (0.2–)0.5–1 mm wide, usually appressed, revolute, subglabrous to scabrid on margins and midrib beneath. Inflorescence loose to somewhat clustered, terminal, racemose, up to 20(–30) cm long; flowers numerous, alternate to whorled; bracteoles linear to subulate, 1.5–2 mm long, ± 0.2 mm wide, subglabrous to minutely scabrid; pedicels 6–13 mm long, rather slender, subglabrous to shortly hispid-pubescent. Calyx brownish-green when fresh, 4–6 mm long, verrucose to hispid or minutely pubescent without; lobes triangular, 1–3 mm long, woolly at margins and inside. Corolla pink to lilac or pale mauve, rarely white, with dark-purple throat rotate; limb (10–)14–15 mm in diameter, caducous. Fertile anther cells more or less ellipsoid, 2.5–2.8 mm long. Capsule oblong to ovoid- or obovoid-oblong, 4.5–5 mm long, 2.5–3.5 mm in diameter, retuse, bearing persistent base of style. Fig. 45, p. 179.

UGANDA. Bunyoro District: Kibangya, Feb. 1943, *Purseglove* 1268!; Elgon, 1905, *Evan James* s.n.!; Masaka District: Kyebe, Buddu, Aug. 1945, *Purseglove* 1783!
KENYA. Trans-Nzoia District: near Kitale, May 1950, *Tweedie* 843!; North Kavirondo District: Kakamega Forest, 11 Apr. 1973, *Hansen* 914!; Masai District: Lolgorien, Sept. 1933, *Napier* 5344!
TANZANIA. Bukoba District: Kakindu, Oct. 1931, *Haarer* 2332!; Ufipa District: Rukwa, Sumbawanga, Kotola Hills, 28 km W of Matai, road to Kasanga, 25 Oct. 1992, *Gereau et al.* 4881!; Songea District: Mpapa Mission, Matengu, 18 Oct. 1956, *Mgaza* 113!

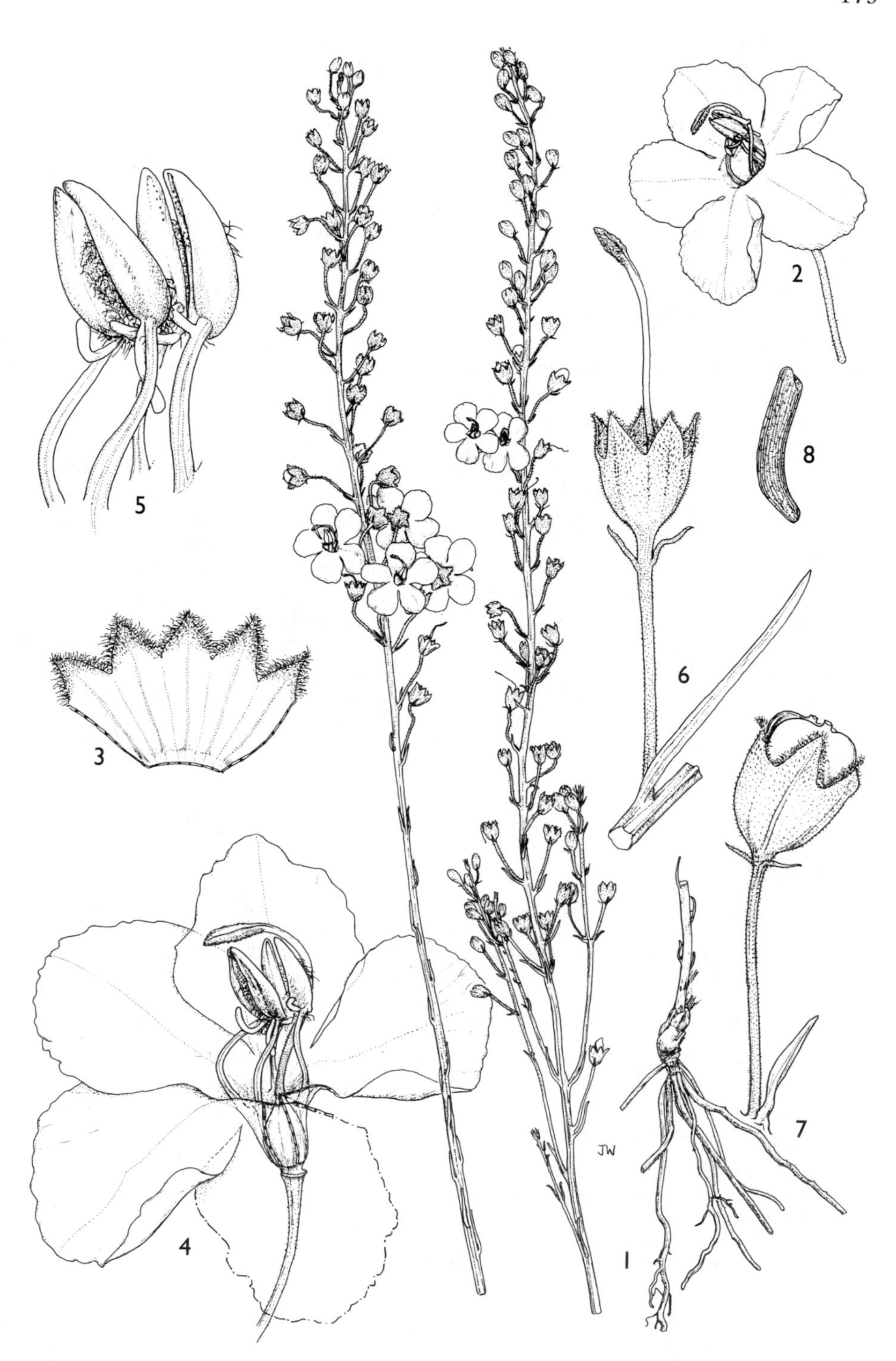

Fig. 45. *SOPUBIA SIMPLEX* — **1**, habit × ⅔; **2**, flower × 2; **3**, calyx opened × 4; **4**, corolla opened × 4; **5**, stamens × 8; **6**, young capsule × 4; **7**, capsule × 4; **8**, seed × 20. 1 from *Bidgood, Mbago & Vollesen* 2587; 2–6 from *Hargreaves* 2000; 7 from *Irwin* 1; 8 from *Gobbo, Sitoni, Kayombo & Mwiga* 564. Drawn by Juliet Williamson.

DISTR. **U** 2–4; **K** 3, 5, 6; **T** 1, 4, 6–8; West Africa, Zambia, Malawi, Mozambique, Zimbabwe
HAB. Mostly in burnt or moist grassland, riversides, swamps or marshes; 600–2300 m
USES. None recorded from our area
CONSERVATION NOTES. Least Concern (LC); widespread

SYN. *Raphidophyllum simplex* Hochst. in Flora 24: 667 (1841)
Sopubia dregeana Benth. in DC., Prodr. 10: 522 (1846) *nom. illegit.* (see Hepper in K.B. 14: 409 (1960))

4. **Sopubia ramosa** (*Hochst.*) *Hochst.* in Flora 27: 27 (1844); Skan in F.T.A. 4(2): 449 (1906); Hepper in F.W.T.A. ed. 2, 2: 369 (1963); Hansen in K.B. 30: 546 (1975); Cribb & Leedal, Mount. Fl. S Tanz.: 118, pl. 28G (1982); Philcox in F.Z. 8 (2): 145 (1990); U.K.W.F.: 258 (1994); Fischer in Fl. Ethiop. & Eritr. 5: 288 (2006). Type: Ethiopia, Shire near Sagal, *Schimper* 507 (K! lecto., chosen by Hansen, 1975)

Erect, perennial, much branched herb or undershrub, 45–100(–150 or more) cm tall, scabrid; stems usually branched above with branches ascending, densely pubescent with pubescence usually in 3 or 4 rows alternating with leaf insertions. Leaves usually in whorls of 3–4, undivided or rarely 3-fid, linear to linear-lanceolate, 15–30(–50) mm long, 1–3.5(–7) mm wide, margins revolute, strigose to scabrid, midrib very prominent. Inflorescence usually terminal, racemose, many-flowered, up to 20(–35) cm long; flowers in whorls of 3–4, frequently closely arranged; bracteoles linear to linear-lanceolate, 2.5–5 mm long, arising 0.3–0.5 mm below base of calyx, long-pubescent to hispid-strigose or subglabrous; pedicels 1.5–4.5 mm long, increasing to 6.5 mm in fruit. Calyx 3.5–4 mm long, increasing to 6.5 mm in fruit, 10-ribbed, shortly scabrid-pubescent; lobes deltate, ± 1.5 mm long, densely tomentose at margins and inside. Corolla pink to lilac or lavender, or occasionally purplish, 7–8 mm long; limb broadly ovoid to subglobose, 9–13.5 mm in diameter, emarginate, glabrous. Fertile anther cells oblong, 2.8–3 mm long. Capsule ovoid to subglobose, 4–5 mm long, 3–3.5 mm in diameter, retuse, glabrous.

UGANDA. Karamoja District: Mt Moroto, W slope, 5 Sept. 1956, *Stella Hardy & Bally* 10732!; Ankole District: Mbarara road near Kalinzu Forest, 18 May 1961, *Symes* 725!; Masaka District: Bugabo, SW of Lake Nabugabo, 1 Feb. 1969, *Lye* 1783!
KENYA. W Suk/Elgeyo Districts: Suk–Cherangani Hills, Oct. 1937, *Graham* in For. Dep. No. 3724!; Machakos/Masai Districts: Chyulu Hills, 21 Apr. 1938, *Bally* 8058!; N Kavirondo District: Yala River area, Kakamega Forest, 10 Dec. 1956, *Verdcourt* 1699!
TANZANIA. Ngara District: Kirushya, Bugufi, 5 Apr. 1960, *Tanner* 4820!; Singida District: Mt Hanang, July 1952, *Brooks* 105!; Songea District: about 19 km E of Songea, 22 June 1956, *Milne-Redhead & Taylor* 10868!
DISTR. **U** 1–4; **K** 2–6; **T** 1–8; West and Central Africa, Sudan, Ethiopia, Angola, Zambia, Malawi, Mozambique, Zimbabwe
HAB. Forest borders and glades, open woodland to grassy hillsides, plains and swampy ground; 0–3000 m
USES. None recorded from our area.
CONSERVATION NOTES. Least Concern (LC); widespread

SYN. *Raphidophyllum ramosum* Hochst. in Flora 24: 668 (1841)
Sopubia trifida D.Don var. *ramosa* (Hochst.) Engl., E.J. 18: 65 (1893); P.O.A. C: 358 (1895)
Sopubia similis Skan in F.T.A. 4(2): 447 (1906). Types: Ethiopia, without precise locality, *Schimper* 695 (K!, holo.)
Sopubia laxior S.Moore in J.B. 49: 186 (1911). Types: Angola, Catemba, near Malange, *Baum* 885 & *Gossweiler* 1095, 1096, 3168 (K!, syn.)

5. **Sopubia ugandensis** *S.Moore* in J.L.S. 37: 192 (1905); Skan in F.T.A. 4(2): 455 (1906); Hansen in K.B. 30 (3): 550 (1975). Type: Uganda, Kigezi District: below Rukiga [Ruchigga], *Bagshawe* 502 (BM!, holo.)

Perennial woody herb to about 55 cm tall; stems erect, branched above, densely silvery-white, crisped pubescent with downwardly directed hairs. Leaves up to 25 mm long, verticillate, trifid with segments about 0.5 mm wide, arising in upper third, glabrous but minutely scabrid on margins. Inflorescence short, densely spicate, 15–30 mm long, to ± 17 mm broad. Flowers mauve. Calyx glabrous without, but sparsely hairy on margins of lobes. Corolla campanulate, tube 4–6 mm long, limb 10–12 mm in diameter, lobes broadly obovate. Capsule not seen.

UGANDA. Kigezi District: below Rukiga [Ruchigga], *Bagshawe* 502!; Ankole District: Karagwe, 1893–94, *G.F. Scott Elliot* 7496! (see note under *S. karaguensis*)
DISTR. **U** 2; known from a single collection from Burundi (*van der Ben* 2006!)
HAB. Montane and submontane grassland; ± 2000 m
USES. None recorded from our area
CONSERVATION NOTES. Not much material from Uganda and that too collected more than 50 years ago; possibly at threat from habitat degradation; assessed here as Vulnerable, VU B1b(iii), with the extent of occurrence to be estimated to be less than 20,000 km² and projected decline in the quality of habitat.

6. **Sopubia karaguensis** *Oliv.* in Trans. Lin. Soc. 29: 123, t. 87, fig. B (1875); De Wild & Durand, Etudes Fl. Katanga: 126 (1903); S. Moore in J.L.S. 37: 191 (1905); Skan in F.T.A. 4(2): 448 (1906); O.J. Hansen in K.B. 30: 552 (1975); Philcox in F.Z. 8 (2): 147 (1990); U.K.W.F.: 258 (1994); Fischer in Fl. Ethiop. & Eritr. 5: 289 (2006). Type: Uganda, Ankole District: Karagwe hilltop, *Grant* 411 (K!, holo.)

Erect, robust perennial or undershrub to 1 m tall; stems up to 6 mm in diameter at base, branched or occasionally simple, branches woody, ascending, pubescent to long-pilose or glabrescent. Leaves opposite to subverticillate, often bearing smaller leaves in their axils, undivided to trifid or 5-fid, 15–40 mm long, 1–1.5 mm wide, margins somewhat thickened, pilose to shortly pubescent, slightly scabrid. Inflorescence 5–20 cm long, loosely spicate or racemose with clearly defined internodes, simple to much branched; flowers many; bracteoles narrowly linear, 6.5–11 mm long, acute, villous; pedicels 0–6.5 mm long. Calyx 6–15 mm long, 4–11 mm wide, densely lanate-tomentose outside and inside; lobes narrowly triangular, 3–13.5 mm long. Corolla pink, mauve to purple; limb 15–22 mm in diameter. Fertile anther thecae cylindric, 3.5–4.5 mm long. Capsule subglobose to broadly ovoid, 5–5.5 mm long, 3.8–5 mm in diameter, glabrous.

1. Calyx less than 11 mm long, woolly 2
   Calyx 11–21 mm long, covered with long, stiff hairs on outside as well as inside, distended .... b. var. *macrocalyx*
2. Calyx less than 8 mm long .... a. var. *karaguensis*
   Calyx 8–11 mm long .... c. var. *welwitschii*

a. var. **karaguensis**

Calyx 6–11 mm long, 4–5 mm wide; calyx lobes 3–9.5 mm long; calyx not apparently inflated or enlarged.

UGANDA. Ankole District: Karagwe hilltop, Dec. 1861, *Grant* 411!
KENYA. Machakos District: Chyulu Mts, above Kibwezi, 3 May 1975, *Friis & Hansen* 2699! & May 1938, *Bally* 8059!; Kisumu-Londiani District: Londiani, *Graham* in For. Dept. 2701!
TANZANIA. Bukoba District: Bugene, Oct. 1931, *Haarer* 2294!; Moshi District: E Kilimanjaro, June 1937, *Haarer* 568! & N Kilimanjaro, 26 Dec. 1932, *Geilinger* 5118!
DISTR. **U** 2; **K** 4–5; **T** 1–2; Zambia, Malawi, Mozambique, Zimbabwe, Swaziland
HAB. Damp ground and grasslands; up to 2400 m
USES. None recorded from our area
CONSERVATION NOTES. Least Concern (LC); widespread

SYN. *S. welwitschii* Engl. var. *micrantha* Engl. in P.O.A. C: 359 (1895). Type: Tanzania, Kilimanjaro, Marangu, *Volkens* 1997 (B†, holo.; BM!, G, iso.)
S. fastigiata Hiern in Fl. Cap. 4, 2: 387 (1904). Type: Swaziland, Piggs Peak, *Galpin* 1337 (K!, holo.)

b. var. **macrocalyx** *O.J.Hansen* in K.B. 30: 553 (1975); Philcox in F.Z. 8 (2): 149 (1990). Type: Zimbabwe, Gwampa Forest Reserve, *Goldsmith* 93/56! (K!, holo.)

Calyx 11–15 mm long or more, 7.5–10 mm wide; calyx lobes 10–13.5 mm long; whole calyx appearing inflated and enlarged.

TANZANIA. Mbeya District: Ruaha National Park, Magangwe Ranger Post, 8 May 1972, *Bjørnstad* 1632!
DISTR. **T** 7; Zambia, Malawi, Mozambique, Zimbabwe
HAB. Grasslands, swamps, and mixed woodland; 1000–1250 m
USES. None recorded from our area
CONSERVATION NOTES. Least Concern (LC)

c. var. **welwitschii** (*Engl.*) *O.J.Hansen* in K.B. 30: 553 (1975). Type: Angola, Huilla, Oiabya, *Welwitsch* 5841 (B†, holo.; K! lecto., chosen by Hansen)

Calyx 8–11 mm long, 6–8 mm wide, calyx lobes 8.5–11 mm long; whole calyx not appearing particularly inflated, but where so, probably due to the density and length of indumentum.

KENYA. Elgeyo District: Kaisungor, Cherangani Hills, Feb. 1965, *Tweedie* 2980!; Laikipia District: Rumuruti, July 1931, *Napier* 1211!; Nakuru District: Ol Joro Orok, Jan. 1932, *Pierce* in *Napier* 1866!
DISTR. **K** 3, 5; Congo-Kinshasa, Angola
HAB. Higher altitude grasslands bordering forests, edges of paths near cultivation; 2100–3000 m
USES. None recorded from our area
CONSERVATION NOTES. Least Concern (LC)

SYN. *S. welwitschii* Engl. In E.J. 18: 66 (1893)

7. **Sopubia lanata** *Engl.* in E.J. 18: 67 (1894); Skan in F.T.A. 4(2): 454 (1906); Hansen in K.B. 30: 553 (1975); Philcox in F.Z. 8(2): 147, t. 48 (1990). Type: Angola, Pungo Andongo, Luxillo stream near Quibinda, *Welwitsch* 5863 (B†, holo.; K!, iso.)

Erect, robust perennial or undershrub up to 1 m or more tall; stems many from a woody base, up to 6.5 mm thick below, simple or branched usually above, densely leafy, densely grey to silvery, appressed lanate almost throughout. Leaves subverticillate, spreading, strict or laterally arcuate, with tufts of smaller leaves in axils, linear to linear-lanceolate, 15–40 mm long, 1–2.5(–5) mm wide, margins entire, slightly revolute, midvein prominent beneath. Inflorescence many-flowered, 4–12(–16) cm long, 1.5–2.5 cm wide, spicate or racemose; flowers closely and tightly arranged so as to obscure the inflorescence axis and base of flowers, or more loosely arranged showing main axis and at least base of lowest flowers; bracteoles linear or linear-lanceolate, 4–10 mm long; pedicels 0–10 mm long. Calyx 6–10 mm long; lobes linear-triangular, 3–5.5 mm long. Corolla pink to lilac or mauve-pink with darker pink or mauve throat; limb 10–16 mm in diameter. Fertile anther thecae cylindric or ellipsoid, 2–3 mm long. Capsule shortly cylindric to subglobose, 4–5 mm long, about 4 mm in diameter, rounded at apex, sparsely pilose.

var. **densiflora** (*Skan*) *O.J.Hansen* in K.B. 30: 554 (1975); Philcox in F.Z. 8(2) 147, tab 48, 1 (1980). Type: Tanzania, Livingstone Range, E of Lake Malawi [Nyasa], *Johnson* s.n. (K! lecto., chosen by Hansen)

Flowers closely and tightly arranged so that the inflorescence axis, calyx, base of flowers, and pedicels if present, are not visible at flowering time.

UGANDA. Kigezi District: Nyakagyeme [Nyakageme], May 1947, *Purseglove* 2411!; Masaka District: Sango Bay region, Oct. 1925, *Maitland* 1068!
TANZANIA. Ngara District: Bushubi, 12 Apr. 1960, *Tanner* 4853!; Mbeya/Ufipa Districts: 75 km NW of Tunduma on Sumbawanga road, 28 May 1990, *Carter et al.* 2516!; Songea District: Miyau, Matengo Hills, 23 May 1956, *Milne-Redhead & Taylor* 10292!
DISTR. **U** 2, 4; **T** 1, 4, 7–8; Congo-Kinshasa, Rwanda, Angola, Zambia, Malawi
HAB. Marshes, swamps, riversides, wet grasslands and rainforest borders; 900–2500 m
USES. None recorded from our area
CONSERVATION NOTES. Least Concern (LC)

SYN. *Sopubia densiflora* Skan in F.T.A. 4(2): 454(1906)

NOTE. The type variety is known from Congo-Kinshasa, Angola and Zambia and is not recorded from the Flora area. It differs from var. *densiflora* in its flowers loosely arranged showing inflorescence axis and calyx and pedicels visible, at least of lower flowers at flowering time.

8. **Sopubia conferta** *S.Moore* in J.L.S. 37: 191 (1905); Skan in F.T.A. 4(2): 455 (1906); Hansen in K.B. 30 (3): 555 (1975). Type: Uganda, Kigezi District: Rukiga [Ruchigga], *Bagshawe* 529 (K! lecto., chosen by Hansen)

Perennial woody herb, 30–90 cm tall; stems 1–many arising from a woody rootstock, indumentum variable but never silvery-appressed hairy. Leaves verticillate, simple or trifid, 15–45 mm long, segments 1–2.5 mm wide, margins outwardly rolled, midvein usually prominent, densely hairy above becoming subglabrous below. Inflorescence terminal, spicate, (1.5–)3–7(–10) cm long, (1.2–)2.1–2.5 cm broad, densely woolly; flowers alternate, congested; bracteoles oblong-lanceolate, 8–20 mm long. Calyx lobes narrowly triangular, ± 4 mm, woolly. Corolla blue-mauve, mauve, pink to pinkish-white, rotate to subcampanulate; limb 10–15 mm in diameter. Fertile anther cells oblong, ± 2 mm long. Capsule ovoid, 4–5 mm long, ± 3 mm in diameter, glabrous.

UGANDA. Ankole District: Ankole, Jan. 1934, *Mainwaring* 5901!; Kigezi District: Rukiga [Ruchigga], *Bagshawe* 529!
TANZANIA. Ngara District: Bushubi, Keza, 15 May 1960, *Tanner* 4965!; Mbulu District: 9.6 km N of Olairobi, on Ngorongoro Crater rim, 28 July 1957, *Bally* 11610! Mpwapwa District: Mpwapwa, Kiboriani Mts, 19 Apr. 1932, *Burtt* 3896!
DISTR. **U** 2, 4; **T** 1, 2, 4, 5, 7; not known elsewhere
HAB. Montane grasslands and light forest, mostly on poor soils; 1500–2200 m
USES. None recorded from our area
CONSERVATION NOTES. Least Concern (LC)

42. **GHIKAEA**

Volkens & Schweinf. in Ghika, Pays des Somalis 214 (1898)

Much branched shrub. Leaves opposite, becoming subopposite on the flowering stems, drying black. Flowers large, showy, solitary, placed above the axil of leaves with 2 small glands between the axil and pedicel; bracteoles 2. Calyx equally 5-lobed; lobes as long as or slightly longer than the tube. Corolla campanulate, tube expanding above. Stamens 4, didynamous, included; adaxial stamens with shorter filaments and bithecal anthers; connective divided with the upper part bearing a single pollen cell, the lower produced into a slender curved spur; abaxial stamens monothecal; connective absent; anthers transverse to filament. Ovary ovoid; style curved distally, exserted; stigma rounded. Capsule loculicidal.

Monotypic; endemic to tropical East Africa.

Ghikaea speciosa (*Rendle*) *Diels* in E. & P. Pf.: 3, 314 (1908); Thulin, Fl. Somal.: 3: 287 (2006); Fischer in Fl. Ethiop. & Eritr. 5: 292 (2006). Type: Ethiopia, Galla Country, Darar to Sheikh Husein (Sheik Husin), Sept. 1894, *Donaldson Smith* s.n. (BM!, holo.)

Shrub to 1.5 m tall, much branched; stems somewhat angular, dark brown with short retrorse hairs, lenticellate. Leaves opposite, becoming subopposite on the flowering stems, ovate to obvate to elliptic, 1–3 × 0.5–2 cm, base tapering into a short petiole, apex acute or obtuse, margins entire, veins impressed, hirsute. Flowers large, up to 4 cm long and across, solitary, placed above the axil of leaves with 2 small glands between the axil and pedicel; bracteoles 2, placed close to the base of the calyx, linear, ± 8 mm long; pedicel 5–6 mm, hirsute. Calyx campanulate, 7–9 mm long, equally 5-lobed; lobes slightly longer than the tube, ovate-lanceolate, 4–5 mm long, acute, hirsute, ciliate on the margins. Corolla reddish purple to violet, campanulate, tube expanding 10–13 mm from base to ± 12 mm across at mouth; lobes 8–10 mm, rounded, spreading, glabrous except on the margins. Stamens with the upper pair of filaments ± 15 mm, connective divided with the upper part bearing a single pollen cell ± 6 mm long, the lower produced into a slender curved spur ± 8 mm long; lower pair of stamens with filaments ± 17 mm. Ovary ovoid, 4–5 mm long, glabrous; style ± 40 mm; stigma rounded. Capsule ovoid, 8–10 mm long, apiculate. Fig. 46, p. 185.

KENYA. Northern Frontier District: Latakwen (Ndoto Mts), Jan. 1959, *Newbould* 3353!; Meru District: Meru National Park, 3 km NW of Tana, 12 Dec. 1969, *Gillett* 18923!; Tana River District: Kora Reserve, 29 Dec. 1982, *van Someren* 878!

DISTR. **K** 1, 3, 4, 7; Ethiopia, Somalia

HAB. Dry grassland, *Acacia–Commiphora* scrub; flowering December to May; 350–1400 m

USES. None recorded on specimens from our area

CONSERVATION NOTES. Least Concern (LC), but may qualify as Near Threatened (NT) due to degradation of habitat

SYN. *Graderia speciosa* Rendle in J.B. 1896: 128 (1896)
Ghikaea spectabilis Volkens & Schweinf. in Ghika, Pays des Somalis, 214 (1898); Engl. in Ann. Ist. Bot. Roma 7: 27 (1897) & E.J. 23: 510, t. 13j, k (1897); Hemsl. & Skan in F.T.A. 4(2): 440 (1906). Type: Somalia, between rivers Jerer & Faf, *Ghika* s.n. (B†, holo.)

NOTE. Interestingly of all the material that I (S.A.G.) have seen with flowers only the one made in June (*Gillett* 13432) is with capsules, but the seeds are not developed.

43. **GRADERIA**

Benth. in DC., Prodr. 10: 521 (1846)

Perennial undershrub, many-stemmed from a woody rootstock. Leaves opposite or alternate. Flowers solitary, axillary. Calyx 5-lobed, campanulate. Corolla tubular, 5-lobed, limb spreading, lobes almost entire. Stamens 4, didynamous, included; filaments filiform, pilose at least towards base, inserted below middle of corolla tube; anthers bithecal, free, thecae divergent, oblong, curved, mucronate at base. Ovary bilocular, compressed; ovules numerous; style slender, glabrous, exserted, incurved above. Capsule compressed perpendicular to septum, loculicidal. Seeds numerous, obovoid-cylindrical.

A genus of 4 species; 3 from Southern Africa and 1 from Socotra.

Graderia scabra (*L.f.*) *Benth.* in DC., Prodr. 10: 521 (1846); Philcox in F.Z. 8(2): 151, t. 49 (1990). Type: South Africa, Cape, *Thunberg* s.n. (LINN, holo. 764.3)

FIG. 46. *GHIKAEA SPECIOSA* — **1**, habit × $^{2}/_{3}$; **2**, corolla opened × 1; **3**, upper anther × 4; **4**, lower anther × 4; **5**, calyx with bracts and style × $1^{1}/_{2}$; **6**, calyx opened with gynoecium × $1^{1}/_{2}$; **7**, capsules × 2; **8**, capsule opened × 2. 1–6 from *Newbould* 3353; 7, 8 from *Gillett* 13432. Drawn by Juliet Williamson.

FIG. 47. *GRADERIA SCABRA* — **1**, habit × $^2/_3$; **2**, flower × 2; **3**, corolla opened × 2; **4**, capsule × 3. 1 from *Methuen* 10; 2, 3 from *Richards* 22573; 4 from *Fries et al.* 3611. Drawn by Christine Grey-Wilson. Reproduced with permission from F.Z.

Erect, ascending or rarely procumbent perennial shrub, 7–30 cm tall, from a woody rootstock, usually branched at or near the base; stems obtusely quadrangular above, leafy, indumentum of large multicellular hairs, slightly herbaceous (at least when young). Leaves opposite or nearly so or alternate, ovate-elliptic or more usually lanceolate. 10–22(–30) mm long, 4–17 mm wide, base subcuneate to subsessile, apex acute, margins incised-pinnatifid with acute teeth or rarely subentire, 3–5-veined with venation prominent beneath, scabrid-hispid. Flowers in axils of upper leaves; bracteoles opposite, sublanceolate to linear, 5–8.5 mm long, 0.6–1 mm wide, 1-veined, inserted at base of calyx; pedicels 0–2 mm long. Calyx campanulate, 7–10 mm long, deeply 5-lobed, 10-veined, scabrid-hispid or pubescent especially on veins; lobes triangular-lanceolate, 4–6 mm long, acute. Corolla purple, mauve or pink, funnel-shaped; tube 10–15 mm long, externally sparsely pubescent; limb 13–16(–22) mm in diameter; lobes broadly rounded, 4–9 mm long, spreading; style filiform, to 14 mm long, persistent in young fruit. Capsule globose, 7 mm long, 6–7 mm wide, strongly longitudinally compressed, appearing broadly unilaterally winged, included in persistent calyx. Fig. 47, p. 186.

TANZANIA. Iringa District: Mufindi, 7 Oct. 1936, *McGregor* 72! & 19 Sept. 1971, *Paget-Wilkes* 995!
DISTR. **T 7**; Malawi, Mozambique, Zimbabwe, Swaziland, South Africa
HAB. Montane grasslands; 1500–2400 m
USES. None recorded from our area
CONSERVATION NOTES. Known from a few specimens from Tanzania; elsewhere of Least Concern (LC)

SYN. *Gerardia scabra* L.f., Suppl.: 279 (1781)
Sopubia scabra (L.f.) G. Don, Gen. Syst. 4: 560 (1837)
Bopusia scabra (L.f.) Presl. in Abh. Bohm. Gesd. Wiss. 3: 521 (1845)

44. MICRARGERIA

Benth. in DC., Prodr. 10: 509 (1846)

Annual herbs; stems erect, branched. Leaves opposite to alternate, linear to filiform, entire or trifid. Flowers small, solitary, axillary or terminally racemose, bi-bracteolate. Calyx campanulate, 5-lobed; lobes ovate to lanceolate, acute or obtuse. Corolla campanulate, tube enlarged above; sometimes incurved; limb ± equally 5-lobed, lobes entire. Stamens 4, didynamous, included; anthers bithecal, apically attached. Ovary 2–4-locular. Capsule globose to subglobose. Seeds numerous, obovoid.

A genus of 4 or 5 species distributed from tropical west, east and southern Africa with 1 species in India.

Micrargeria filiformis (*Schum. & Thonn.*) *Hutch. & Dalz.* in F.W.T.A. 2: 223 (1931); Hepper in F.W.T.A. ed. 2, 2: 366 (1963); Philcox in F.Z. 8(2): 151, t. 50 (1990); U.K.W.F.: 259 (1994). Type: Ghana, Pramforam and Ningo, *Thonning* 366 (C, syn.; P isosyn.)

Slender herb, 35–80 cm tall; stem much-branched above, rarely simple, minutely scabrid. Leaves opposite, sessile, filiform to narrowly linear, 0.7–3.5(–5) mm long, 0.5–1.3 mm wide, margins entire, erect to spreading. Inflorescence of open racemes; flowers subtended by reduced leaf-like bracts; bracteoles 2, linear to subulate, 1.3–3.2 mm long; pedicels 2–4 mm long. Calyx 2.5–4.5 mm long; lobes broadly deltate, 0.8–1.4 mm long. Corolla yellow with purple centre, pinkish-white or lilac to reddish-purple, campanulate, 8.5–12 mm long, narrowed at base into a

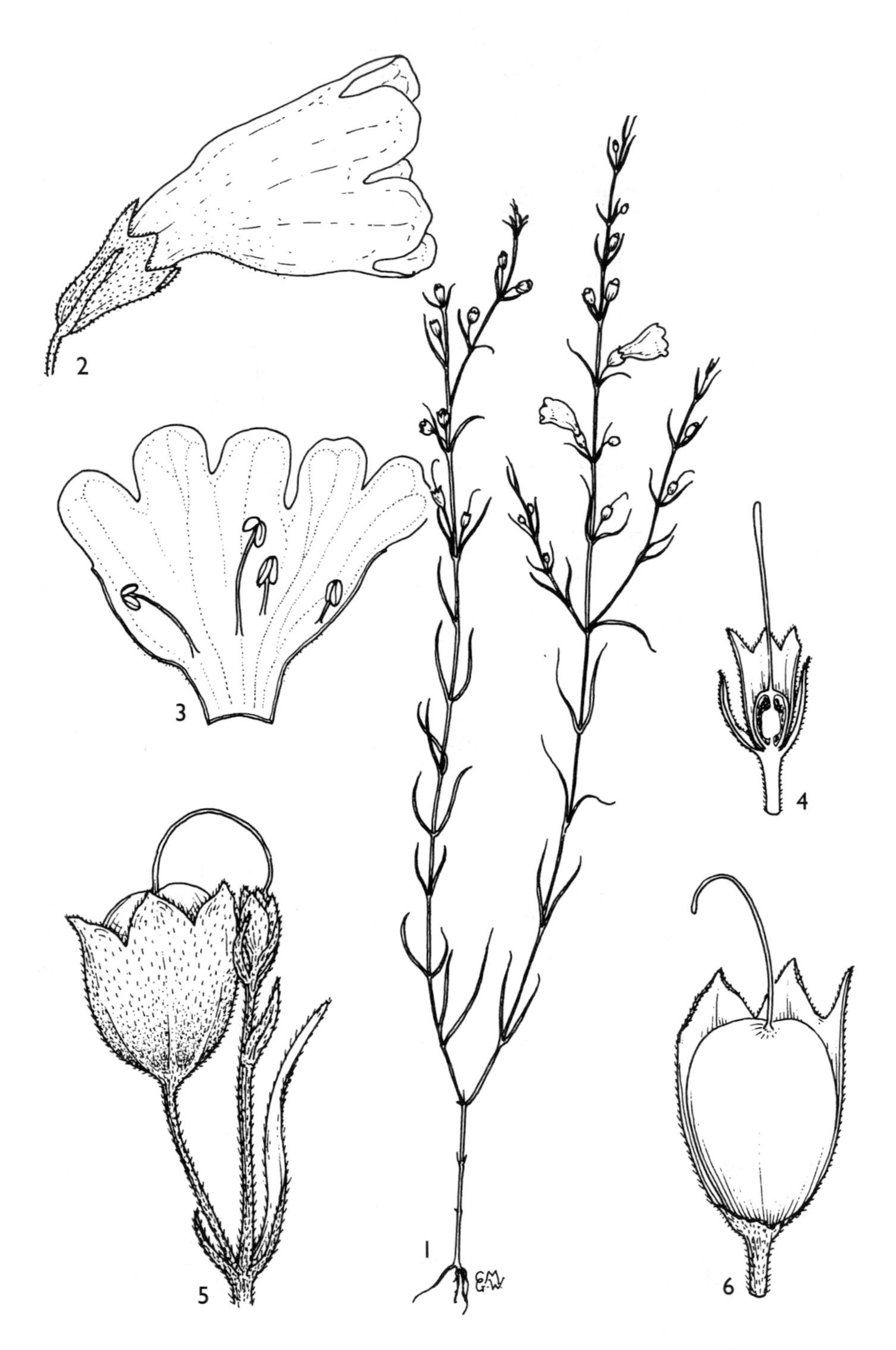

FIG. 48. *MICRARGERIA FILIFORMIS* — **1**, habit × ²⁄₃; **2**, flower × 4; **3**, corolla opened × 4; **4**, flower dissected × 4; **5**, capsule × 8; **5**, capsule with part calyx removed × 8. All from *Philcox, Drummond & Pope* 9086. Drawn by Christine Grey-Wilson. Reproduced with permission from F.Z.

short tube; tube whitish; lobes rounded, externally pubescent, finely ciliate. Filaments reddish purple; anther thecae subequal. Capsule dark brown to black, subglobose, 2.5–3.5 long, 2.8–3.5 mm in diameter, slightly longitudinally compressed, glabrous. Fig. 48, p. 188.

UGANDA. Mt Elgon N, Kapchorwa, 7 Sept. 1954, *Lind* 472!; Bunyoro/Mubende District: Ngusi R., April 1906, *Bagshawe* 964!; Masaka District: 4–5 km N of Lake Nabugabo, 17 April 1969, *Lye* 2555!

KENYA. Trans-Nzoia District: Darrads farm, Oct. 1967, *Tweedie* 3497!; Kwale District: N of Kwale, 13 Aug. 1937, *Greenway* 4963!; Lamu coast, 6 Oct. 1947, *Joy Adamson* in Bally B 6141!

TANZANIA. Tanga District: Msubugwe Forest Reserve, 12 Sept. 1955, *Tanner* 2202; Singida District: 40 km from Sekenke on the Singida road, 1 May 1962, *Polhill & Paulo* 2268!; Rufiji District: Mafia Is., 2 Apr. 1933, *Wallace* 701!

DISTR. **U** 3, 4; **K** 3, 7; **T** 3–8; ?**Z**; Sudan, Angola, Zambia, Malawi, Mozambique, Zimbabwe; Madagascar

HAB. Swamps, damp and wet areas in grassland and seepages in woodland; 0–2000 m

USES. None recorded from our area

CONSERVATION NOTES. Least Concern (LC); widespread and common

SYN. *Gerardia filiformis* Schum. & Thonn., Beskr. Guin. Pl.: 272 (1827)
Sopubia filiformis (Schum. & Thonn.) G.Don, Gen. Syst. 4: 560 (1837)
Sopubia filiformis Hiern, Cat. Afr. Pl. Welw. 776 (1898) *non* (Schum. & Thonn.) G.Don., *nom. illegit.* Type: Angola, Huilla–Morino, de Morro de Momino, *Welwitsch* 5830 (LISC, syn.; BM, isosyn.) and Lopollo, *Welwitsch* 5831 (LISC, syn.; BM, isosyn.)
Gerardianella scopiformis Klotzsch in Peters, Reise Mossamb., Bot.: 229, t. 36 (1861). Type: Mozambique, Querimba, *Peters* s.n. (B†, holo.)
Sopubia scopiformis (Klotzsch) Vatke in Linnaea 43: 313 (1882)
Micrargeria scopiformis (Klotzsch) Engl., P.O.A. C. 359 (1895); Hemsl. & Skan in F.T.A. 4(2): 457 (1906)

NOTE. Recorded to occur in Zanzibar, but I (S.A.G.) have not seen any material from there.

45. GERARDIINA

Engl. in E.J. 23: 507, t. 10 g–m (1897)

Annual or perennial herbs, erect; stems trailing, rooting at lower nodes then erect. Leaves opposite, entire. Inflorescence racemose. Flowers opposite, bracteate, pedicellate. Calyx 5-lobed, campanulate, lobes subequal, shorter than tube. Corolla campanulate, oblique, subrotate, 5-lobed; lobes subequal, rounded. Stamens 4, didynamous with anterior pair about 2–2.5 times longer than posterior pair; anterior filaments long pilose above middle; anthers bithecal; thecae divergent, subequal. Ovary bilocular; ovules numerous. Capsule ovoid, equalling to somewhat longer than calyx. Seeds numerous, narrowly cylindric, usually tapering.

Two species from tropical southeast Africa.

Gerardiina angolensis *Engl.* in E.J. 23: 507, t. 10 g–m (1897); Hiern, Cat. Afr. Pl. Welw. 1: 770 (1898) & in Fl. Cap. 4(2): 378 (1904); Cribb & Leedal, Mount. Fl. S Tanz.: 119, pl. 29B (1982); Philcox in F.Z. (8)2: 156, t. 52 (1990). Type: Angola, Malange district, by the Lopollo and Humpata streams, *Welwitsch* 5846 (B†, holo.; K!, iso.)

Herb 35–80 cm tall; stems simple, subquadrangular, minutely hispid to subglabrous, sulcate, blackening on drying. Leaves erect or appressed to stem, linear-lanceolate, (2.5–)6–11 cm long, 0.3–1.9 cm wide, sessile, decurrent at base, apex obtuse or rarely subacute, margins entire, 3-5-veined, scabrid above, subglabrous beneath, except scabrid on prominent veins, much exceeding internodes. Racemes

FIG. 49. *GERARDIINA ANGOLENSIS* — **1**, habit × ²⁄₃; **2**, flower × 1; **3**, corolla opened × 2; **4**, flower dissected showing gynoecium × 2; **5**, capsule × 2. 1 from *Milne-Redhead* 3181; 2–5 from *Richards* 1148. Drawn by Christine Grey-Wilson. Reproduced with permission from F.Z.

(5–)10–20 cm long; flowers numerous; bracts cordate to broadly ovate, 5–8.5 mm long, somewhat connate, glabrous; pedicels 4–14 mm long, glabrous. Calyx 5–8 mm long, obscurely veined, glandular-punctate; lobes broadly ovate-deltate, 1.5–3 mm long, obtuse, glabrous. Corolla pink to purple or blue, or occasionally white, 18–30 mm long, externally glabrous, pubescent within especially at base; lobes subequal, occasionally upper 2 clearly smaller. Style 16–18 mm long, arcuate above, equalling or slightly exceeding longer stamens, base persistent in fruit. Capsule ovoid, 7–8 mm long, 5–6.5 mm in diameter. Fig. 49, p. 190.

TANZANIA. Rungwe District: Kyimbila, N of Lake Nayasa, 2 Feb. 1914, *Stolz* 2505!; Iringa District: Dabaga Highlands, Kibengu, 30 km S of Dabaga, 17 Feb. 1962, *Polhill & Paulo* 1517!; Njombe District: Mwakete–Njombe, 17 Jan. 1957, *Richards* 7873!

DISTR. **T** 7; only collected from the above locations in Tanzania; Congo-Kinshasa, Burundi, Angola, Zambia, Malawi, Mozambique, Zimbabwe, South Africa

HAB. River and steam banks, wet grasslands, grassy hillsides, and open forests; 350–2500 m

USES. None recorded from our area

CONSERVATION NOTES. Least Concern (LC), widespread but known from only a few locations in Tanzania

46. BARTSIA

L., Sp. Pl. 2: 2: 602 (1753), *nom. conserv.*

Annual or perennial herbs, densely pubescent, often glandular. Leaves opposite, sessile, with revolute, crenate, rarely entire margins. Flowers solitary, axillary, shortly pedicellate. Calyx campanulate, unequally 4-lobed, the median clefts usually deeper than lateral ones. Corolla yellow (in Flora area), white, red or violet-brown, 2-lipped, tubular, the tube longer than the calyx, galeate. Stamens 4, inserted at the top of the corolla tube, included; anthers bithecal, transverse, bearded. Ovary oblong to subglobose; style entire, dilated at apex. Capsule oblong. Seeds numerous, testa reticulate, with longitudinal wings, the wings striate.

55 species mainly S and C America; a few in tropical Africa, Europe and N America.

1. Leaf margin revolute; corolla tube curved distally; lobes of lower lip with entire or crenulate margins; anthers apiculate .... 2. *B. decurva*
   Leaf margin not revolute; corolla tube almost straight; margin of lower lip with erose or entire margins; anthers acute or mucronate 2
2. Corolla tube 15–23 mm long; margin of lower lip erose; anthers acute at apex; seeds broadly winged .... 1. *B. longiflora*
   Corolla tube 6–7 mm long; margin of lower lip entire; anthers mucronate at apex; seeds ± winged .... 3. *B. trixago*

1. **Bartsia longiflora** *Benth.* in DC., Prodr. 10: 545 (1846); Molau in Opera Bot. 102: 23–25 (1990); U.K.W.F.: 262 (1994); K.T.S.L.: 589, fig., map (1994); Wood, Handb. Fl. Yemen: 266 (1997); Fischer in Fl. Ethiop. & Eritr. 5: 307 (2006). Type: Ethiopia, Mt Kubbi, *Schimper* 418 (K!, holo.; B, BM!, BR, G, GOET, HAL, HBG, JE, L, M, MPU, OXF, P, S, UPS, Z, iso.)

Erect perennial herb or undershrub, up to 60 cm tall; stems sparsely branched, pubescent, glandular, viscid. Leaves dense on the distal part of the stems, oblong to elliptic, (5–)15–55 mm long, (3–)4–14 mm broad, cuneate at base, apex obtuse, margins crenate, pubescent, ± glandular. Inflorescence spicate, the flowers spreading; bracts somewhat divaricate, similar to leaves, glandular-pubescent; pedicels 2–5 mm.

Calyx green, 10–15 mm, unequally 5-lobed to 1/3 (lateral clefts) or 2/3 (median clefts) of its length; lobes ovate-oblong, acute to obtuse, median one or both lobes with small irregular lobes on one margin, glandular-pubescent. Corolla lemon yellow, often with reddish veins, tubular; tube straight, 15–23 mm, glandular-pubescent outside and inside, expanding above into 5 lobes; lip of 3 lobes, rounded, deflexed, 7–8 mm, margins erose, dorsally glabrous; galea 7–8 mm, hairy at the top. Stamens attached to the top of the corolla tube; filaments 4–5 mm, with small scale-like papillae along the length; anther ± 3 mm, at right angles to the filament, bearded along the anther suture, acute at apex. Ovary oblong, ± 5 mm; style ± 30 mm, curved distally, puberulous; stigma clavate. Capsule ovoid, 10–12 mm long, 5–7 mm in diameter, short to long villous to ± glabrescent. Seeds 1–2 mm long, broadly winged.

Molau (op. cit.) recognizes two subspecies based on the length of the corolla, and the length:width ratio of leaves. Whereas the length of the corolla tube mostly holds true in Flora area, the leaf length:width ratio is variable. I have followed Molau in the treatment of the subspecies, as subsp. *macrophylla* is found only in a small part of the Flora area (Ruwenzori and Virunga Ranges of Uganda) extending to Congo-Kinshasa and Rwanda.

subsp. **longiflora**

Leaves mostly 3–5 times as long as wide; calyx densely glandular-puberulous; corolla tube 20–30 mm long; corolla more than twice as long as the calyx. Fig. 50, p. 193.

UGANDA. Mbale District: Bulambuli, Bugishu, 4 Sept. 1932, *A.S. Thomas* 538!; Mt Elgon, Sasa trail on W approach, 23 Sept. 1997, *Lye & Pócs* in Lye 23072! & W slope above Butadiri, 2 Dec. 1967, *Hedberg* 4459!
KENYA. Northern Frontier District: Mt Nyiru, S end of Lake Rudolf, June 1935, *Jex-Blake* 35!; Naivasha District: S Kinangop, Chamya valley, 4 Sept. 1965, *Gillett* 16888!; Nanyuki District: W slope of Mt Kenya, Naro–Moru track, 1 Oct. 1967, *Hedberg* 4269!
TANZANIA. Arusha District: Mt Meru, western slopes above Olkakola estate, 31 Oct. 1948, *Hedberg* 1553!; Mt Meru, Arusha National Park, 6 Oct. 1977, *Raynal* 19469! & Kilimanjaro, Mweika route, 9 Oct. 1966, *V.C. Gilbert* G91!
DISTR. **U** 3; **K** 1, 3, 4; **T** 2; Ethiopia; Yemen
HAB. Heath and afroalpine vegetation zone, grassland; 700–3500 m
USES. None recorded on specimens from our area
CONSERVATION NOTES. Least Concern (LC); widespread in the mountains of E Tropical Africa

SYN. *Bartsia similis* Hemsl. in F.T.A. 4(2): 461 (1906). Type: Tanzania, Kilimanjaro, Kifinika Volcano, Sept. 1893, *Volkens* 926 (K!, holo.; BM!, E!, G, iso.)
B. keniensis Standl. in Smithson. Misc. Coll. 68(5): 16 (1917). Type: Kenya, W slopes of Mt Kenya, Sept. 1909, *Mearns* 1487 (US, holo.)
B. macrophylla Hedberg subsp. *gughensis* Cuf. in Senck. Biol. 46: 93 (1965). Type: Ethiopia, Gamu-Gofa, Mt Dita, *Kuls* 745 (FR, holo.)

subsp. **macrophylla** (*Hedberg*) *Hedberg* in Norw. J. Bot. 26: 7 (1979); Molau in Opera Bot. 102: 24 (1990). Type: Uganda, Ruwenzori, Bujuku Valley near Bigo Camp, *Hedberg* 349 (UPS holo.; BR, EA, K!, LD, S, iso.)

Leaves mostly 3 times as long as wide; calyx glandular-villous; corolla tube 12–16 mm long; corolla at most twice as long as the calyx.

UGANDA. Ruwenzori, SW ridge of Stanley, 5 Aug. 1953, *Osmaston* 3262!; Kigezi District: Mt Muhavura, 1 Oct. 1946, *J. Williams* in *Bally* 5063!; Virunga Range, between Muhavura and Mgahinga, 30 Oct. 1954, *Stauffer* 662!
DISTR. **U** 2; Congo-Kinshasa, Rwanda
HAB. Heath and afroalpine vegetation zone; 2400–3800 m
USES. None recorded on specimens from our area
CONSERVATION NOTES. Least Concern (LC); a localized endemic, not common but not at threat in our area

SYN. *Bartsia macrophylla* Hedberg, A.V.P.: 174 (1957)

FIG. 50. *BARTSIA LONGIFLORA* SUBSP. *LONGIFLORA* — **1**, habit × ²⁄₃; **2**, **3**, flower front and back views × 1; **4**, calyx × 3; **5**, calyx opened with gynoecium × 1½; **6**, corolla opened × 2; **7**, stamen × 8; **8**, capsule × 4; **9**, capsule dehisced × 4. 1–3, 5–7 from *O. Hedberg* 4269; 4 from *Croat* 28323; 8 from *Tothill* 2256; 9 from *Thomas* 538. Drawn by Juliet Williamson.

2. **Bartsia decurva** *Benth* in DC., Prodr. 10: 545 (1846); Molau in Opera Bot. 102: 25 (1990); U.K.W.F.: 262 (1994); K.T.S.L.: 589 (1994); Fischer in Fl. Ethiop. & Eritr. 5: 307 (2006). Type: Ethiopia, Begemder, Simien range, Mt Selki [Silke], *Schimper* 1329 (K!, holo.; BM!, BR, G, GH, HAL, L, M, MO, NY, OXF, P, iso.)

Erect perennial herb or undershrub, up to 50 cm; stems simple or branched, hirsute or shortly villous, usually glandular. Leaves mostly crowded on the distal part of the stems (the ones below falling soon), erect or patent, ovate to ovate-lanceolate or lanceolate, 6–35 mm long, 2–10 mm broad, rounded to ± cordate at base, apex obtuse, margins revolute, crenate with 6–12 small lobes along each side, puberulous to villous, often puberulous on the veins only below, ± glandular. Inflorescence spicate with the flowers spreading; bracts somewhat erect or ascending, similar to leaves, glandular-pubescent; pedicels 1–3 mm, glandular-hirsute. Calyx green, 6–12 mm, unequally 5-lobed to $^1/_3$ (lateral clefts) or $^2/_3$ (median clefts) of its length; lobes ovate, obtuse, entire, glandular-pubescent. Corolla yellow, often shaded with purple, tubular; tube curved, 13–22 mm, glandular-pubescent inside and outside, tube expanding above into 5 lobes; lip of 3 lobes, rounded, deflexed, 5–8 mm, margin entire, dorsally glabrous, sparsely pilose on the inside; galea 4–6 mm, pilose within. Stamens attached to the top of the corolla tube; filaments 4–5 mm; anther ± 3 mm, at right angles to the filament, distinctly apiculate at apex. Ovary oblong, ± 5 mm; style ± 25 mm, curved distally, puberulous; stigma capitate or clavate. Capsule ovoid, 8–15 mm long, 5–7 mm in diameter, short to long villous. Seeds ± 2 mm long, ± broadly winged.

UGANDA. Mt Elgon, Sasa trail, 23 Oct. 1996, *Wesche* 09! & N side, 6 June 1949, *Osmaston* 4011! & Jacksons summit, April 1930, *Liebenberg* 1649!
KENYA. Mt Kenya, above Meru hut, 18 Jan. 1985, *Townsend* 2255!; Aberdares National Park, Muirs Massif, 18 March 1987, *Beentje* 3238!; North Nyeri District: Mt Kenya around end of Sirimon track, 4 April 1975, *Hepper & Field* 4863!
TANZANIA. Kilimanjaro, Maurdi crater, 15 Oct. 1993, *Grimshaw* 93/948! & Shira Mt, Feb. 1928, *Haarer* 1115! & by path from Bismarck hut to Peter's hut, 25 Feb. 1953, *G.H.S. Wood* 922!
DISTR. **U** 3; **K** 3, 4; **T** 2; Ethiopia
HAB. Afroalpine vegetation zone, short grassland, rocky places, stream beds; 3000–3700 m
USES. None recorded on specimens from our area
CONSERVATION NOTES. Least Concern (LC); localised but common

SYN. *Bartsia kilimandscharica* Engl., Hochgebirgsfl. Afrika: 384 (1892). Type: Tanzania, Mt Kilimanjaro, between Muebach and vegetation limit on Kibo, *Meyer* 245 (B, lecto., chosen by Molau, 1990)
B. keniensis R.E.Fr. in Acta Horti Berg. 8: 68 (1924). Type: Kenya, Mt Kenya, *Fries & Fries* 1311 (UPS, holo.; S, Z, iso.), *non* Standl. (1917)
B. macrocalyx R.E.Fr. in Acta Horti Berg. 8: 69 (1924). Type: Uganda/Kenya, Mt Elgon, *Lindblom* s.n. (S, holo.)

3. **Bartsia trixago** *L.*, Sp. Pl. 602 (1753); U.K.W.F.: 262 (1994); Fischer in Fl. Ethiop. & Eritr. 5: 307 (2006). Type: Locality and collector not stated, Herb. Burser XIV(1): 36 (UPS, lecto.), designated by Molau in Opera Bot. 102 : 27 (1990)

Erect annual herb, to 50 cm, semi-parasitic; stems simple or sparsely branched above, glandular-hairy. Leaves present throughout the stem, sessile, lanceolate to linear-lanceolate, 10–30(–48) mm long, 1–5(–14) mm broad, base ± cordate, apex obtuse, margins coarsely serrate with 5–7 teeth on either side, glandular-pilose. Inflorescence terminal, spicate; flowers sessile to very shortly pedicellate, spreading; bracts similar to leaves but smaller in size, glandular-pubescent. Calyx green, campanulate, 5–6 mm, unequally 5-lobed, cleft into 2 lobes; lobes broadly triangular, obtuse, glandular-pilose. Corolla white to pinkish, 2-lipped, tubular; tube straight above the calyx, 6–7 mm, glandular-pubescent ouside, expanding above into 5 lobes; lip of 3 lobes, rounded, deflexed, ± 7 mm, margin erose, dorsally glabrous; galea ±

7 mm. Stamens attached to the top of the corolla tube; filaments 5–6 mm; anther ± 2 mm, at right angles to the filament, mucronate at apex, bearded along the anther suture. Ovary oblong, ± 4 mm, pilose; style ± 10 mm, curved slightly distally, puberulous; stigma clavate. Capsule ovoid, 8–9 mm long, 5–6 mm in diameter, pilose. Seeds ± 0.5 mm, longitudinally and transversely striate, ± narrowly winged.

KENYA. Elgeyo District: Cherangani, E side of Kaisungur, Dec. 1971, *Tweedie* 4201! & 4 Jan. 1971, *Mabberley* 589! & W Aberdares, Geta Valley, Jan. 1932, *Dale* 2683!

DISTR. **K** 3; Ethiopia, South Africa (Cape of Good Hope); widespread in Southern Europe, Canary Islands, Azores, S America

HAB. Alpine vegetation zone, short grassland, rocky places; 2000–3700 m

USES. None recorded on specimens from our area

CONSERVATION NOTES. Least Concern (LC); very localised but not uncommon

SYN. *Bellardia trixago* (L.) All., Fl. Pedem. 1: 61 (1785); Tutin in Fl. Eur. 3: 269 (1972); Wood, Handb. Fl. Yemen: 266 (1997)

NOTE. Although Qaiser (in Jafri & El-Gadi, *Fl. Libya* 88: 40. 1982) indicated 758.3 (LINN) as type, it carries no annotation by Linnaeus to link it with this name, and also came from Eastern Asia which is at variance with the stated Italian provenance of the species. It is not original material for the name (see Linnaean Plant Typification Project).

47. HEDBERGIA

Molau in Nord. Journ. Bot. 8: 194 (1988)

Perennial herbs, sometimes shrubby, erect to scrambling, branched below, becoming tufted, pubescent. Leaves opposite, sessile, crenate or serrate, decreasing in size above. Flowers solitary, axillary, shortly pedicellate, ebracteolate. Calyx 4-lobed. Corolla subrotate, 5-lobed; limb weakly bilabiate but not separated into galea and lip. Stamens 4, didynamous; anther thecae 2, equal, parallel, mucronate. Style simple, filiform; stigma entire. Capsule ovoid. Seeds numerous, white, longitudinally winged.

Monotypic genus of tropical Africa.

Molau described this genus as distinct from *Bartsia* L. by reason of its 5-lobed, subrotate, not markedly bilabiate, and non-galeate corolla.

Hedbergia abyssinica (*Benth.*) *Molau* in Nord. Journ. Bot. 8: 195 (1988); Philcox in F.Z. 8(2): 156 (1990); U.K.W.F.: 262, pl. 113 (1994); Fischer in Fl. Ethiop. & Eritr. 5: 305 (2006). Type: Ethiopia, Mt Scholoda, *Schimper* I: 356 (K! lecto., chosen by Hedberg et al. 207 (1980) (see under synonomy); BM!, BR, G, HAL, HBG, L, LE, M, MO, OXF, P!, S, W, iso.)

Perennial suffrutescent herb to 3 m tall; stems several arising from a woody rootstock, erect, ascending or scrambling, much branched especially within inflorescence, rarely almost simple, usually more or less densely pubescent with straight or hooked, patent or retrorse glandular hairs. Leaves sessile or subsessile, thick, rigid, spreading, elliptic to lanceolate, 3–12(–30) mm long, 1–2(–4) mm wide, cuneate or rounded at base, apex acute or obtuse, minutely short-hispid to subglabrous, margins shallowly or deeply serrate or crenate, more or less reflexed. Inflorescence a usually much branched many-flowered raceme; flowers, opposite to subopposite; pedicels 2–5 mm long. Calyx 5–7.5 mm long, shortly hispid; lobes oblong, equal to or slightly shorter than the tube, slightly bilabiate with lobes at times irregularly crenate-dentate. Corolla white to pale pink or at times purple-white to pink or magenta, obliquely campanulate; tube curved, shorter than limb;

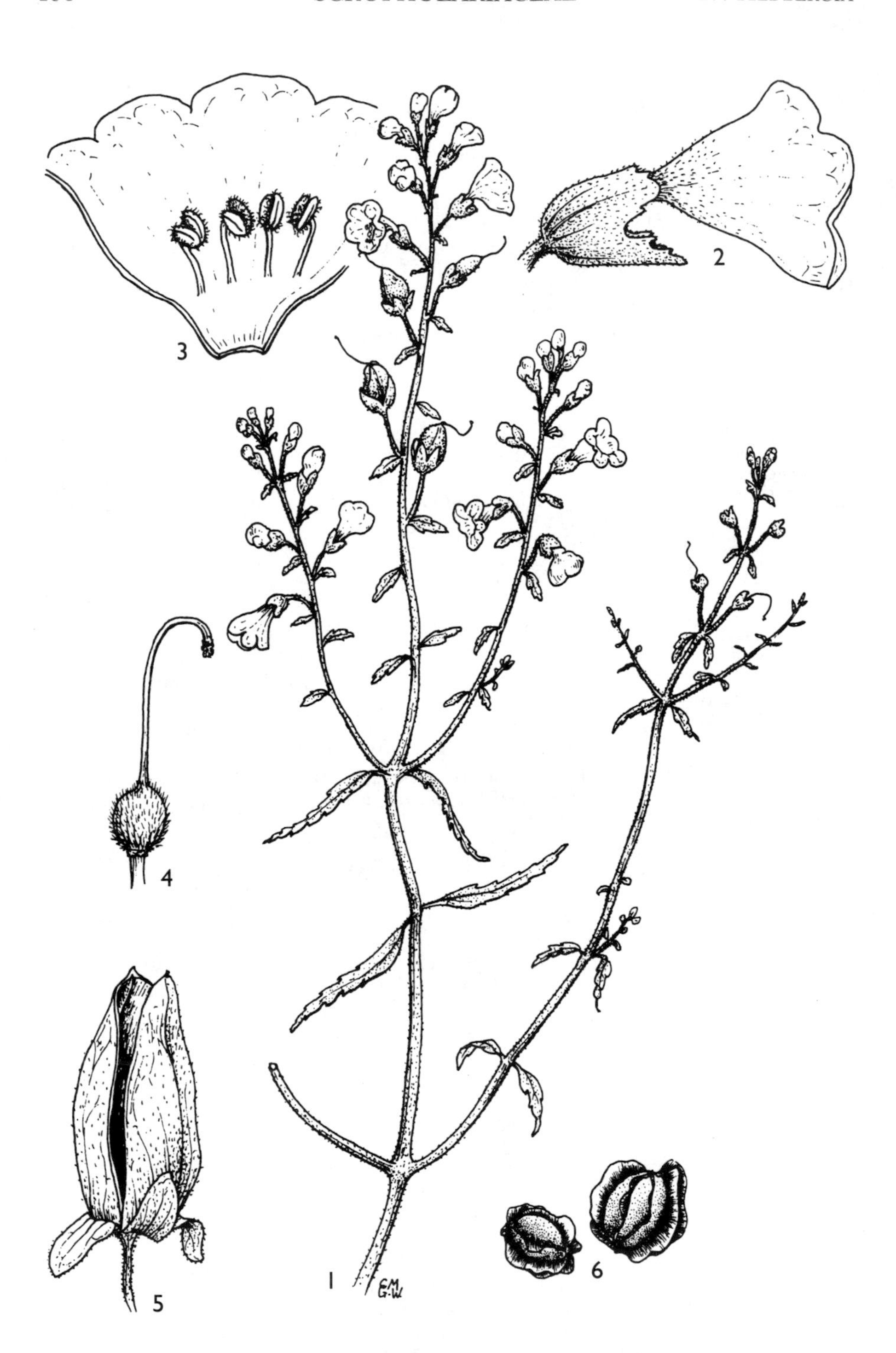

FIG. 51. *HEDBERGIA ABYSSINICA* — **1**, habit × ⅔; **2**, flower × 3; **3**, corolla opened × 3; **4**, gynoecium × 3; **5**, capsule dehisced × 3; **6**, seeds × 12. All from *White* 2748. Drawn by Christine Grey-Wilson. Reproduced with permission from F.Z.

limb almost equally 5-lobed; style 5.5–7 mm long, minutely pubescent or glabrous. Capsule ovoid to subglobose, (5–)6.5–13 mm long, (4.5–)5–6.5 mm broad, short- to long-pilose. Seeds slightly curved, longitudinally ridged with 6–14 ridges. Fig. 51, p. 196.

UGANDA. Mt Elgon, Feb. 1955, *Hedberg et al.* 2101!; Kigezi District: Rubaya, July 1946, *Purseglove* 2096!

KENYA. Trans-Nzoia District: W of Kipsane Hill, 25 Dec. 1967, *Gillett* 18454!; Narok District: Nasampolai valley, 9 Dec. 1972, *Greenway & Kanuri* 15065!; Aberdare Mts, W part of Nyeri track, 17 July 1948, *Hedberg* 1953!

TANZANIA. Meru District: Mt Meru, S side, 23 Apr. 1968, *Greenway & Kanuri* 13509!; Kilimanjaro, 18 Feb. 1934, *Schlieben* 4802!; Rungwe District: Kiware R., Lower Fishing Camp, 10 Aug. 1949, *Greenway* 8403!

DISTR. **U** 2; **K** 2–4, 6; **T** 2, 7; Nigeria, Cameroon, Congo-Kinshasa, Sudan, Ethiopia, Zambia, Malawi

HAB. Open grasslands, secondary bushland or forest margins; 2000–2400 m

USES. None recorded on specimens from our area

CONSERVATION NOTES. Least Concern (LC); widespread

SYN. *Bartsia abyssinica* Benth. in DC., Prodr. 10: 545 (1846); Engl., Hochgebirgsfl.: 384 (1892); Skan in F.T.A. 4(2): 460 (1906); Hedberg et al. in Bot. Not. 133: 207 (1980); Cribb & Leedal, Mount. Fl. S Tanz.: 117, pl. 28A, B (1982)

Alectra abyssinica (Benth.) A.Rich., Tent. Fl. Abyss. 2: 118 (1851)

A. petitiana A.Rich., Tent. Fl. Abyss. 2: 118 (1851). Type: Ethiopia, Ouodgerate, *Petit* s.n. (P, holo.; K!, iso.)

Bartsia mannii Skan in F.T.A. 4(2): 459 (1906). Type: Cameroon, Cameroon Mt, *Mann* 1986 (K!, lecto., chosen by Hedberg et al. 210 (1980); GH, iso.)

B. petitiana (A.Rich.) Skan in F.T.A. 4(2): 460 (1906)

B. elgonensis R.E. Fries. in Acta Hort. Berg. 8: 67 (1924). Type: Kenya/Uganda, Mt Elgon, *Snowden* 450 (K!, holo.; BM!, iso.)

B. nyikensis R.E. Fries in Acta Hort. Berg. 8: 66 (1924). Type: Malawi, Nyika Plateau, *McClounie* 60 (K!, holo.)

B. abyssinica Benth. var. *nyikensis* (R.E. Fries) Hedberg et al. in Bot. Not. 133, 211 (1980)

B. abyssinica Benth. var. *petitiana* (A.Rich.) Hedberg et al. in Bot. Not. 133: 211 (1980)

48. HEBENSTRETIA

L., Sp. Pl. 2: 629 (1753) & Gen Pl.: ed. 5, 277 (1754); Roessler in Mitt. Bot. Staatss. München 15: 18–89 (1979)

Annual or perenial herbs or small shrubs. Leaves alternate, or the lower opposite or clustered in axils appearing fascicled, narrow. Flowers sessile, in short or long dense or paniculate spikes, bracteate. Bracts usualy imbricate, exceeding the calyx. Calyx 1 lobed, spathaceous, entire or emarginate, membranous or hyaline, frequently biveined. Corolla tubular, divided almost to the base forming a large, flattened or concave 4(–5)-lobed limb. Stamens 4, didynamous, inserted on limb below lobes; anthers oblong or linear. Ovary bilocular; style entire. Capsule separating into 2 cocci at maturity.

A genus of 24 species distributed in tropical and southern Africa.

Hebenstretia angolensis *Rolfe* in J.B. 24: 174 (1886) as "*Hebenstreitia*"; A.V.P.: 167 (1957); Roessler in Mitt. Bot. Staatss. München 15: 77 (1979); Blundell, Wild. Flow. E Afr.: 377, pl. 115 (1987); Philcox in F.Z. 8(2): 163, t. 56 (1990); U.K.W.F.: 262, pl. 113 (1994); Fischer, F.A.C. Scrophulariaceae: 16, pl. 3 (1999) & in Fl. Ethiop. & Eritr. 5: 254 (2006). Type: Angola, Humpata stream, *Welwitsch* 4786 (K!, lecto.) & Jan. 1860, *Welwitsch* 4787 (K!, isosyn.)

Perennial herb to 50 cm; stems reddish-brown, erect, branched; branches leafy, glabrous to minutely pubescent with ± retrorse hairs. Leaves often fascicled with smaller leaves in the axils, linear, 10–30 mm long, 1–1.5 mm wide, apex acute, margins entire, glabrous, single-veined. Inflorescence spicate, 3–7 cm long; bracts ovate, 3–4 mm long, acuminate to apiculate, margins hyaline, glabrous. Calyx ovate, 3–4 mm long, obtuse or emaginate, hyaline, glabrous. Corolla white with an orange to reddish brown throat, 8–12 mm long; tube 2.5–4 mm, slit to about $^2/_3$ its length; limb to 8 mm, glabrous; lobes 1–1.5 mm, oblong. Anthers ± 1 mm; filaments short. Ovary ovoid, ± 4 mm long, glabrous; style shorter than the tube, bent apically at ± right angles and exserting through the slit in the tube; stigma elongate. Capsule oblong, 3–5 mm long, glabrous, separating into 2 cocci at maturity. Fig. 52, p. 199.

UGANDA. Karamoja District: Mt Morongole, 11 Nov. 1939, *A.S. Thomas* 3298!; Ruwenzori, 2 April 1932, *Oliver* s.n.!; Mt Elgon, Sasa Trail, 14 Nov. 1996, *Wesche* 245!

KENYA. Northern Frontier District: Mt Nyiru, 30 March 1995, *Bytebier et al.* 162!; Trans-Nzoia District: Cherangani Mts, 10 Feb. 1957, *Bogdan* 4390!; Mt Kenya, 17 Aug. 1931, *Slade* 17!

TANZANIA. Arusha District: below (N of) Losirwa village, 17 Feb. 2001, *Simon, Festo & Minde* 795!; Mpanda District: Mahali Mts, 27 Aug. 1958, *Jefford & Newbould* 1797!; Singida District: Mt Hanang, July 1952, *Brooks* s.n.!

DISTR. **U** 1–3; **K** 1–6; **T** 1, 2, 4, 5, 7; Congo-Kinshasa, Sudan, Eritrea, Ethiopia, Angola, Zambia, Malawi, Mozambique, Zimbabwe, South Africa, Lesotho

HAB. Submontane grasslands, mountain slopes, amongst rocks; 1500–3600 m

USES. None recorded on specimens from our area; recorded to be poisonous to cattle (fide *Adamson* 554)

CONSERVATION NOTES. Least Concern (LC); widespread and common

SYN. *H. dentata* var. *integrifolia* sensu Hiern in Cat. Afr. Pl. Welw. 1(4): 825 (1900), *non* Choisy
H. bequaertii De Wild. in Rev. Zool. Afr. 8, Suppl. Bot.: 41 (1920). Type: Congo-Kinshasa, Ruwenzori, *Bequaert* 4504 (BR)
H. dentata sensu Hedberg, A.V.P.: 167 (1957)

NOTE. Rolfe describes the species by syntypes *Welwitsch* 4786 & Jan. 1860, *Welwitsch* 4787 (both at K). Hedberg (op. cit.) designated *Welwitsch* 4786 as the holotype. This was noted by Roessler who changed it to lectotype.

49. **SELAGO**

L., Sp. Pl. 2: 629 (1753); Choisy in Mém. Phys. Hist. Nat. 2(2): 71 (1823) & Mém. Fam. Sélag.: 28 (1823); Hilliard, The Tribe Selagineae (Scrophulariaceae): 48 (1999)

Walafrida E.Mey., Comm. 272 (1838); Choisy in DC., Prodr. 12: 21 (1848); Philcox in F.Z. 8(2): 165 (1990) pro parte

Perennial shrubs and herbs, rarely annual herbs; stems erect, decumbent or prostrate. Leaves usually alternate, sometimes opposite near base, often in fascicles. Inflorescence a congested or elongate spike or raceme, simple, corymbose or paniculate. Bracts adnate to the pedicel and/or calyx tube. Calyx glabrous, (2–)3–5-lobed, (3- or 5-lobed in Flora area); lobes ± equal or very unequal with the dorsal lobes small and the anterior lobe deeply cleft, the calyx appearing 2- or 3-lobed, persistent. Corolla tubular, tube campanulate or cylindric, glabrous; limb 2-lipped, 5-lobed, glabrous inside. Stamens 4, didynamous; posterior pair decurrent down the tube, anterior pair inserted in throat; anthers monothecal. Ovary bilocular; style filiform; stigma clavate. Capsule loculicidal, separating into 2 cocci, included in calyx. Seeds elongate, 1 in each locule.

About 190 species distributed mainly in Africa, especially southern Africa; one species in Madagascar.

FIG. 52. *HEBENSTRETIA ANGOLENSIS* — **1**, habit × $^2/_3$; **2**, flower × 3; **3**, sepals × 3; **4**, corolla opened × 3; **5**, gynoecium × 3; **6**, mature capsules × 4. 1 from *Adamson* 374; 2–5 from *Salubeni* 720; 6 from *Brass* 16788. Drawn by Christine Grey-Wilson. Reproduced with permission from F.Z.

1. Inflorescence of congested racemes or glomerules forming narrow elongated panicles; calyx 5-lobed 2
   Inflorescence of congested spikes forming glomerules arranged in rounded-topped corymbose panicles; calyx 3–5-lobed 3
2. Leaves spreading; bracts up to 0.8 mm wide .... 1. *S. nyasae*
   Leaves sharply ascending; bracts up to 1.2 mm wide .... 2. *S. thyrsoidea*
3. Calyx 3-lobed .... 5. *S. goetzei*
   Calyx 5-lobed 4
4. Perennial herbs; bracts glabrous, villous on margins only; corolla tube 1.5–1.8 mm long; style with globular glands only .... 4. *S. viscosa*
   Annual herbs; bracts villous; corolla tube 1–2 mm long; style with small globular glands and acute hairs .... 3. *S. thomsonii*

1. **Selago nyasae** *Rolfe* in F.T.A. 5: 270 (1900); Hilliard, Selagineae: 224 (1999). Type: Tanzania, Njombe/Rungwe District: higher plateau north of Lake Nyasa, *Thomson* s.n. (K!, lecto., chosen by Hilliard, 1999)

Small shrub, to 1 m tall with a woody root-stock, branching from the base; branches dark reddish brown, often virgate, pubescent. Leaves alternate, strongly fascicled, spreading, linear to narrowly oblanceolate, 5–25 mm long, ± 1 mm wide, apex acute, margins entire, midrib channelled above, glabrous, glandular-punctate. Inflorescence of few-flowered congested racemes in narrow branched panicles; bracts oblong to linear-oblong, adnate to the calyx tube and pedicel, 2.5–3 mm long, glabrous or with few hairs on the lower margins; pedicels up to 1 mm long. Calyx obliquely campanulate, 1–2 mm long, 5-lobed to half the length; tube minutely puberulous and glandular-punctate; lobes oblong, ± equal, obtuse, margins pale, villous on the margins. Corolla funnel-shaped; tube 2–3 mm long; limb pale to violet-blue, ± 3 mm across; lobes elliptic to oblong, 1.5–2.5 mm long, the two anterior ones obtuse. Stamens with filaments ± 2 mm long; anthers exserted. Ovary ± 1 mm; style and stigma ± 3.5 mm long, glabrous or minutely puberulous. Cocci subrotund, ± 1 mm long, reddish brown, smooth, glabrous.

TANZANIA. Njombe Distrct: Kitulo [Elton] Plateau, May 1975, *Hepper, Field & Mhoro* 5348!; Mbeya District: Kitulo [Elton] Plateau, May 1957, *Richards* 9698! & between Mt Islinga & Sanshashi Hill, May 1987, *Iversen et al.* 87722!

DISTR. **T** 7; only known from the high plateau of Tanzania

HAB. Montane grassland and heath with tussock grass and composites; 1800–2800 m

USES. None recorded on our specimens

CONSERVATION NOTES. Least concern (LC); localized but not uncommon and not at any threat as far as is known

SYN. *S. thyrsoidea* Bak. var. *nyikensis* (Rolfe) Brenan in Mem. N.Y. Bot. Gard. 9: 32 (1954) pro parte. Type: Malawi, Nyika Plateau near Mwanemba, *McClounie* 39 (K!, lecto.)

S. longithyrsa Gilli in Ann. Naturhist. Mus. Wien. 77: 46 (1973). Type: Tanzania, Rungwe District: Kyimbila, *Stolz* 2077 (W, holo; C, MO, PRE, S, SAM, Z, iso.)

NOTE. The separation of *S. thyrsoidea* Bak. and *S. nyasae* Rolfe has been dealt with in detail by Hilliard (op. cit.). Following Hilliard, all our material earlier identified as *S. thyrsoidea* Bak. var. *nyikensis* from the Southern Highlands of Tanzania, especially the Kitulo Plateau is referable to *S. nyasae*. The flowers are recorded to be mildly fragrant (fide *Greenway* 8425).

2. **Selago thyrsoidea** *Baker* in K.B. 1898: 159 (1898); Rolfe in F.T.A. 5: 270 (1900); Brenan in Mem. N.Y. Bot. Gard. 9: 31 (1954), pro parte; Philcox in F.Z. 8(2): 171, t. 58 (1990); Hilliard, Selagineae: 222 (1999). Type: Malawi, Nyika Plateau, *Whyte* 144 (K!, holo.)

Small shrub, to 1 m tall with a woody root-stock, branching from the base; branches dark reddish brown, often virgate, pubescent. Leaves alternate, fascicled, ascending, linear to narrowly oblanceolate, 5–25 mm long, ± 1 wide, apex acute, entire, midrib channelled above, glabrous, glandular-punctate. Inflorescence of few-flowered congested racemes in simple or in thyrsoid panicles; bracts oblong-lanceolate, adnate to the calyx tube and pedicel, 2.5–3 mm long, often incurved and concave, pubescent; pedicels ± 1 mm long. Calyx obliquely campanulate, 1–2 mm long, lobed to half the length; lobes oblong, obtuse, margins pale, villous on the margins. Corolla funnel-shaped, tube 2.5–3 mm long; limb pale to violet-blue, ± 3 mm across; lobes elliptic, 1.5–2.5 mm long, the two anterior ones obtuse. Stamens with filaments ± 2 mm long; anthers exserted. Ovary ± 1 mm; style and stigma ± 3.5 mm long, glabrous or minutely puberulous. Cocci reddish brown, ± 1 mm in diameter, smooth, glabrous.

TANZANIA. Iringa District: near Ludewa, Itimbo, 6 Nov. 1987, *Mwasumbi et al.* 13410!
DISTR. **T** 7; Malawi
HAB. Montane grassland and heath and wet meadow; ± 2000 m
CONSERVATION NOTES. Known in the Flora area from two collections, but possibly of Least Concern (LC) throughout its range

SYN. *S. nyikensis* Rolfe in K.B. 1908: 261 (1908). Type: Malawi, Nyika Plateau near Mwanemba, *McClounie* 39 (K!, lecto.)
S. thyrsoidea Bak. var. *nyikensis* (Rolfe) Brenan in Mem. N.Y. Bot. Gard. 9: 32 (1954) pro parte excl. *S. nyasae* Rolfe

NOTE. Hilliard (op. cit. p. 222–223) cites a single collection from Tanzania, Ukinga Mts, NNE of Lake Malawi (*Goetze* 961), but states that it differs from the typical species in its longer calyx with acute anterior lobes and a longer corolla, and whose status needs confirmation. I have seen this collection, it agrees with all characters of this species and I am convinced that it is *S. thyrsoidea*. It is either very rare or more likely undercollected in our area.

3. **Selago thomsonii** *Rolfe* in J.L.S. 21: 402 (1885) & in F.T.A. 5: 270 (1900); Brenan in Mem. N.Y. Bot. Gard. 9: 30 (1954); Hilliard, Selagineae: 237 (1999); Fischer, F.A.C. Scrophulariaceae: 18 (1999). Type: Tanzania, Kilimanjaro, *Thomson* 35 (K!, lecto., chosen by Hilliard, 1999)

Annual herb with the main stem arising from a simple taproot; stems reddish or pale brown, simple below but soon branching, erect or decumbent, 10–50 cm long, leafy throughout, pubescent with down-curving and spreading hairs. Leaves opposite and alternate, densely fascicled, spreading, linear, 5–20 mm long, 0.5–2 mm wide, base tapering, apex acute to obtuse, margins entire or with 1–2 teeth in the apical part, ± thickened and ± revolute, hairy on surfaces and margins, glandular-punctate. Inflorescence of congested racemes forming small round-topped corymbose panicles; bracts adnate to pedicel and about half of calyx tube, oblong to elliptic, 2–3 mm, obtuse, concave, villous on margins and back, sometime only sparsely so; pedicels up to 1 mm long. Calyx 1.5–2 mm long, 5-lobed to about half the length, rarely 3-lobed, the two dorsal lobes larger than the other three, oblong to triangular, margins pale, hairy, glandular. Corolla tube 1–2 mm long, limb ± 3 mm across the lateral lobes; lobes violet to purplish, rarely white, elliptic to oblong-elliptic, ± 1 mm long. Filaments 1–2 mm long; anthers shortly exserted. Ovary ovoid, ± 0.5 mm; style and stigma ± 2 mm long with small rounded glands and acute hairs. Cocci reddish brown, subrotund, ± 1.5 mm in diameter, smooth. Fig. 53: 1–7, p. 202

KENYA. West Suk District: Cherangani Hills, Jan. 1971, *Tweedie* 3879!; Elgeyo District: Marakwet Hills, *Dale* 3395!; Masai District: Nasampolai valley, Aug. 1972, *Greenway & Kanuri* 15045!
TANZANIA. Kilimanjaro, track to Shira Plateau, Feb. 1969, *Richards* 24009!; Lushoto District: Western Usambaras, 1.6 km NW of Kwai, June 1953, *Drummond & Hemsley* 2900!; Kilosa District: Mamiwa Forest Reserve, Aug. 1972, *Mabberley & Alehe* 1498!

FIG. 53. *SELAGO THOMSONII* — **1**, habit × $^{2}/_{3}$; **2**, lower stem with roots × $^{2}/_{3}$; **3**, inflorescence × 6; **4**, flower with bract × 10; **5**, corolla opened × 12; **6,** capsule with calyx and bract × 12; **7**, cocci with calyx removed × 12. *SELAGO GOETZEI* — **8**, inflorescence × 5. 1, 4–6 from *Grimshaw* 93/665; 2 from *Baldock* 10; 3, 7 from *Newbould* 3429; 8 from *Richards* 9651. Drawn by Juliet Williamson.

Distr. **K** 1–4, 6; **T** 2, 3, 5–7; not found elsewhere
Hab. Montane grassland, stony hillsides with scattered undergrowth, damp open places, periodically burnt secondary bush; 2000–3000 m
Uses. None recorded on our specimens
Conservation notes. Least concern (LC); common and widespread

Syn. *S. johnstonii* Rolfe in Trans. Linn. Soc. Ser 2, 2: 344 (1887) & in F.T.A. 5: 269 (1900). Type: Tanzania, Kilimanjaro, *Johnston* 147 (K!, holo.)
S. holstii Rolfe in F.T.A. 5: 269 (1900), pro parte. Type: Tanzania, Lushoto District: Usambaras, Kwa Mshuza, *Holst* 9088 (K!, lecto., chosen by Hilliard; COI, E, S, Z, isolecto.)

Note. This is the most widespread of all species of *Selago* in the Flora area. According to Hillaird (op.cit.) it is distinguished from the others in its annual life-cycle, but I (S.A.G.) have seen material of this species that is apparently perennial (also recorded to be such on the label), but nevertheless agrees with all other characters of *S. thomsonii* (e.g. Tanzania: W Usambaras, *Brenan & Greenway* 8308; Lushoto, *Drummond & Hemsley* 2121 & *Shabani* 1222; Kilimanjaro, *Leippert* 6090; Kenya: Nasampolai, *Greenway & Kanuri* 15045).

I am following Hilliard for *Selago*, but am not fully convinced of her treatment of the group of species which includes this taxon, as there are many sheets that show intermediary characters especially those for *S. thomsonii* and *S. caerulea* Rolfe (not recorded in Flora area). There is also a dearth of material from the southern highlands of Tanzania which adds to the lack of information for this species group.

As treated here *S. thomsonii* is, to a large extent, geographically isolated from the others in the Flora area (all other species found in **T** 7), it is also the northernmost representative of the genus in eastern and southern Africa.

4. **Selago viscosa** *Rolfe* in F.T.A. 5: 267 (1900); Hilliard, Selagineae: 232 (1999). Type: Tanzania, lower plateau N of Lake Nyasa, *Thomson* s.n. (K!, holo.)

Perennial herb up to 1 m with stems arising from a woody root-stock; stems reddish or pale brown, erect or decumbent, simple or branched, leafy throughout, pubescent with down-curving hairs. Leaves opposite and alternate, densely or loosely fascicled, spreading, oblanceolate to linear-lanceolate to linear, 5–25 mm long, 0.5–6 mm wide, base tapering, apex acute to obtuse, margins entire, midrib channelled above, mainly the margins and midline hairy, ± sparsely hairy on surfaces, glandular-punctate; fascicled leaves smaller above, 5–15 mm long. Inflorescence of few-flowered racemes forming lax corymbs arranged in round-topped corymbose panicles; bracts adnate to pedicel and part of calyx tube, oblong to oblanceolate, 2–3 mm long, obtuse, concave, glabrous, villous on margins; pedicels up to 1 mm long. Calyx 1.5–2 mm long, 5-lobed to about half its length, the two dorsal lobes larger than the other three, oblong to triangular, margins pale, hairy. Corolla tube 1.5–1.8 mm long, limb 3–4 mm across the lateral lobes; lobes pale blue to purplish, elliptic to oblong-elliptic, ± 1 mm long. Filaments 1–2 mm long; anthers exserted. Ovary ovoid, ± 0.5 mm; style and stigma ± 1.5 mm long with small rounded glands. Cocci reddish brown, subrotund, ± 1.5 mm in diameter, smooth.

Tanzania. Iringa Distrct: Dabaga highlands, 55 km SE of Iringa, Jan. 1971, *Mabberley* 646! & Muengo [Munaga], near Ulaya Hill, Aug. 1952, *Carmichael* 107!; Njombe District: by Hagafilo R., 11 km S of Njombe, July 1956, *Milne-Redhead & Taylor* 10778!
Distr. **T** 7; only known from the high plateau of SW Tanzania and adjoining Nykia plateau in Malawi (see also Hilliard)
Hab. Montane grassland, mountain slopes, by sides of streams; 1750–2400 m
Uses. None recorded on our specimens.
Conservation notes. Least concern (LC); localized but not uncommon

Syn. *S. maclouniei* Rolfe in K.B. 1908: 262 (1908). Type: Malawi, [Nyika Plateau], Panda Peak, *McClounie* 139 (K!, lecto. chosen by Hilliard)
S. thomsonii Rolfe var. *caerulea* (Rolfe) Brenan in Mem. N.Y. Bot. Gard. 9: 30 (1954) pro parte. Type of *S. caerulea* Rolfe, Malawi, summit of Nyika plateau, *Whyte* 145 (K!, holo; E, SAM, iso.)

S. laticorymbosa Gilli in Ann. Naturhist. Mus. Wien 77: 45 (1973). Type: Tanzania, Njombe District: near Njombe, *Gilli* 506 (W, holo.; K! iso.)

5. **Selago goetzei** *Rolfe* in E.J. 30: 402 (1901); Hilliard, Selagineae: 234 (1999). Type: Tanzania, Poroto Mts, Usafwa [Usafa], *Goetze* 1043 (K!, holo.; BR, E, Z, iso.)

Perennial herb up to 45 cm, with stems arising from a woody root-stock; stems dark reddish brown, erect or decumbent and rooting near the base, simple or branching from the base, leafy throughout, villous. Leaves alternate, fascicled, ascending or spreading, lanceolate to linear-lanceolate, 8–28 mm long, 1–3 mm wide, apex acute, entire or with 3 teeth in the apical half, channelled above, villous to sparse on surfaces, margins villous, glandular-punctate. Inflorescence of few-flowered congested racemes forming glomerules arranged in pyramidal or flat-topped panicles; bracts adnate to pedicel and calyx tube, concave, villous on margins, glandular-punctate in the upper half; pedicels up to 1 mm long. Calyx mostly 3-lobed, ± 3 mm long; lobes broadly triangular, ± 1.5 mm, acute, dorsal lobe much narrower and smaller to 1 mm, villous on margins, glandular. Corolla tube 3–4.5 mm long, limb ± 4.5 mm across; lobes white to mauve to pale purple, oblong-elliptic, 1–2 mm, obtuse, anterior lobes longer. Stamens with filaments 1–2 mm long; anthers exserted. Ovary ovoid, ± 0.5 mm; style and stigma 3–4 mm long with small globular glands and few short hairs. Cocci yellowish brown, ovoid, 1–1.5 mm, smooth. Fig. 53: 8, p. 202.

TANZANIA. Rungwe District: Isongole Usafwa, Mbeya–Tukuyu road, May 1975, *Hepper, Field & Mhoro* 5363! & Kyimbila, N of Lake Nyasa, Feb. 1915, *Stolz* 2545!; Iringa District: Imagi Mt, ± 48 km E of Iringa, Dec. 1961, *Richards* 15637!

DISTR. **T** 7; only known from the high plateau of Tanzania, in the mountains N of Lake Malawi, around Tukuyu, Mbeya and Rungwe (see also Hilliard, op. cit.)

HAB. Montane grassland, slopes, pastures, harvested fields; 1800–2400 m

USES. None recorded on our specimens

CONSERVATION NOTES. Least concern (LC); localized but not uncommon

SYN. *Walafrida goetzei* (Rolfe) Brenan in Mem. N.Y. Bot. Gard. 9: 33 (1954); Philcox in F.Z. 8(2): 167 (1990)

EXCLUDED SPECIES

Schweinfurthia pedicellata (*T. Anders.*) *Balf.* in Trans. Roy. Soc. Edinb. 31: 201 (1888).

There is a specimen of this species at K (Tanzania, Lushoto District: Amani, 1929, *Toms* 6) which has been annotated by Hepper as it being an unlikely locality for the species. The specimen is a duplicate sent from EA. Toms was a gardener at Amani and it is possible that the plant was a weed in the extensive cultivations of introduced plants there. The species has not been recorded or collected since from there or anywhere else in the Flora area.

GEOGRAPHICAL DIVISIONS OF THE FLORA

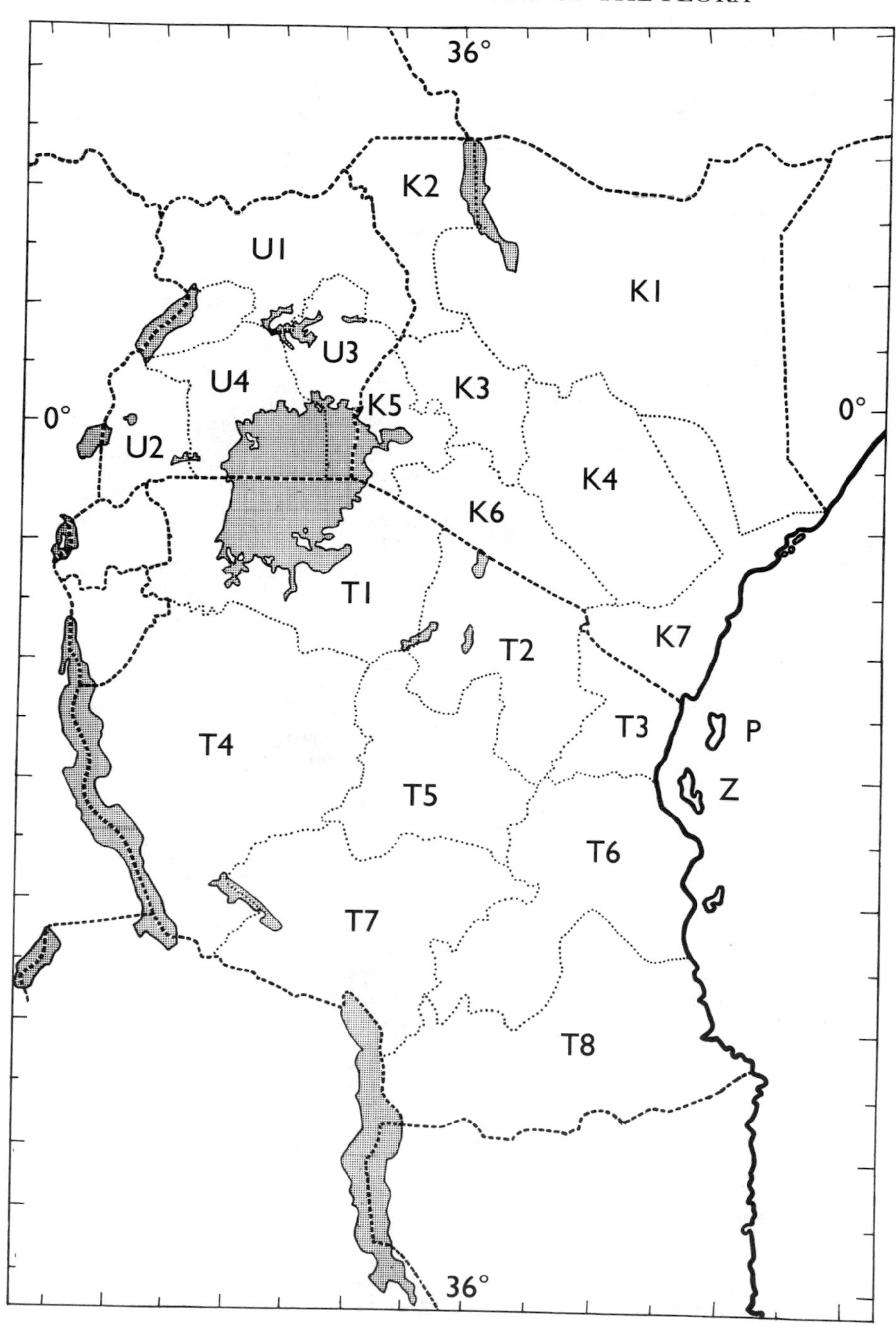

INDEX TO SCROPHULARIACEAE

New names validated in this part

Lindernia hepperi (*Eb.Fisch.*) *Philcox*, **comb. nov.**
Lindernia lindernioides (*E.A.Bruce*) *Philcox*, **comb. nov.**
Lindernia serpens *Philcox*, **nom. nov.**
Lindernia ugandensis (*Skan*) *Philcox*, **comb. nov.**

First published in 2008 by
Royal Botanic Gardens, Kew
Richmond, Surrey, TW9 3AB, UK
www.kew.org

ISBN 978 1 84246 192 1

British Library Cataloguing in Publication Data
A catalogue record for this book is available from the British Library

Design and typesetting by Margaret Newman,
Kew Publishing, Royal Botanic Gardens, Kew.

For information or to purchase all Kew titles please visit
www.kewbooks.com or email publishing@kew.org

LIST OF ABBREVIATIONS

A.V.P. = O. Hedberg, Afroalpine Vascular Plants; **B.J.B.B.** = Bulletin du Jardin Botanique de l'Etat, Bruxelles; Bulletin du Jardin Botanique Nationale de Belgique; **B.S.B.B.** = Bulletin de la Société Royale de Botanique de Belgique; **C.F.A.** = Conspectus Florae Angolensis; **E.J.** = A. Engler, Botanische Jahrbücher für Systematik, Pflanzengeschichte und Pflanzengeographie; **E.M.** = A. Engler, Monographieen Afrikanischer Pflanzen-Familien und Gattungen; **E.P.** = A. Engler, Das Pflanzenreich; **E.P.A.** = G. Cufodontis, Enumeratio Plantarum Aethiopiae Spermatophyta; in B.J.B.B. 23, Suppl. (1953) et seq.; **E. & P. Pf.** = A. Engler & K. Prantl, Die Natürlichen Pflanzenfamilien; **F.A.C.** = Flore d'Afrique Centrale (*formerly* F.C.B.); **F.C.B.** = Flore du Congo Belge et du Ruanda-Urundi; Flore du Congo, du Rwanda et du Burundi; **F.E.E.** = Flora of Ethiopia & Eritrea; **F.D.-O.A.** = A. Peter, Flora von Deutsch-Ostafrika; **F.F.N.R.** = F. White, Forest Flora of Northern Rhodesia; **F.P.N.A.** = W. Robyns, Flore des Spermatophytes du Parc National Albert; **F.P.S.** = F.W. Andrews, Flowering Plants of the Anglo-Egyptian Sudan *or* Flowering Plants of the Sudan; **F.P.U.** = E. Lind & A. Tallantire, Some Common Flowering Plants of Uganda; **F.R.** = F. Fedde, Repertorium Speciorum Novarum Regni Vegetabilis; **F.S.A.** = Flora of Southern Africa; **F.T.A.** = Flora of Tropical Africa; **F.W.T.A.** = Flora of West Tropical Africa; **F.Z.** = Flora Zambesiaca; **G.F.P.** = J. Hutchinson, The Genera of Flowering Plants; **G.P.** = G. Bentham & J.D. Hooker, Genera Plantarum; **G.T.** = D.M. Napper, Grasses of Tanganyika; **I.G.U.** = K.W. Harker & D.M. Napper, An Illustrated Guide to the Grasses of Uganda; **I.T.U.** = W.J. Eggeling, Indigenous Trees of the Uganda Protectorate; **J.B.** = Journal of Botany; **J.L.S.** = Journal of the Linnean Society of London, Botany; **K.B.** = Kew Bulletin, *or* Bulletin of Miscellaneous Information, Kew; **K.T.S.** = I. Dale & P.J. Greenway, Kenya Trees and Shrubs; **K.T.S.L.** = H.J. Beentje, Kenya Trees, Shrubs and Lianas; **L.T.A.** = E.G. Baker, Leguminosae of Tropical Africa; **N.B.G.B.** = Notizblatt des Botanischen Gartens und Museums zu Berlin-Dahlem; **P.O.A.** = A. Engler, Die Pflanzenwelt Ost-Afrikas und der Nachbargebiete; **R.K.G.** = A.V. Bogdan, A Revised List of Kenya Grasses; **T.S.K.** = E. Battiscombe, Trees and Shrubs of Kenya Colony; **T.T.C.L.** = J.P.M. Brenan, Check-lists of the Forest Trees and Shrubs of the British Empire no. 5, part II, Tanganyika Territory; **U.K.W.F.** = A.D.Q. Agnew (or for ed. 2, A.D.Q. Agnew & S. Agnew), Upland Kenya Wild Flowers; **U.O.P.Z.** = R.O. Williams, Useful and Ornamental Plants in Zanzibar and Pemba; **V.E.** = A. Engler & O. Drude, Die Vegetation der Erde, IX, Pflanzenwelt Afrikas; **W.F.K.** = A.J. Jex-Blake, Some Wild Flowers of Kenya; **Z.A.E.** = Wissenschaftliche Ergebnisse der Deutschen Zentral-Afrika-Expedition 1907–1908, 2 (Botanik).

FAMILIES OF VASCULAR PLANTS REPRESENTED IN THE FLORA OF TROPICAL EAST AFRICA

The family system used in the Flora has diverged in some respects from that now in use at Kew and the herbaria in East Africa. The accepted family name of a synonym or alternative is indicated by the word "see". Included family names are referred to the one used in the Flora by "in" if in accordance with the current system, and "as" if not. Where two families are included in one fascicle the subsidiary family is referred to the main family by "with".

PUBLISHED PARTS

Foreword and preface
*Glossary
Index of Collecting Localities

Acanthaceae
Part 1
*Actiniopteridaceae
*Adiantaceae
Aizoaceae
Alangiaceae
Alismataceae
*Alliaceae
*Aloaceae
*Amaranthaceae
*Amaryllidaceae
*Anacardiaceae
*Ancistrocladaceae
Anisophyllaceae — as Rhizophoraceae
Annonaceae
*Anthericaceae
Apiaceae — see Umbelliferae
Apocynaceae
*Part 1
*Aponogetonaceae
Aquifoliaceae
*Araceae
Araliaceae
Arecaceae — see Palmae
*Aristolochiaceae
Asparagaceae
*Asphodelaceae
Aspleniaceae
Asteraceae — see Compositae
Avicenniaceae — as Verbenaceae
*Azollaceae

*Balanitaceae
*Balanophoraceae
*Balsaminaceae
Basellaceae
Begoniaceae
Berberidaceae
Bignoniaceae
Bischofiaceae — in Euphorbiaceae
Bixaceae
Blechnaceae
*Bombacaceae
*Boraginaceae
Brassicaceae — see Cruciferae
Brexiaceae
Buddlejaceae — as Loganiaceae
*Burmanniaceae
*Burseraceae
Butomaceae
Buxaceae

Cabombaceae
Cactaceae
Caesalpiniaceae — in Leguminosae
*Callitrichaceae
Campanulaceae
Canellaceae
Cannabaceae
Cannaceae — with Musaceae
Capparaceae
Caprifoliaceae
Caricaceae
Caryophyllaceae
*Casuarinaceae
Cecropiaceae — with Moraceae
*Celastraceae
*Ceratophyllaceae
Chenopodiaceae
Chrysobalanaceae — as Rosaceae
Clusiaceae — see Guttiferae
Cobaeaceae — with Bignoniaceae
Cochlospermaceae

Colchicaceae
Combretaceae
Compositae
*Part 1
*Part 2
Part 3
Connaraceae
Convolvulaceae
Cornaceae
Costaceae — as Zingiberaceae
*Crassulaceae
*Cruciferae
Cucurbitaceae
Cupressaceae
Cyanastraceae — in Tecophilaeaceae
Cyatheaceae
Cycadaceae
*Cyclocheilaceae
*Cymodoceaceae
Cyphiaceae — as Lobeliaceae

*Davalliaceae
*Dennstaedtiaceae
*Dichapetalaceae
Dilleniaceae
Dioscoreaceae
Dipsacaceae
*Dipterocarpaceae
Dracaenaceae
Dryopteridaceae
Droseraceae

*Ebenaceae
Elatinaceae
Equisetaceae
Ericaceae
*Eriocaulaceae
*Eriospermaceae
*Erythroxylaceae
Escalloniaceae
Euphorbiaceae
*Part 1
*Part 2

Fabaceae — see Leguminosae
Flacourtiaceae
Flagellariaceae
Fumariaceae

*Gentianaceae
Geraniaceae
Gesneriaceae
Gisekiaceae — as Aizoaceae
*Gleicheniaceae
Goodeniaceae
Gramineae
Part 1
Part 2
*Part 3
Grammitidaceae
Gunneraceae — as Haloragaceae
Guttiferae

Haloragaceae
Hamamelidaceae
*Hernandiaceae
Hippocrateaceae — in Celastraceae
Hugoniaceae — in Linaceae
*Hyacinthaceae
*Hydnoraceae
*Hydrocharitaceae
*Hydrophyllaceae
*Hydrostachyaceae
Hymenocardiaceae — with Euphorbiaceae
Hymenophyllaceae
Hypericaceae — see also Guttiferae
Hypoxidaceae

Icacinaceae
Illecebraceae — as Caryophyllaceae
*Iridaceae
Irvingiaceae — as Ixonanthaceae
Isoetaceae
*Ixonanthaceae

Juncaceae
Juncaginaceae

Labiatae
Lamiaceae — see Labiatae
*Lauraceae
Lecythidaceae
Leeaceae — with Vitaceae
Leguminosae
Part 1, Mimosoideae
Part 2, Caesalpinioideae
Part 3 }
Part 4 } Papilionoideae
Lemnaceae
Lentibulariaceae
Liliaceae (*s.s.*)
Limnocharitaceae — as Butomaceae
Linaceae
*Lobeliaceae
Loganiaceae
*Lomariopsidaceae
*Loranthaceae
Lycopodiaceae
*Lythraceae

Malpighiaceae
Malvaceae
Marantaceae
*Marattiaceae
*Marsileaceae
Melastomataceae
*Meliaceae
Melianthaceae
Menispermaceae
*Menyanthaceae
Mimosaceae — in Leguminosae
Molluginaceae — as Aizoaceae
Monimiaceae
Montiniaceae
*Moraceae
*Moringaceae
Muntingiaceae — with Tiliaceae
*Musaceae
*Myricaceae
*Myristicaceae
*Myrothamnaceae
*Myrsinaceae
*Myrtaceae

*Najadaceae
Nectaropetalaceae — in Erythroxylaceae
*Nesogenaceae
*Nyctaginaceae
*Nymphaeaceae

Ochnaceae
Octoknemaceae — in Olacaceae
Olacaceae
Oleaceae
*Oleandraceae
Oliniaceae
Onagraceae
*Ophioglossaceae
Opiliaceae
Orchidaceae
Part 1, Orchideae
*Part 2, Neottieae, Epidendreae
*Part 3, Epidendreae, Vandeae
Orobanchaceae
*Osmundaceae
Oxalidaceae

*Palmae
Pandaceae — with Euphorbiaceae
*Pandanaceae

Papaveraceae
Papilionaceae — in Leguminosae
*Parkeriaceae
Passifloraceae
Pedaliaceae
Periplocaceae — see Apocynaceae (Part 2)
Phytolaccaceae
*Piperaceae
Pittosporaceae
Plantaginaceae
Plumbaginaceae
Poaceae — see Gramineae
Podocarpaceae
Podostemaceae
Polemoniaceae — see Cobaeaceae
Polygalaceae
Polygonaceae
*Polypodiaceae
Pontederiaceae
*Portulacaceae
Potamogetonaceae
Primulaceae
*Proteaceae
*Psilotaceae
*Ptaeroxylaceae
*Pteridaceae

*Rafflesiaceae
Ranunculaceae
Resedaceae
Restionaceae
Rhamnaceae
Rhizophoraceae
Rosaceae
Rubiaceae
Part 1
*Part 2
*Part 3
*Ruppiaceae
*Rutaceae

*Salicaceae
Salvadoraceae
*Salviniaceae
Santalaceae
*Sapindaceae
Sapotaceae
*Schizaeaceae
Scrophulariaceae
Scytopetalaceae
Selaginellaceae
Selaginaceae — in Scrophulariaceae
*Simaroubaceae
*Smilacaceae
Sonneratiaceae
Sphenocleaceae
Strychnaceae — in Loganiaceae
*Surianaceae
Sterculiaceae

Taccaceae
Tamaricaceae
Tecophilaeaceae
Ternstroemiaceae — in Theaceae
Tetragoniaceae — in Aizoaceae
Theaceae
Thelypteridaceae
Thismiaceae — in Burmanniaceae
Thymelaeaceae
*Tiliaceae
Trapaceae
Tribulaceae — in Zygophyllaceae
*Triuridaceae
Turneraceae
Typhaceae

Uapacaceae — in Euphorbiaceae
Ulmaceae
*Umbelliferae
*Urticaceae

Vacciniaceae — in Ericaceae
Valerianaceae
Velloziaceae
*Verbenaceae
*Violaceae
*Viscaceae
*Vitaceae
*Vittariaceae

*Woodsiaceae

*Xyridaceae

*Zannichelliaceae
*Zingiberaceae
*Zosteraceae
*Zygophyllaceae

FORTHCOMING PARTS

Acanthaceae
Part 2
Apocynaceae
Part 2
Asclepiadaceae — see Apocynaceae
Commelinaceae
Cyperaceae
Solanaceae

Editorial adviser, National Museums of Kenya: Quentin Luke
Adviser on Linnaean types: C. Jarvis

Parts of this Flora, unless otherwise indicated, are obtainable from:
Royal Botanic Gardens, Kew, Richmond, Surrey TW9 3AB, England. www.kew.org or www.kewbooks.com

*** only available through CRC Press at:**
UK and Rest of World (except North and South America):
CRS Press/ITPS,
Cheriton House, North Way, Andover, Hants SP10 5BE.
e: uk.tandf@thomsonpublishingservices. co.uk

North and South America:
CRC Press,
2000NW Corporate Blvd, Boco Raton, FL 33431-9868,
USA.
e: orders@crcpress.com

Information on current prices can be found at www.kewbooks.com or www.tandf.co.uk/books/